Lecture Notes in Mathematics   2138

More information about this series at http://www.springer.com/series/304

Ihsen Yengui

Constructive Commutative Algebra

Projective Modules Over Polynomial Rings
and Dynamical Gröbner Bases

Springer

Ihsen Yengui
Fac. of Science, Dept. of Mathematics
University of Sfax
Sfax, Tunisia

ISSN 0075-8434 ISSN 1617-9692 (electronic)
Lecture Notes in Mathematics
ISBN 978-3-319-19493-6 ISBN 978-3-319-19494-3 (eBook)
DOI 10.1007/978-3-319-19494-3

Library of Congress Control Number: 2015956600

Mathematics Subject Classification (2010): 13Cxx, 13Pxx, 14Qxx, 03Fxx

Springer Cham Heidelberg New York Dordrecht London

Springer International Publishing AG Switzerland is part of Springer Science+Business Media
(www.springer.com)

Contents

Chapter 1
Introduction

Constructive algebra can be seen as an abstract version of computer algebra. In computer algebra, on the one hand, one attempts to construct efficient algorithms for solving concrete problems given in an algebraic formulation, where a problem is understood to be concrete if its hypotheses and conclusion have computational content. Constructive algebra, on the other hand, can be understood as a "prepro-cessing" step for computer algebra that leads to general algorithms, even if they are sometimes not efficient. In constructive algebra, one tries to give general algorithms for solving "virtually any" theorem of abstract algebra.

Therefore, a first task in constructive algebra is to define the computational con-tent hidden in hypotheses that are formulated in a very abstract way. For example, what is a good constructive definition of a local ring (i.e., a ring with a unique max-imal ideal), a valuation ring (i.e., a ring in which all elements are comparable under division), an arithmetical ring (i.e., a ring which is locally a valuation ring), a ring of Krull dimension $\leq n$ (i.e., a ring in which every chain $\mathfrak{p}_0 \subset \mathfrak{p}_1 \subset \cdots \subset \mathfrak{p}_k$ of prime ideals has length $k \leq n$), and so on? A good constructive definition must be equiv-alent to the usual definition within classical mathematics; it must have computational content; and it must be fulfilled by "usual" objects that satisfy the definition.

As a typical example, let us consider the classical theorem "any polynomial P in $\mathbf{K}[X]$ is a product of irreducible polynomials ($\mathbf{K}$ a field)". This leads to an interesting problem: it seems like no general algorithm that produces the irreducible factors. What, then, is the constructive content of this theorem? A possible answer is as follows: when performing computations with P, proceed as if its decomposition into irreducible polynomials were known (at the beginning, proceed as if P were irreducible). When something strange happens (e.g., when the gcd of P and another polynomial Q is a strict divisor of P), use this fact to improve the decomposition of P. This trick was invented in Computer Algebra as the D5-philosophy [46, 51, 125], and later taken up in the form of the dynamical proof method in algebra [42]. It indeed enables one to carry out computations inside the algebraic closure $\widetilde{\mathbf{K}}$ of

© Springer International Publishing Switzerland 2015
I. Yengui, *Constructive Commutative Algebra*, Lecture Notes in Mathematics 2138,
DOI 10.1007/978-3-319-19494-3_1

$\mathbf{K}$ even if it is not possible to effectively construct $\widetilde{\mathbf{K}}$, for in general this would require transfinite methods as Zorn's Lemma. The foregoing has been referred to as "dynamical evaluation" of the algebraic closure.

From a logical point of view, the "dynamical evaluation" gives a constructive substitute for two highly nonconstructive tools of abstract algebra: the Law of Excluded Middle and Zorn's Lemma. For instance, these tools are required in order to "construct" the complete prime factorization of an ideal in a Dedekind domain (i.e., in a Noetherian domain which is locally a valuation domain), while the dynamical method reveals the computational content of this "construction". We refer to [42] for more details on the dynamical proof method in algebra, including a wealth of examples. It is worth mentioning Schuster's new approach with Open Induction [157, 158]. As compared with the dynamical proof method, Schuster's approach is somewhat closer both to everyday mathematical practice, and to the established proof-theoretic methods for the extraction of algorithms, and thus computer programs, from formalized proofs. The approach with Open Induction therefore is a direct competitor of the dynamical proof method, both regarding objectives and techniques.

Following this "dynamical" philosophy, the main goal is to find the constructive content hidden in abstract proofs of concrete theorems in Commutative Algebra and especially well-known theorems concerning finitely-generated projective modules (i.e., the images of idempotent matrices) over polynomial rings and syzygy module (i.e., the relations module) of multivariate polynomials with coefficients in a valuation ring, or, more generally, in an arithmetical ring (a ring which is locally a valuation ring). As explained above, the general method consists in replacing some abstract ideal objects whose existence is based on the excluded middle principle and the axiom of choice by incomplete specifications of these objects.

The constructive rewriting of "abstract local-global principles" is very important. In classical proofs using this kind of principle, the argument is "let us see what happens after localization at an arbitrary maximal ideal of $\mathbf{R}$". From a computational point of view, maximal ideals are too abstract objects, particularly if one wishes to deal with a general commutative ring. In the constructive rereading, the argument is "let us see what happens when the ring is a residually discrete local ring", i.e., if $\forall x, (x \in \mathbf{R}^{\times} \text{ or } \forall y (1 + xy) \in \mathbf{R}^{\times})$. If a constructive proof is obtained in this particular case, the process can be completed by "dynamically evaluating an arbitrary ring $\mathbf{R}$ as a residually discrete local ring". For example, in these lecture notes, Dedekind domains will behave dynamically as valuation domains.

Dynamical methods were used successfully in order to find constructive substitutes to very elegant abstract theorems such as Quillen Patching, Quillen Induction and Lequain-Simis. In Chap. 2, the problem of freeness of projective modules over polynomial rings originally raised by Serre [160] in 1955 is approached constructively. Serre remarked that it was not known whether there exist finitely-generated projective modules over $\mathbf{A} = \mathbf{K}[X_1, \ldots, X_n]$, $\mathbf{K}$ a field, which are not free. This remark turned into the so-called "Serre's conjecture" or "Serre's problem", stating that indeed there were no such modules. Proved independently by Quillen [145] and Suslin [164] in 1976 (see also [151]), it became known subsequently as the Quillen-Suslin theorem. Quillen's and Suslin's proofs had a big effect on the subsequent development of the study of projective modules. The book of Lam [92] is a nice exposition of Serre's conjecture. It was known [161] well before Serre's conjecture was settled that finitely-generated projective modules over $\mathbf{A}$ are *stably free*, i.e, every finitely-generated projective $\mathbf{A}$-module P is isomorphic to the kernel of an $\mathbf{A}$-epimorphism T from $\mathbf{A}^m$ onto $\mathbf{A}^s$. In that situation the matrix T is unimodular: that is, the maximal minors of T generate the unit ideal in $\mathbf{A}$. The fact that P is free is nothing but the fact that the matrix T can be completed (we say that it is *completable*) to an invertible matrix by adding a suitable number of new rows. For $s = 1$, we speak of a *unimodular row* $(b_1, \ldots, b_m) \in \mathbf{A}^{1 \times m}$ ($^t(b_1, \ldots, b_m)$ is called a *unimodular vector*), i.e., such that $\langle b_1, \ldots, b_m \rangle = \mathbf{A}$.

Constructive versions of Quillen and Suslin proofs of Serre's problem, simple and constructive proofs of some subsequent developments in the theory of projective modules over polynomial rings, are presented, as well as recent progress on the Hermite ring Conjecture using tools from Constructive Algebra.

The Hermite Ring Conjecture [91]. *For any ring* $\mathbf{R}$, *all finitely-generated stably free modules over* $\mathbf{R}[X]$ *are extended from* $\mathbf{R}$ *(i.e., isomorphic to* $N \otimes_{\mathbf{R}} \mathbf{R}[X]$ *for some* $\mathbf{R}$*-module* N*).*

The following new conjectures about unimodular completion arising from the constructive approach to the unimodular completion problem will be presented:

The One-Dimension Conjecture. *For any ring* $\mathbf{R}$ *of Krull dimension* ≤ 1, *and* $k \in \mathbb{N}$, *all finitely-generated stably free modules over* $\mathbf{R}[X_1, \ldots, X_k]$ *are free. In other words, if* $\mathbf{R}$ *is a of Krull dimension* ≤ 1, *then for any* $k \in \mathbb{N}$, $\mathbf{R}[X_1, \ldots, X_k]$ *is Hermite.*

(The One-dimension Conjecture is known to be true for Noetherian rings [92] and in the univariate case [182].)

The One Square Conjecture. *The unimodular vector* $^t(x_1, x_2, x_3)$ *is not completable (i.e., can not be a first column of an invertible matrix) over the ring* $\mathbb{F}_2[x_1, x_2, x_3, y_2, y_3]$ *modulo* $x_1^2 + x_2 y_2 + x_3 y_3 = 1$.

Or more generally:

The One Square Conjecture, bis. *If* $\mathbf{K}$ *is a field then the unimodular vector* $^t(x_1, x_2, x_3)$ *is not completable over the ring* $\mathbf{K}[x_1, x_2, x_3, y_2, y_3]$ *modulo* $x_1^2 + x_2 y_2 + x_3 y_3 = 1$.

The Sum of Squares Conjecture. *Let* $\mathbf{K}$ *be a field,* $n \geq 2$, *and* $m \geq 0$ *with* $n + m \geq 3$. *The unimodular vector* $^t(x_1, \ldots, x_n, y_1, \ldots, y_m)$ *is elementarily completable (i.e., is the first column of an elementary matrix) over the ring* $\mathbf{K}[x_1, \ldots, x_n, y_1, \ldots, y_m, z_1, \ldots, z_m]$ *modulo* $x_1^2 + \cdots + x_n^2 + y_1 z_1 + \cdots + y_m z_m = 1$ *if and only if* -1 *is the sum of* $n - 1$ *squares in* $\mathbf{K}$.

Besides deciphering theorems such as Quillen Patching, Quillen Induction and Lequain-Simis which at some step use a localization at a "generic" maximal ideal, a method for deciphering constructively abstract proofs which pass modulo a "generic" maximal ideal to prove that a given ideal contains 1 will be presented in Chap. 2. As academic example, a lemma of Suslin (Theorem 57), which played a central role in his elementary proof of Serre's problem [165] and which is the only nonconstructive step in his solution, will be studied. This is the second main aspect of use of maximal ideals in commutative algebra besides localization. This lemma says that for a commutative ring $\mathbf{R}$ if $\langle v_1(X), \ldots, v_n(X) \rangle = \mathbf{R}[X]$ where v_1 is monic and $n \geq 3$, then there exist $\gamma_1, \ldots, \gamma_\ell \in \mathrm{E}_{n-1}(\mathbf{R}[X])$ (the subgroup of $\mathrm{SL}_{n-1}(\mathbf{R}[X])$ generated by elementary matrices) such that $\langle \mathrm{Res}(v_1, e_1.\gamma_1{}^t(v_2, \ldots, v_n)), \ldots, \mathrm{Res}(v_1, e_1.\gamma_\ell{}^t(v_2, \ldots, v_n)) \rangle = \mathbf{R}$ (here Res means resultant and $e_1.u$ means the first component of u). By the constructive proof of the above-mentioned lemma, Suslin's proof of Serre's problem becomes fully constructive. Moreover, the new method used for treating this academic example may be a model for miming constructively abstract proofs in which one works modulo each maximal ideal to prove that a given ideal contains 1. The Concrete local-global principle developed in Sect. 2.1.3 cannot be used here since the proof we want to decipher constructively, instead of passing to the localizations at each maximal ideal, passes to the residue fields modulo each maximal ideal.

The Quillen-Suslin theorem finds natural applications either in signal processing (see, e.g., the work of Park [131, 132, 133, 134]) or in mathematical systems theory and control theory (see, e.g., the work of Fabiańska and Quadrat [62]). There are several papers [29, 62, 64, 93, 94, 97, 110, 124, 135, 186] in the literature proposing algorithms for the Quillen-Suslin theorem but the first full implementation (a MAPLE package QuillenSuslin by Fabiańska [61]) is only available recently. An implementation of the Quillen-Suslin theorem in MACAULAY2 [72] is underway [115]. Also, a recent implementation of the Quillen-Suslin theorem has been developed in the homalg package of GAP 4.5. All these implementations rely on the Logar-Sturmfels paper [97], namely, rely on the facts that for a discrete field $\mathbf{K}$, the ring $\mathbf{K}[X_1, \ldots, X_k]$ is Noetherian and has an effective Nullstellensatz. As a matter of fact, roughly speaking, in order to eliminate one variable, say X_k, via comaximal resultants $r_1, \ldots, r_m$ (i.e., such that $1 \in \langle r_1, \ldots, r_m \rangle$, see the above-mentioned lemma of Suslin), they compute a maximal ideal $\mathfrak{M}$ of $\mathbf{K}[X_1, \ldots, X_{k-1}]$ containing the ideal $\langle r_1, \ldots, r_j \rangle$ generated by the current list $[r_1, \ldots, r_j]$ of resultants, and then, enter a "local loop" by localizing $\mathbf{K}[X_1, \ldots, X_{k-1}]$ at $\mathfrak{M}$ in order to find a new resultant $r_{j+1} \notin \langle r_1, \ldots, r_j \rangle$. The fact that $\mathbf{K}[X_1, \ldots, X_{k-1}]$ is Noetherian ensures the termination of this search for comaximal resultants. In Sect. 2.2, we will give a method avoiding this heavy use of maximal ideals and Noetherianity by means of efficient elementary operations in the case where the base ring is an infinite field (which is often the case in concrete applications) [124]. Indeed, in that case, we know in advance the comaximal resultants we need.

It is worth pointing out that in concrete applications in systems theory (e.g., in signal processing and control theory), most of the arising polynomial matrices are actually multivariate Laurent polynomial matrices (partly due to the time-delay). For example, various signal processing problems can be understood in terms of multi-input multi-output MIMO systems which are characterized by their transfer matrices whose entries are in $\mathbb{C}[X_1^{\pm}, \ldots, X_k^{\pm}]$ [133]. An efficient algorithm [4] for unimodular completion over $\mathbf{K}[X_1^{\pm}, \ldots, X_k^{\pm}]$), where $\mathbf{K}$ is an infinite field, will be presented (see Exercise 375 and its correction). Contrary to the paper [133], this algorithm for unimodular completion over multivariate Laurent polynomials does not convert the given Laurent polynomial vector to a "regular" polynomial vector, it eliminates all the variables at one time (contrary to the polynomial case), and does not use maximal ideals nor Noetherianity.

Recently Gröbner bases techniques in multivariate polynomial rings over $\mathbb{Z}/m\mathbb{Z}$ and $(\mathbb{Z}/p^{\alpha}\mathbb{Z}) \times (\mathbb{Z}/p^{\alpha}\mathbb{Z})$ (in particular $\mathbb{Z}/2^{\alpha}\mathbb{Z}$ and $(\mathbb{Z}/2^{\alpha}\mathbb{Z}) \times (\mathbb{Z}/2^{\alpha}\mathbb{Z})$) have attracted some attention due to their potential applications in formal verification of data paths [21, 74, 162], and coding theory [25, 127, 128, 129, 130, 144] (see also the recent Ph.D. thesis of Wienand [177]). Also, many authors [5, 54, 71, 136, 154, 169, 178] have been interested in computing Gröbner bases over $\mathbb{Z}/p^{\alpha}\mathbb{Z}$ (where p is a "lucky" prime number) [71], because modular methods give a satisfactory way to avoid intermediate coefficients swell with Buchberger's algorithm for computing Gröbner bases over the rational numbers [5]. In Chap. 3, a new approach for the construction of Gröbner bases over valuation rings (possibly with zero-divisors) is presented. Recall that, for us, a valuation ring is a ring equipped with a divisibility test and in which every two elements are comparable under division. The ring $\mathbb{Z}/p^{\alpha}\mathbb{Z}$, where p is a prime number and $\alpha \geq 2$, is a typical example of a valuation ring with zero-divisors. The dynamical method will be used to extend this Gröbner bases construction to arithmetical rings (possibly with zero-divisors). For us, arithmetical rings are strongly discrete rings (i.e., rings equipped with a membership test to finitely-generated ideals) which are locally valuation rings ($\mathbb{Z}/m\mathbb{Z}$ is a typical example of an arithmetical ring). This gives birth to the notion of "dynamical Gröbner bases" [77, 181]. It is a new alternative for computation with multivariate polynomials over rings. Contrary to the methods that have been proposed, which suggest that for (Noetherian) rings the analog of Gröbner bases over field should be computed [3, 23, 73, 137, 171], a dynamical substitute is proposed. Instead of a Gröbner basis describing the situation globally, use a finite number of Gröbner bases, not over the base ring, but over comaximal localizations of this ring. At each localization, the computation behaves as if a valuation ring were present. In a nutshell, it is somewhat like Serre's method in "Corps locaux" [160] but follows the lazy fashion of computer algebra [8, 11, 16, 34, 35, 38, 39, 40, 41, 42, 46, 47, 51, 55, 99, 102, 103, 104, 138, 180, 181, 182]. Borrowing words from [137], the difference between our approach and classical approaches is well illustrated by the following example: a Gröbner basis of the ideal $\langle 2X_1, 3X_2 \rangle$ in $\mathbb{Z}[X_1, X_2]$ is $\{2X_1, 3X_2\}$ according to Trinks [171], $\{2X_1, 3X_2, X_1X_2\}$ according to Buchberger [23], and

$$\{(\mathbb{Z}[\frac{1}{2}, X_1, X_2], \{X_1, 3X_2\}), (\mathbb{Z}[\frac{1}{3}, X_1, X_2], \{2X_1, X_2\})\}$$

for us. In fact, the ring of integers $\mathbb{Z}$ is more than an arithmetical ring, it is a Bezout domain of Krull dimension 1. We will prove that, over such rings, when computing dynamical Gröbner bases, one can avoid branching.

An essential property of a Dedekind domain is that its integral closure in a finite algebraic extension of its quotient field remains a Dedekind domain. This property is difficult to capture from an algorithmic point of view if one requires complete prime factorization of ideals (see [120]). Besides, even if such factorization is possible in theory, one rapidly encounters impracticable methods that involve huge complexities such as factorizing the discriminant. In [24], Buchmann and Lenstra proposed to compute inside the rings of integers without using a $\mathbb{Z}$-basis. An important algorithmic fact is that it is always easier to obtain partial factorization for a family of natural integers, i.e., a decomposition of each of these integers into a product of factors picked in a family of pairwise coprime integers [16, 17]. This just is the strategy adopted when computing dynamical Gröbner bases, the use of which provides a way to overcome such difficulties.

Another feature of the use of dynamical Gröbner bases is that it enables to easily resolve the delicate problem caused by the appearance of zero divisors as leading coefficients. Cai and Kapur concluded their paper [28] by mentioning the open question of how to generalize Buchberger's algorithm for Boolean rings (see also [85], in which Boolean rings are used to model propositional calculus). As a typical example of a problematical situation, Cai and Kapur studied the case where the base ring is $\mathbf{A} = \mathbb{F}_2[a,b]$ with $a^2 = a$ and $b^2 = b$. In that case, the method they proposed does not work due to the fact that an annihilator of $ab + a + b + 1 \in \mathbf{A}$ can be either a or b; thus, there may exist noncomparable multiannihilators for an element in $\mathbf{A}$. Dynamical Gröbner bases allow one to fairly overcome this difficulty. As a matter of fact, in this specific case, a computation of a dynamical Gröbner base made up of four Gröbner bases on localizations of $\mathbf{A}$ will be conducted. For $x, y \in \mathbf{A}$, denoting $\mathbf{A}_{x,y} := \mathbf{A}[\frac{1}{x}, \frac{1}{y}]$, this can be represented by the following tree:

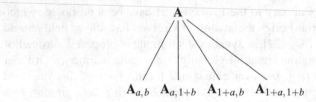

So, computing a dynamical Gröbner basis over $\mathbf{A}$ amounts to computing four (classical) Gröbner bases over the field with two elements $\mathbb{F}_2$ (possibly, in a parallel way) which correspond to $(a,b) = (1,1)$, $(1,0)$, $(0,1)$ or $(0,0)$. Of course, at each leaf of the tree above, the problem Cai and Kapur pointed disappears completely. Thus, by systematizing the dynamical construction above, it is directly shown that dynamical Gröbner bases could be a satisfactory solution to this open problem.

It is true that most of the examples given in these lecture notes are over $\mathbb{Z}/n\mathbb{Z}$, over rings of integers having a $\mathbb{Z}$-basis, or over quotients of polynomial rings, and that such problems can be treated directly in most software systems such as MAGMA [116] and SINGULAR [44] without using a dynamical approach. Dynamical Gröbner bases are potentially more appropriate for dealing with Dedekind domains which are intractable to this type of computer algebra software. However, the computations are restricted to small, simple examples because most of the work is done by hand or using nonoptimized implementations. For lack of an optimized implementation of dynamical Gröbner bases, a practical comparison with other methods is impossible. A serious analysis of improvements to the dynamical method proposed is therefore outside the scope of this course. No doubt, almost all the improvements that have been made in the case where the base ring is a field will prove to be easily adaptable to the dynamical context. Our goal is simply to introduce the main lines of the computation of dynamical Gröbner bases over Dedekind rings, with the hope that in the future dynamical Gröbner bases will be implemented in one of the available computer algebra systems. Of course, in such case, one must take into account the considerable number of optimizations that have been made in recent years for the purpose of speeding up Buchberger's algorithm in case where the base ring is a field (the faster version was given in [63]). The interested reader can refer to [73] for a modern introduction to this subject.

Another important issue raised in Chap. 3 is the *"Gröbner Ring Conjecture"*:

The Gröbner Ring Conjecture. *Let* $\mathbf{V}$ *be a valuation domain. Then for each* $n \in \mathbb{N}$ *and each finitely-generated ideal* I *of* $\mathbf{R}[X_1, \dots, X_n]$, *fixing a monomial order on* $\mathbf{R}[X_1, \dots, X_n]$, *the ideal* $\langle \mathrm{LT}(f), f \in I \rangle$ *of* $\mathbf{R}[X_1, \dots, X_n]$ *generated by the leading terms of the elements of* I *is finitely-generated, if and only if* $\mathbf{V}$ *has Krull dimension* ≤ 1.

First, a positive answer to this conjecture will be presented in case of one variable, and then this positive answer will be extended to the multivariate case with a lexicographic monomial order. Contrary to the common idea that Gröbner bases can be computed exclusively on Noetherian ground, the computation of (finite) Gröbner bases over $\mathbf{R}[X_1, \dots, X_n]$, where $\mathbf{R}$ is a Prüfer domain (i.e., an arithmetical domain), has nothing to do with Noetherianity. It is only related to the fact that the Krull dimension of $\mathbf{R}$ is ≤ 1. This opens the doors to a wider class of rings over which Gröbner bases can be computed (the class of Prüfer domains of Krull dimension ≤ 1 instead of that of Dedekind domains), at least for the lexicographic monomial order.

In the literature, there is no well-established terminology for *Dedekind rings*. In this book, we follow the definition adopted in [120]: they are arithmetical coherent (i.e., the relations module between finitely many elements in the ring is finitely-generated) Noetherian rings (i.e., rings in which every nondecreasing sequence $(I_n)_{n \in \mathbb{N}}$ of finitely-generated ideals pauses, that is, there exists $n \in \mathbb{N}$ such that $I_n = I_{n+1}$). In addition, in order for our theorems to have an algorithmic content, we will demand that the arithmetic character and coherence are explicit. Let us finally point out that, in [108], the authors further demand that the ring is reduced.

The computation of syzygies (that is, relations between the generators of a module) and the submodule membership problem are central to homological algebra and represent the two principal tools required for solving linear systems over rings. The first is used for testing particular solutions and the second for solving the homogeneous associated system. These two major problems have been chosen to illustrate our dynamical computation with multivariate polynomials over Dedekind rings. The resolution of a finitely-generated module is nothing but the computation of the iterated syzygies of its presentation matrix. It is worth mentioning that in the examples given in this document we are restricted to the computation of the first syzygy because the computation is done by hand or using nonoptimized implementations, as explained above. The method used for the computation of syzygies over multivariate polynomials with coefficients in a field is not the optimal one. As a matter of fact, the algorithms implemented in computer algebra systems that compute such syzygies (SINGULAR for example) are largely inspired by Schreyer's original proof [155, 156]. Moreover, by performing reductions between the generators, one can obtain a more balanced presentation of the syzygy module. Here, it is emphasized that the classical approach can be adapted to the dynamical setting; thorough optimization of the approach remains to be done.

It is folklore [68, 75] that if V is a valuation domain without any restriction on its Krull dimension (i.e., even if Gröbner bases can not be computed over V), then $V[X_1, \ldots, X_n]$ is coherent: that is, the syzygy modules of finitely-generated ideals of $V[X_1, \ldots, X_n]$ are all finitely-generated. In general, however, there is no algorithm for this remarkable result, and it seems difficult to find one even for small polynomials. An exception is the case of coherent archimedean valuation rings (valuation domains of Krull dimension ≤ 1, or coherent zero-dimensional valuation rings with zero divisors), which can be done via Gröbner bases computation [77] (see Chap. 3).

The main objective of the fourth chapter is to give a general algorithm for computing a finite generating set for the syzygies of any finitely-generated ideal of $V[X_1, \ldots, X_k]$ (V a valuation domain) which neither relies on Noetherianity nor on Krull dimension. In fact, the presented algorithm computes a finite generating set for the V-saturation of any finitely-generated submodule S of $V[X_1, \ldots, X_k]^n$, i.e.,

$$\{s \in V[X_1, \ldots, X_k]^n \mid \alpha s \in S \text{ for some } \alpha \in V \setminus \{0\}\} = (S \otimes_V K) \cap V[X_1, \ldots, X_k]^m,$$

where K is the quotient field of V. This algorithm is based on a notion of "echelon form" which ensure its correctness. The proposed algorithm terminates when two (Hilbert) series on the quotient field of V and the residue field of V coincide. Computing syzygies over $V[X_1, \ldots, X_k]$ is one important application of the saturation algorithm. In the univariate case [48, 113], precise complexity bounds will be given.

The present notes are self-contained. All the proofs are constructive. The text is based on four lectures on Constructive Algebra the author gave on the occasion of the CIMPA Summer School held at Yaoundé (Cameroun) from the 24th of August to the 4th of September 2009, and also, on the occasion of the Summer School and Conference on "Mathematics, Algorithms and Proofs" held at the ICTP (Trieste, Italy) from the 11th to the 29th of August 2008.

The undefined terminology is standard as in [19, 53, 66, 90, 92] for commutative algebra, [12, 153] for algebraic K-theory, [108, 109, 120] for constructive algebra, and [3, 14, 43, 73, 88, 176] for computer algebra.

Chapter 2
Projective Modules Over Polynomial Rings

2.1   Quillen's Proof of Serre's Problem

2.1.1   Finitely-Generated Projective Modules

Definition 1. Let **R** be a ring.

(1) Let M be an **R**-module and $k \in \mathbb{N}$. We say that M is a *free of rank k* if it is isomorphic to $\mathbf{R}^k$.

(2) Let N be a submodule of an **R**-module M. We say that N is a *direct summand* in M if there exists a submodule L of M such that $M = L \oplus N$. The interesting situation is when N if a finite-rank free module (see the following proposition).

Proposition 2. (Splitting off) *Let **R** be a ring, M an **R**-module, and $v_1, \ldots, v_n \in M$. Then, the following two assertions are equivalent:*

(1) *The submodule $N = \langle v_1, \ldots, v_n \rangle$ of M is free with basis $(v_1, \ldots, v_n)$ and is a direct summand in M.*

(2) *There exists an alternating n-linear form $\varphi : M^n \to \mathbf{R}$ such that $\varphi(v_1, \ldots, v_n) = 1$.*

Proof. "(1) $\Rightarrow$ (2)" Write $N \oplus L = M$ for some submodule L of M, and denote by $\pi : M = N \oplus L \to N$ the projection on N parallel to L. Denoting by $(v_1^*, \ldots, v_n^*)$ the dual basis of $(v_1, \ldots, v_n)$, i.e., $v_j^*(\sum_{i=1}^n \lambda_i v_i) = \lambda_j$ with $\lambda_i \in \mathbf{R}$, it suffices to take

$$\varphi(u_1, \ldots, u_n) = \det \left((v_j^*(\pi(u_i))_{1 \le i, \, j \le n}) \right).$$

"(2) $\Rightarrow$ (1)" Consider the **R**-linear map $p : M \to M$ with

$$p(u) = \sum_{j=1}^n \varphi(v_1, \ldots, v_{j-1}, u, v_{j+1}, \ldots, v_n) v_j.$$

© Springer International Publishing Switzerland 2015
I. Yengui, *Constructive Commutative Algebra*, Lecture Notes in Mathematics 2138,
DOI 10.1007/978-3-319-19494-3_2

It is clear that $p(v_i) = v_i$ for all $1 \leq i \leq n$, and that $\text{Im}(p) \subseteq N = \langle v_1, \ldots, v_n \rangle$. It follows that $p^2 = p$ (p is a projection), $\text{Im}(p) = N$, and N is a direct summand in M ($M = N \oplus \text{Im}(\text{Id}_M - p)$).

Now, let us prove that $v_1, \ldots, v_n$ are **R**-linearly independent. For this, if $u = \sum_{j=1}^{n} \lambda_j v_j = 0$ with $\lambda_j \in \mathbf{R}$, then $\varphi(v_1, \ldots, v_{j-1}, u, v_{j+1}, \ldots, v_n) = \lambda_j = 0$ for $1 \leq j \leq n$. $\qquad\qquad\qquad\qquad\qquad\qquad\qquad\qquad\qquad\qquad\qquad\qquad\qquad\qquad\quad\square$

In case $n = 1$, we say that $v_1 \in M$ is *unimodular* if there exists a linear form $\varphi : M \to \mathbf{R}$ such that $\varphi(v_1) = 1$. For example, if $a_1, b_1 \ldots, a_k, b_k \in \mathbf{R}$ with $a_1 b_1 + \cdots + a_k b_k = 1$, then $v = (a_1, \ldots, a_k)$ is unimodular in $\mathbf{R}^k$. The corresponding linear form is

$$\varphi : \mathbf{R}^k \to \mathbf{R}; \quad (x_1, \ldots, x_k) \mapsto b_1 x_1 + \cdots + b_k x_k.$$

Definition 3. Let P be a module over a ring $\mathbf{R}$. We say that P is a *projective* **R**-module if any surjective **R**-module homomorphism $\alpha : M \to P$ has a right inverse $\beta : P \to M$; or equivalently, if it is isomorphic to a direct summand in a free **R**-module. It is finitely-generated and projective if and only if it is isomorphic to a direct summand in $\mathbf{R}^n$ for some n.

Definition 4. Let $\mathbf{R}$ be a ring. The isomorphism classes of finitely-generated projective modules over $\mathbf{R}$ form an abelian monoid $\text{Proj}\,\mathbf{R}$ with $\oplus$ as the addition operation and with the 0-module as the identity element. As $\text{Proj}\,\mathbf{R}$ need not be a group, it is therefore convenient to consider the group that it generates, namely its *group* $K_0(\mathbf{R})$. For example, if $\mathbf{R}$ is a Bezout domain then every finitely-generated projective module over $\mathbf{R}$ is free, and so $\text{Proj}\,\mathbf{R} \cong \mathbb{N}$ and $K_0(\mathbf{R}) \cong \mathbb{Z}$.

Example 5.

(i) Every free module is projective.

(ii) Suppose that m and n are coprime natural numbers. Then as abelian groups (and also as $(\mathbb{Z}/mn\mathbb{Z})$-modules), we have $\mathbb{Z}/mn\mathbb{Z} \cong \mathbb{Z}/n\mathbb{Z} \oplus \mathbb{Z}/m\mathbb{Z}$. Thus, $\mathbb{Z}/m\mathbb{Z}$ is a projective $(\mathbb{Z}/mn\mathbb{Z})$-module which is not free as it contains fewer than mn elements.

(iii) An ideal I of an integral domain $\mathbf{R}$ is projective if and only if it is invertible. Integral domains in which every ideal is invertible are known as Dedekind domains, and they are important in number theory. For example, the ring of integers in any algebraic number field is a Dedekind domain. So, by considering a Dedekind domain which is not a PID, one can find an example of a projective module (an invertible ideal) which is not free (not principal).

(iv) Let e be a nontrivial idempotent of a ring $\mathbf{R}$, i.e., $e^2 = e$, $e \neq 0$, and $e \neq 1$ (for example, one can consider the generic case $\mathbf{R} = \mathbf{K}[u]/\langle u^2 - u \rangle = \mathbf{K}[\bar{u}]$ and $e = \bar{u}$ where $\mathbf{K}$ is a field). As $\mathbf{R} \cong e\mathbf{R} \oplus (1-e)\mathbf{R}$, $e\mathbf{R}$ is a projective $\mathbf{R}$ module. Note that $e\mathbf{R}$ is not free since $(1-e)(e\mathbf{R}) = 0$ while $(1-e)\mathbf{R} \neq 0$.

Definition and Proposition 6. (Finitely-Presented Modules)

(1) Let $\mathbf{R}$ be a ring. A *finitely-presented* $\mathbf{R}$-module is an $\mathbf{R}$-module given by a finite number of generators and relations. Thus, it is a finitely-generated $\mathbf{R}$-module having a finitely-generated relations module. Equivalently, it is an $\mathbf{R}$-module isomorphic to the cokernel of a linear application

$$\gamma : \mathbf{R}^m \to \mathbf{R}^q.$$

The matrix $G \in \mathbf{R}^{q \times m}$ of γ has as columns a generating set of the relations module between the generators g_i which are the images of the canonical basis by the epimorphism $\pi : \mathbf{R}^q \to M$. The matrix G is called a presentation matrix of the module M for the generating system $(g_1, \ldots, g_q)$. We have:

- $[g_1 \cdots g_q] G = 0$, and
- each relation between the g_i's is a linear combination of the columns of G, i.e., if $[g_1 \cdots g_q] C = 0$ with $C \in \mathbf{R}^{q \times 1}$ then there exists $C' \in \mathbf{R}^{m \times 1}$ such that $C = GC'$.

For example, a free module of rank k (i.e., isomorphic to $\mathbf{R}^k$) is finitely-presented. Its presentation matrix is a column matrix formed by k zeroes. More generally, if P is a finitely-generated projective module then, as it is isomorphic to the image of an idempotent matrix $F \in \mathbf{R}^{n \times n}$ for some $n \in N^*$ (see Remark 9 below) and $\mathbf{R}^n = \operatorname{Im}(F) \oplus \operatorname{Im}(\mathrm{I}_n - F)$, we get $P \cong \operatorname{Coker}(\mathrm{I}_n - F)$ and thus P is finitely-presented.

(2) The definition above can be rephrased as follows: An $\mathbf{R}$-module M is finitely-presented if there is an epimorphism $\pi : \mathbf{R}^q \to M$ for some $q \in \mathbb{N}^*$ (and thus, $\mathbf{R}^q / \operatorname{Ker}(\pi) \cong M$) whose kernel $\operatorname{Ker}(\pi)$ is finitely-generated. The module M is specified using finitely many generators (the images of the q generators of $\mathbf{R}^q$) and finitely many relations (the generators of $\operatorname{Ker}(\pi)$).

For example, for $a \in \mathbf{R}$, the ideal $a\mathbf{R}$ is finitely-presented if and only if the annihilator $\operatorname{Ann}(a) := \{b \in \mathbf{R} \mid ba = 0\}$ of a is finitely-generated. The epimorphism π corresponds to the multiplication by a and its kernel is $\operatorname{Ann}(a)$. More generally, a finitely-generated ideal $\langle a_1, \ldots, a_n \rangle$ of $\mathbf{R}$ is finitely-presented if and only if the syzygy module

$$\operatorname{Syz}(a_1, \ldots, a_n) := \{(b_1, \ldots, b_n) \in \mathbf{R}^n \mid b_1 a_1 + \cdots + b_n a_n = 0\}$$

is finitely-generated.

(3) *Two matrices $G \in \mathbf{R}^{q \times m}$ and $H \in \mathbf{R}^{r \times m}$ present the same module, i.e., their cokernels are isomorphic, if and only if the following two matrices are equivalent:*

$$\begin{pmatrix} G & 0_{q,r} & 0_{q,q} & 0_{q,n} \\ 0_{r,m} & \mathrm{I}_r & 0_{r,q} & 0_{r,n} \end{pmatrix},$$

$$\begin{pmatrix} 0_{q,m} & 0_{q,r} & \mathrm{I}_q & 0_{q,n} \\ 0_{r,m} & 0_{r,r} & 0_{r,q} & H \end{pmatrix}.$$

Definition and Proposition 7. (Coherent Rings and Modules) Let $\mathbf{R}$ be a ring.

(1) An $\mathbf{R}$-module M is said to be *coherent* if for any $(v_1, \ldots, v_n) \in M^n$, the syzygy module

$$\mathrm{Syz}(v_1, \ldots, v_n) := \{(b_1, \ldots, b_n) \in \mathbf{R}^n \mid b_1 v_1 + \cdots + b_n v_n = 0\}$$

is finitely-generated.

The ring $\mathbf{R}$ is said to be coherent if it is coherent as an $\mathbf{R}$-module.

(2) *Every finite direct sum of coherent R-modules is a coherent $\mathbf{R}$-module.*

(3) *An R-module M is coherent if and only if*

    (i) *The intersection of any two finitely-generated submodules of M is a finitely-generated $\mathbf{R}$-module.*

    (ii) *The annihilator $\{b \in \mathbf{R} \mid bv = 0\}$ of any $v \in M$ is a finitely-generated ideal of $\mathbf{R}$.*

(4) $\mathbf{R}$ is said to be *stably coherent* if $\mathbf{R}[X_1, \ldots, X_k]$ is coherent for any k.

Definition 8.   (1) A ring $\mathbf{R}$ is said to be *discrete* if there is an algorithm deciding if $x = 0$ or $x \neq 0$ for an arbitrary element of $\mathbf{R}$.

(2) A ring is said to be *strongly discrete* if it is equipped with a membership test for finitely-generated ideals.

(2) A ring $\mathbf{R}$ is said to be *a domain* or an *integral ring* if any element in $\mathbf{R}$ is either null or regular (with an explicit "or"). In other words, if the annihilator of any element in $\mathbf{R}$ is either $\langle 1 \rangle$ or $\langle 0 \rangle$ (with an explicit "or").

(3) The total ring of fractions of an integral ring is a discrete field: any element is either null or invertible (with an explicit "or").

(4) A ring $\mathbf{R}$ is called *without zero-divisors* if for all $x, y \in \mathbf{R}$, $(xy = 0 \Rightarrow x = 0$ or $y = 0)$, with an explicit "or".

In a ring without zero-divisors, the annihilator of an element is an idempotent ideal. If this annihilator is finitely-generated, then it is generated by an idempotent (say, by e), and, thus, it is equal to $\langle 0 \rangle$ or $\langle 1 \rangle$ $(e(1 - e) = 0 \Rightarrow e = 0 \text{ or } e = 1)$.

It follows that, constructively, *a ring is integral if and only if it is without zero-divisors and the annihilator of any element is finitely-generated.* In particular, a coherent ring is integral if and only if it is without zero-divisors.

Remark 9.

(i) **Projective modules via idempotent matrices [153]:** There is another approach to finitely-generated projective modules which is more concrete and therefore more convenient for our constructive approach. If P is a finitely-generated projective **R**-module, we may assume (replacing P by an isomorphic module) that $P \oplus Q = \mathbf{R}^n$ for some n, and we consider the idempotent matrix M of the **R**-module homomorphism p from $\mathbf{R}^n$ to itself which is the identity on P and 0 on Q written in the standard basis. So, P can be seen (up to isomorphism) as the image of an idempotent matrix M. Conversely, different idempotent matrices can give rise to the same isomorphism class of projective modules. As a matter of fact, if M and N are idempotent matrices over a ring **R** (of possibly different sizes m and n, respectively), the corresponding finitely-generated projective modules are isomorphic if and only if it is possible to enlarge the sizes of M and N (by adding zeros in the lower right-hand corner) so that they have the same size $s \times s$ and conjugate under the group $\mathrm{GL}_s(\mathbf{R})$.

In more details ([153, Lemma 1.2.1] or [108, Lemma V.2.10]), if the isomorphism from $\mathrm{Im}(M)$ to $\mathrm{Im}(N)$ is coded by U and its inverse is coded by U', we obtain a matrix

$$A = \begin{pmatrix} \mathrm{I}_m - F & -U' \\ U & \mathrm{I}_n - G \end{pmatrix} = \begin{pmatrix} \mathrm{I}_m & 0 \\ U & \mathrm{I}_n \end{pmatrix} \begin{pmatrix} \mathrm{I}_m & -U' \\ 0 & \mathrm{I}_n \end{pmatrix} \begin{pmatrix} \mathrm{I}_m & 0 \\ U & \mathrm{I}_n \end{pmatrix}$$

in $\mathrm{GL}_{n+m}(\mathbf{R})$ with $\begin{pmatrix} 0_m & 0 \\ 0 & G \end{pmatrix} = A \begin{pmatrix} F & 0 \\ 0 & 0_n \end{pmatrix} A^{-1}$ (with usual block matrix notation).

The matrix $\begin{pmatrix} 0_m & 0 \\ 0 & G \end{pmatrix}$ is obviously conjugate to $\begin{pmatrix} G & 0 \\ 0 & 0_m \end{pmatrix}$ by a permutation matrix.

We will embed $\mathrm{M}_n(\mathbf{R})$ in $\mathrm{M}_{n+1}(\mathbf{R})$ by $M \mapsto \begin{pmatrix} M & 0 \\ 0 & 0 \end{pmatrix}$, $\mathrm{GL}_n(\mathbf{R})$ in $\mathrm{GL}_{n+1}(\mathbf{R})$ by the group homomorphism $M \mapsto \begin{pmatrix} M & 0 \\ 0 & 1 \end{pmatrix}$, so that we can define by $\mathrm{M}(\mathbf{R})$ (resp., $\mathrm{GL}(\mathbf{R})$) as the infinite union of the $\mathrm{M}_n(\mathbf{R})$ (resp., $\mathrm{GL}_n(\mathbf{R})$). Denoting by $\mathrm{Idem}(\mathbf{R})$ the set of idempotent matrices in $\mathrm{M}(\mathbf{R})$, $\mathrm{Proj}\,\mathbf{R}$ may be identified with the set of conjugation orbits of $\mathrm{GL}(\mathbf{R})$ on $\mathrm{Idem}(\mathbf{R})$. The monoid operation is induced by $(M,N) \mapsto \begin{pmatrix} M & 0 \\ 0 & N \end{pmatrix}$ and $\mathrm{K}_0(\mathbf{R})$ is the Groethendieck group of this monid. Denoting by $M = (m_{i,j})_{i,j \in I}$ and $N = (n_{k,\ell})_{k,\ell \in J}$, the Kronecker product $M \otimes N := (r_{(i,k),(j,\ell)})_{(i,k),(j,\ell) \in I \times J}$, where $r_{(i,k),(j,\ell)} = m_{i,j} n_{k,\ell}$, corresponds to the tensor product $\mathrm{Im}\,M \otimes \mathrm{Im}\,N$.

(ii) **Projective modules via Fitting ideals [108, 109]:** The theory of Fitting ideals of finitely-presented modules is an extremely efficient computing machinery from a theoretical constructive point of view. Recall that if G is a presentation matrix of a module T given by q generators related by m relations, the Fitting ideals of T are the ideals

$$\mathscr{F}_n(T) := \mathscr{D}_{q-n}(G),$$

where for any integer k, $\mathscr{D}_k(G)$ denotes the determinantial ideal of G of order k, that is, the ideal generated by all the minors of G of size k, with the convention that for $k \leq 0$, $\mathscr{D}_k(G) = \langle 1 \rangle$, and for $k > \min(m,n)$, $\mathscr{D}_k(G) = \langle 0 \rangle$. It is worth pointing out that the Fitting ideals of a finitely-presented module T don't depend on its presentation matrix G and that one has

$$\langle 0 \rangle = \mathscr{F}_{-1}(T) \subseteq \mathscr{F}_0(T) \subseteq \mathscr{F}_q(T) = \langle 1 \rangle.$$

Projectivity can be tested via the Fitting ideals as follows: *a finitely-presented* **R**-*module is projective if and only if its Fitting ideals are principal generated by idempotent elements* (see Theorems V.6.1 and V.8.14 in [108]).

Recall that a ring **R** is *local* if it satisfies:

$$\forall x \in \mathbf{R}, \quad x \in \mathbf{R}^\times \vee 1 - x \in \mathbf{R}^\times.$$

Theorem 10. *If* **R** *is a local ring, then every finitely-generated projective* **R**-*module is free. In particular,* $K_0(\mathbf{R}) \cong \mathbb{Z}$ *(since* $\operatorname{Proj} \mathbf{R} \cong \mathbb{N}$*) with generator the isomorphism class of a free module of rank* 1 *(*$\cong \mathbf{R}$*).*

Proof. Let $F = (f_{i,j})_{1 \leq i,j \leq m}$ be an idempotent matrix with coefficients in a local ring **R**. Let us prove that F is conjugate to a standard projection matrix

$$I_{r,m} := \begin{pmatrix} I_r & 0_{r,m-r} \\ 0_{m-r,r} & 0_{m-r,m-r} \end{pmatrix}.$$

Two cases may arise:

- If $f_{1,1}$ is invertible, then one can find $G \in \operatorname{GL}_m(\mathbf{R})$ such that

$$GFG^{-1} = \begin{pmatrix} 1 & 0_{1,m-1} \\ 0_{m-1,1} & F_1 \end{pmatrix},$$

  where F_1 is an idempotent matrix of size $(m-1) \times (m-1)$, and an induction on m applies.

- If $1 - f_{1,1}$ is invertible, then one can find $H \in \operatorname{GL}_m(\mathbf{R})$ such that

$$HFH^{-1} = \begin{pmatrix} 0 & 0_{1,m-1} \\ 0_{m-1,1} & F_2 \end{pmatrix},$$

  where F_2 is an idempotent matrix of size $(m-1) \times (m-1)$, and again an induction on m applies.

$\square$

Example 11. ($K_0(\mathbb{Z}/m\mathbb{Z})$ via the Chinese Remainder Theorem)

(a) If $\mathbf{R} = \mathbf{R}_1 \times \mathbf{R}_2$ is a product of two rings $\mathbf{R}_1$ and $\mathbf{R}_2$, it is easy to see $\mathrm{Idem}(\mathbf{R}) = \mathrm{Idem}(\mathbf{R}_1) \times \mathrm{Idem}(\mathbf{R}_2)$, $\mathrm{GL}(\mathbf{R}) = \mathrm{GL}(\mathbf{R}_1) \times \mathrm{GL}(\mathbf{R}_2)$, and thus

$$\mathrm{Proj}(\mathbf{R}) \cong \mathrm{Proj}(\mathbf{R}_1) \times \mathrm{Proj}(\mathbf{R}_2) \ \& \ K_0(\mathbf{R}) \cong K_0(\mathbf{R}_1) \times K_0(\mathbf{R}_2).$$

(b) Let $m \in \mathbb{N} \setminus \{0, 1\}$ and suppose that we know the prime factorization $m = p_1^{\alpha_1} \cdots p_\ell^{\alpha_\ell}$ of m, where $\ell, \alpha_i \in \mathbb{N}^*$ and the p_i's are pairwise different prime numbers. By the Chinese remainder theorem, we have the ring isomorphism

$$\mathbb{Z}/m\mathbb{Z} \cong (\mathbb{Z}/p_1^{\alpha_1}\mathbb{Z}) \times (\mathbb{Z}/p_2^{\alpha_2}\mathbb{Z}) \times \cdots \times (\mathbb{Z}/p_\ell^{\alpha_\ell}\mathbb{Z}).$$

Now, as $\mathbb{Z}/p_i^{\alpha_1}\mathbb{Z}$ is a local ring, we know by Theorem 10 that $K_0(\mathbb{Z}/p_i^{\alpha_1}\mathbb{Z}) \cong \mathbb{Z}$ and, thus, by virtue of (a), we have

$$K_0(\mathbb{Z}/m\mathbb{Z}) \cong \mathbb{Z}^\ell.$$

The following theorem gives a local characterization of projective modules.

Theorem 12. *An* $\mathbf{R}$*-module* P *is projective if and only if there exist comaximal elements* $s_1, \ldots, s_k \in \mathbf{R}$ *(i.e., satisfying* $\langle s_1, \ldots, s_k \rangle = \mathbf{R}$*) such that for each* $1 \leq i \leq k$, $P_{s_i} := P \otimes \mathbf{R}[\frac{1}{s_i}]$ *is a free* $\mathbf{R}[\frac{1}{s_i}]$*-module.*

Definition 13. (Extended Modules) A module M over $\mathbf{R}[X_1, \ldots, X_n] = \mathbf{R}[\underline{X}]$ is said to be *extended from* $\mathbf{R}$ (or simply, *extended*) if it is isomorphic to a module $N \otimes_\mathbf{R} \mathbf{R}[\underline{X}]$ for some $\mathbf{R}$-module N. Necessarily

$$N \simeq \mathbf{R} \otimes_{\mathbf{R}[\underline{X}]} M \text{ through } \rho : \mathbf{R}[\underline{X}] \to \mathbf{R}, f \mapsto f(0),$$

i.e., $N \simeq M/(X_1 M + \cdots + X_n M)$. In particular, if M is finitely-presented, denoting by $M^0 = M[0, \ldots, 0]$ the $\mathbf{R}$-module obtained by replacing the X_i by 0 in a relation matrix of M, then M is extended if and only if

$$M \simeq M^0 \otimes_\mathbf{R} \mathbf{R}[\underline{X}],$$

or equivalently, if the matrices M and M^0 are equivalent (once properly enlarged, see Definition and Proposition 6) using invertible matrices with entries in $\mathbf{R}[\underline{X}]$.

If M is given as the image of an idempotent matrix $F = F(X_1, \ldots, X_n)$, then M is extended if and only if F is conjugate to $F(0, \ldots, 0)$.

Definition 14. (Finitely-Generated Projective Modules of Constant Rank)

(i) **Classical approach [92]:** The rank of a nonzero free module $\mathbf{R}^m$ is defined by $\mathrm{rk}_\mathbf{R}(\mathbf{R}^m) = m$. If P is a finitely-generated projective module, as it is locally free (i.e., $P_\mathfrak{p} := P \otimes_\mathbf{R} R_\mathfrak{p}$ is a free $R_\mathfrak{p}$-module for any $\mathfrak{p} \in \mathrm{Spec}(\mathbf{R})$, where $\mathrm{Spec}(\mathbf{R})$ denotes the set of prime ideals of $\mathbf{R}$), we define the (rank) map $\mathrm{rk}(P) : \mathrm{Spec}(\mathbf{R}) \to \mathbb{N}$ by $\mathrm{rk}(P)(\mathfrak{p}) = \mathrm{rk}_{R_\mathfrak{p}}(P_\mathfrak{p})$. The map $\mathrm{rk}(P)$ is locally constant. Especially, if $\mathrm{Spec}(\mathbf{R})$ is connected, i.e., if $\mathbf{R}$ is not a direct product of nontrivial rings (or equivalently, if $\mathbf{R}$ has no nontrivial idempotents), then $\mathrm{rk}(P)$ is constant.

(ii) **Constructive approach [108, 109]:** Roughly speaking, if $\varphi : P \to P$ is an endomorphism of a finitely-generated projective **R**-module P, then supposing that $P \oplus Q$ is isomorphic to a free module, then the determinant of $\varphi_1 := \varphi \oplus \mathrm{Id}_Q$ depends only on φ; it is called the determinant of φ. Now, let us consider the **R**$[X]$-module $P[X] := P \otimes_{\mathbf{R}} \mathbf{R}[X]$. The polynomial $R_P(X) := \det(X\mathrm{Id}_P)$ is called the rank polynomial of the module P. If P is free of rank k, then clearly $R_P(X) = X^k$. Moreover, $R_{P \oplus Q}(X) = R_P(X)R_Q(X)$, $R_P(X)R_P(Y) = R_P(XY)$, and $R_P(1) = 1$, in such a way the coefficients of $R_P(X)$ form a fundamental system of orthogonal idempotents ($\sum e_i = 1$ and $e_i e_j = 0$ for $i \neq j$).

Now, this terminology being established, a finitely-generated projective **R**-module P is said to have rank equal to h if $R_P(X) = X^h$. If we don't specify h, we say that P has a constant rank.

For any finitely-generated projective **R**-module P, denoting by $R_P(X) = \sum_{h=0}^{n} r_h X^h$ (as said above, the r_h's form a fundamental system of orthogonal idempotents), we have $P = \bigoplus_{h=0}^{n} r_h P$ as **R**-modules, and each module $r_h P$ is a constant rank projective **R**$/\langle 1 - r_h \rangle$-module of rank h (recall that $\mathbf{R}/\langle 1 - r_h \rangle \cong \mathbf{R}[\frac{1}{r_h}]$).

(iii) **Projective modules of rank one:** To any ring **R**, we can associate its *Picard group* Pic **R**, i.e., the group of projective **R**-modules of rank one equipped with tensor product as group operation. The inverse of P is its dual $P^\star$. If $P \simeq \mathrm{Im} M$ then $P^\star \simeq \mathrm{Im}\,^t M$. In particular, if M is a rank one idempotent matrix, then $M \otimes {}^t M$ is an idempotent matrix whose image is a rank one free module.

In case **R** is an integral domain or a Noetherian ring, Pic **R** is isomorphic to the *class group* of **R**, the group of invertible fractional ideals in the field of fractions of **R**, modulo the principal ideals. So, this generalizes to an arbitrary ring the class group introduced originally by Kummer.

Example 15. Let e be a nontrivial idempotent of a ring **R**, i.e., $e^2 = e$, $e \neq 0$, and $e \neq 1$, and consider the projective **R**-module $P = e\mathbf{R}$. We have $R_P(X) = 1 - e + eX$. The module P hasn't a constant rank: it is of rank 1 over the ring $\mathbf{R}[\frac{1}{e}] \cong \mathbf{R}/\langle 1 - e \rangle$ and of rank 0 over the ring $\mathbf{R}[\frac{1}{1-e}] \cong \mathbf{R}/\langle e \rangle$.

2.1.2   Finitely-Generated Stably Free Modules

Definition 16. An **R**-module P is said to be *finitely-generated stably free* (of rank $n - m$) if $P \oplus \mathbf{R}^m \cong \mathbf{R}^n$ for some m, n. This amounts to saying that P is isomorphic to the kernel of an epimorphism $f : \mathbf{R}^n \to \mathbf{R}^m$. If M is the $m \times n$ matrix associated with f, then M is right invertible, i.e., there exists an $n \times m$ matrix N such that $MN = I_m$. Conversely, the kernel of any right invertible matrix defines a stably free module. So, the study of finitely-generated stably free **R**-modules becomes equivalent to the study of right invertible rectangular matrices over **R**.

Example 17.

(i) Every free module is stably free.

(ii) Every stably free module is projective. The converse does not hold. To see this, it suffices to consider a nonprincipal ideal in a Dedekind domain (for example, the ideal $\langle 3, 2 + \sqrt{-5} \rangle$ in the Dedekind domain $\mathbb{Z}[\sqrt{-5}]$). It is a rank one projective module (as it is an invertible ideal) but not a stably free module since as will be seen in Theorem 25, stably free modules of rank one are free.

Note that for a ring $\mathbf{R}$, the fact that every projective $\mathbf{R}$-module is stably free is equivalent to the fact that $K_0(\mathbf{R}) = \mathbb{Z}$.

(iii) Let $\mathbf{R} = \mathbb{R}[X_1, X_2, X_3]/\langle X_1^2 + X_2^2 + X_3^2 - 1 \rangle = \mathbb{R}[x_1, x_2, x_3]$ be the affine coordinate ring of the real 2-sphere, and consider the syzygy module

$$T = \mathrm{Syz}(x_1, x_2, x_3) := \{(y_1, y_2, y_3) \in \mathbf{R}^3 \mid x_1 y_1 + x_2 y_2 + x_3 y_3 = 0\}.$$

As (x_1, x_2, x_3) is unimodular (see Definition 19 below), we have $T \oplus \mathbf{R} \cong \mathbf{R}^3$, and, thus, T is a rank 2 stably free module. It is well-known that T is not free for topological reasons but this remarkable fact is still lacking a simple algebraic proof.

The following gives a criterion for the freeness of finitely-generated stably free modules in matrix terms.

Proposition 18. *For any right invertible $r \times n$ matrix M with entries in a ring $\mathbf{R}$, the (stably free) solution space of M is free if and only if M can be completed to an invertible matrix by adding a suitable number of new rows.*

Proof. "$\Rightarrow$" Write $MM' = I_r$ with $M' \in \mathbf{R}^{n \times r}$. As $\mathbf{R}^n = \mathrm{Ker}\, M \oplus \mathrm{Im}\, M'$, considering a matrix S' whose columns form a basis of $\ker M$, the matrix $A' = (S' \ M')$ has as columns a basis of $\mathbf{R}^n$ and, thus, it is invertible. Necessarily, its inverse has the form $\begin{pmatrix} S \\ M \end{pmatrix}$. "$\Leftarrow$" Let $A = \begin{pmatrix} S \\ M \end{pmatrix} \in \mathrm{GL}_n(\mathbf{R})$ and denote by $A' = A^{-1}$. Writing A' in the form $A' = (S' \ M')$, as $MS' = 0_{r, n-r}$ and $MM' = I_r$, we have $\mathrm{Im}\, S' \subseteq \mathrm{Ker}\, M$, $\mathrm{Ker}\, M \oplus \mathrm{Im}\, M' = \mathbf{R}^n = \mathrm{Im}\, S' \oplus \mathrm{Im}\, M'$, and hence $\mathrm{Ker}\, M = \mathrm{Im}\, S'$ is free. $\qquad \square$

Definition 19. Let $b_1, \ldots, b_n \in \mathbf{R}$. Recall that a row $(b_1, \ldots, b_n)$ is said to be *unimodular* (or that ${}^t(b_1, \ldots, b_n)$ is a unimodular vector) if the row matrix $(b_1, \ldots, b_n)$ is right invertible, i.e., if $\langle b_1, \ldots, b_n \rangle = \mathbf{R}$. The set of such unimodular rows will be denoted by $\mathrm{Um}_n(\mathbf{R})$ (in order to lighten the notation, we use the same notation for unimodular vectors).

If a unimodular row over $\mathbf{R}$ can be completed to an invertible matrix (i.e., can be written as the first row of an invertible matrix with entries in $\mathbf{R}$), we say that it is *completable* over $\mathbf{R}$. For example, every unimodular row (a, b) of length 2 is completable. As a matter of fact, writing $ac + bd = 1$, the matrix $\begin{pmatrix} a & b \\ -d & c \end{pmatrix}$ has determinant 1.

The following gives a criterion for the freeness of all finitely-generated stably free modules over a ring $\mathbf{R}$ in terms of unimodular rows. It is a consequence of Proposition 18.

Proposition 20. *For any ring $\mathbf{R}$, the following are equivalent:*

 (i) *Any stably free module is free.*

 (ii) *Any unimodular row over $\mathbf{R}$ is completable.*

Definition 21. Rings satisfying the above equivalent properties will be called *Hermite rings*.

The following proposition gives a more precise formulation of Proposition 20.

Proposition 22. *For any ring $\mathbf{R}$ and integer $d \geq 0$, the following are equivalent:*

 (i) *Any finitely-generated stably free module of rank $> d$ is free.*

 (ii) *Any unimodular row over $\mathbf{R}$ of length $q \geq d + 2$ is completable.*

 (iii) *For any unimodular row v over $\mathbf{R}$ of length $q \geq d + 2$, there exists $G \in \mathrm{GL}_q(\mathbf{R})$ such that $vG = (1, 0, \ldots, 0)$.*

 (iv) *For $q \geq d + 2$, $\mathrm{GL}_q(\mathbf{R})$ acts transitively on $\mathrm{Um}_q(\mathbf{R})$.*

 (v) *For any unimodular row v over $\mathbf{R}$ of length $q \geq d + 2$, we have $\mathbf{R}^q \cong \mathbf{R}v \oplus \mathbf{R}^{q-1}$.*

Proof. Assertions (iii) and (iv) as well as assertions (ii) and (v) are obviously equivalent. Assertions (iii) and (ii) are equivalent since for a unimodular row $v \in \mathbf{R}^{1 \times q}$, saying that there exists $G \in \mathrm{GL}_q(\mathbf{R})$ such that $vG = (1, 0, \ldots, 0)$ amounts to saying that v is the first row of G^{-1}.

"(i) $\Rightarrow$ (v)" Let v be a unimodular row $v \in \mathbf{R}^{1 \times q}$ with $q \geq d + 2$. Then, by Proposition 2, we can write $\mathbf{R}^q = \mathbf{R}v \oplus M$, where M is stably free of rank $q - 1 > d$, and, thus, free.

"(v) $\Rightarrow$ (i)" Let M be a stably free $\mathbf{R}$-module of rank $n > d$. We can write $N = M \oplus \mathbf{R}v_1 \oplus \cdots \oplus \mathbf{R}v_r$, where $N \cong \mathbf{R}^{n+r}$. If $r = 0$, there is nothing to do. Else, v_r is a unimodular row in N. By hypothesis, $L/\mathbf{R}v_r \cong \mathbf{R}^{n+r-1}$, and, thus, $M \oplus \mathbf{R}v_1 \oplus \cdots \oplus \mathbf{R}v_{r-1} \cong \mathbf{R}^{n+r-1}$. The result follows by induction on r.

$\square$

In fact, when studying finitely-generated stably free modules, one has only to care about stably free modules of rank ≥ 2, since as will be seen in Theorem 25, stably free of rank 1 are free.

Notation 23. Let $\mathbf{R}$ be ring and $A \in \mathbf{R}^{n \times m}$ an $n \times m$ matrix with entries in $\mathbf{R}$. Denote by $A_1, \ldots, A_m$ the columns of A, so that we can write $A = [A_1, \ldots, A_m]$. If $I = (i_1, \ldots, i_r)$ is a sequence of natural numbers with $1 \leq i_1 < \cdots < i_r \leq m$, we denote by A_I the matrix $[A_{i_1}, \ldots, A_{i_r}]$.

Binet-Cauchy Formula 24. *Let $\mathbf{R}$ be ring and consider two matrices $M \in \mathbf{R}^{s \times r}$ and $N \in \mathbf{R}^{r \times s}$, $r \leq s$. Then*

$$\det(MN) = \sum_I \det(M_I)\det(N_I),$$

where I runs through all sequences of natural numbers $(i_1, \ldots, i_r)$ with $1 \leq i_1 < \cdots < i_r \leq s$.

Theorem 25. *For any ring $\mathbf{R}$, any stably free $\mathbf{R}$-module of rank 1 is free ($\cong \mathbf{R}$).*

Proof. Let P be a stably free $\mathbf{R}$-module of rank 1 (i.e., $P \oplus \mathbf{R}^{n-1} \cong \mathbf{R}^n$ for some $n \geq 2$) represented as the solution space of a right invertible $(n-1) \times n$ matrix M. That is, $P = \mathrm{Ker}\,M$ and $\exists N \in \mathbf{R}^{n \times (n-1)}$ such that $MN = \mathrm{I}_{n-1}$. Proving that P is free is nothing else than proving that M can be completed to an invertible matrix (see Proposition 18). This clearly amounts to proving that the maximal minors $b_1, \ldots, b_n$ are comaximal, i.e., $1 \in \langle b_1, \ldots, b_n \rangle$. As a matter of fact, if $a_1 b_1 + \cdots + a_n b_n = 1$ then M can be completed to a matrix of determinant 1 by adding a last row $[a_1, \ldots, a_n]$ with appropriate signs. Thus, our task is reduced to prove that $1 \in \langle b_1, \ldots, b_n \rangle$.

Classical approach: Let $\mathfrak{m}$ be a maximal ideal of $\mathbf{R}$. Then, modulo $\mathfrak{m}$, we have $\bar{M}\bar{N} = \mathrm{I}_{n-1}$. Since $\bar{M}$ is right invertible, it has rank $n-1$ and can be completed by linear algebra to an invertible matrix $M_{\mathfrak{m}} \in \mathrm{GL}_n(\mathbf{R}/\mathfrak{m})$. Thus, $\det M_{\mathfrak{m}} \neq \bar{0}$ and a fortiori $\langle b_1, \ldots, b_n \rangle \nsubseteq \mathfrak{m}$.

Constructive approach: Reasoning modulo $\langle b_1, \ldots, b_n \rangle$, the fact that $\bar{M}\bar{N} = \mathrm{I}_{n-1}$ together with the Binet-Cauchy Formula 24 give that $\bar{1} = \bar{0}$. Thus, $1 \in \langle b_1, \ldots, b_n \rangle$. This an example of a nontrivial use of trivial rings [150]. $\qquad\square$

Remark 26. It is worth pointing out that there is no analogue to Theorem 25 for projective modules. As a matter of fact, for any domain $\mathbf{R}$, all finitely-generated projective $\mathbf{R}[X]$-modules of rank one are extended from $\mathbf{R}$ if and only if $\mathbf{R}$ is semi-normal, that is, each time $b^2 = c^3$ in $\mathbf{R}$, there exists $a \in \mathbf{R}$ such that $a^3 = b$ and $a^2 = c$ (this is the Traverso-Swan theorem which has been treated recently constructively by Coquand [32] followed by Lombardi and Quitté [107] and also by Barhoumi and Lombardi [8, 9]; see Exercise 381). If $\mathbf{R}$ is a domain which is not seminormal then one can explicitly construct a rank one projective $\mathbf{R}$-module which is not free (see Schanuel's example which will be given in Question (6.b) of Exercise 381).

Definition and Proposition 27. (Finite Free Resolution) Let **R** be a ring.

(1) A *complex* $\mathscr{F}$ of **R**-modules is a sequence of modules F_i and maps $\varphi_i : F_i \to F_{i-1}$ such that $\varphi_i \circ \varphi_{i+1} = 0$ for all i. The module $H_i := \mathrm{Ker}(\varphi_i)/\mathrm{Im}(\varphi_{i+1})$ is called the *homology* of this complex at F_i.

If $H_i = 0$ for all i, we say the complex $\mathscr{F}$ is *exact*. For example, if U is a submodule of a module M then the complex $0 \to U \xrightarrow{i} M \xrightarrow{\pi} M/U \longrightarrow 0$ is exact, where i is inclusion and π is the canonical projection.

For $a \in \mathbf{R}$, the homology of the complex $0 \longrightarrow \mathbf{R} \xrightarrow{\varphi_a} \mathbf{R}$ (called Koszul complex of length 1), where $\varphi_a(x) = ax$, is the annihilator $\mathrm{Ann}(a) := \{b \in \mathbf{R} \mid ba = 0\}$ of a.

A *finite free resolution* of length n of a module M is a complex

$$0 \longrightarrow \mathbf{R}^{r_n} \xrightarrow{\varphi_n} \cdots \xrightarrow{\varphi_2} \mathbf{R}^{r_1} \xrightarrow{\varphi_1} \mathbf{R}^{r_0} \longrightarrow 0$$

which is exact except at $\mathbf{R}^{r_0}$ and such that $M = \mathrm{Coker}(\varphi_1)$ and $r_i \in \mathbb{N}^*$.

(2) **Hilbert Syzygy Theorem:** *If **K** is a field then every finitely-generated module over* $\mathbf{K}[X_1, \ldots, X_k]$ *has a finite free resolution.*

Note that such finite-free resolution can be computed effectively using Gröbner bases [15, 53, 122].

(3) **Serre's Theorem:** *Any projective module with a finite free resolution is stably free.*

To see this, if

$$0 \longrightarrow \mathbf{R}^{r_n} \xrightarrow{\varphi_n} \cdots \xrightarrow{\varphi_2} \mathbf{R}^{r_1} \xrightarrow{\varphi_1} \mathbf{R}^{r_0} \xrightarrow{\varphi_0} P \longrightarrow 0$$

is a free resolution of the projective module P, then

$$P \text{ projective} \Rightarrow \mathrm{Ker}(\varphi_0) \text{ projective} \Rightarrow \mathrm{Im}(\varphi_1) = \mathrm{Ker}(\varphi_0) \text{ projective} \Rightarrow \cdots$$

$$P \oplus \mathbf{R}^{r_1} \oplus \mathbf{R}^{r_3} \oplus \cdots \cong \mathbf{R}^{r_0} \oplus \mathbf{R}^{r_2} \oplus \cdots \quad \& \quad \mathrm{rk}(P) = \sum_{i=0}^{n} (-1)^i r_i.$$

(4) Combining (2) and (3) we get that:

*If **K** is a field then every finitely-generated projective module over* $\mathbf{K}[X_1, \ldots, X_k]$ *is stably free.*

Readers interested in new constructive techniques in finite free resolutions can refer to the nice paper [36] in which Coquand and Quitté greatly simplify the main proofs given in Northcott's book [126] "Finite Free Resolutions" without any use of minimal prime ideals. It is well-known that the existence of minimal prime ideals is equivalent to the axiom of choice.

2.1.3 Concrete Local-Global Principle

We explain here how the constructive deciphering of classical proofs in commutative algebra using a local-global principle works. This section comes essentially from [106].

2.1.3.1 From Local to Quasi-Global

The classical reasoning by localization works as follows. When the ring is local, a property P is satisfied by virtue of a quite concrete proof. When the ring is not local, the same property remains true (from a classical nonconstructive point of view) as it suffices to check it locally.

When carefully examining the first proof, some computations come into view. These computations are feasible thanks to the following principle:

$$\forall x \in \mathbf{R}, \quad x \in \mathbf{R}^{\times} \lor x \in \text{Rad}(\mathbf{R}).$$

This principle is in fact applied to elements coming from the proof itself. In case of a nonnecessarily local ring, we repeat the same proof, replacing at each disjunction "x is a unit or x is in the radical" in the passage of the proof we are considering, by the consideration of two rings $\mathbf{T}_x := \mathbf{T}[\frac{1}{x}]$ and $\mathbf{T}_{1+x\mathbf{T}}$ (the localization of $\mathbf{T}$ at the monoid $1 + x\mathbf{T}$), where $\mathbf{T}$ is the "current" localization of the ring $\mathbf{R}$ we start with. When the initial proof is completely unrolled, we obtain a finite number (since the proof is finite) of localizations $\mathbf{R}_{S_i}$, for each of them the property is true. Moreover, the corresponding Zariski open subsets U_{S_i} cover $\text{Spec}(\mathbf{R})$ implying that the property P is true for $\mathbf{A}$, and this time in an entirely explicit way.

It is worth pointing out that, in order to roll out the method described above, one needs Lemma 34 which guarantees that an element remains in the radical once it is forced into being in.

Definition 28. (Constructive Definition of the Radical)
Recall that a ring $\mathbf{R}$ is said to be *discrete* if there is an algorithm deciding if $x = 0$ or $x \neq 0$ for an arbitrary element of $\mathbf{R}$.

Constructively, the *radical* $\text{Rad}(\mathbf{R})$ of a ring $\mathbf{R}$ is the set of all $x \in \mathbf{R}$ such that $1 + x\mathbf{R} \subseteq \mathbf{R}^{\times}$, where $\mathbf{R}^{\times}$ is the group of units of $\mathbf{R}$. A ring $\mathbf{R}$ is *local* if it satisfies:

$$\forall x \in \mathbf{R}, \quad x \in \mathbf{R}^{\times} \lor 1 + x \in \mathbf{R}^{\times}. \tag{2.1}$$

It is *residually discrete local* if it satisfies:

$$\forall x \in \mathbf{R}, \quad x \in \mathbf{R}^{\times} \lor x \in \text{Rad}(\mathbf{R}). \tag{2.2}$$

From a classical point of view, we have (2.1) ⇔ (2.2), but the constructive meaning of (2.2) is stronger than that of (2.1). Constructively a *discrete field* is defined as a ring in which each element is zero or invertible, with an explicit test for the "or". A *Heyting field* (or a field) is defined as a local ring whose Jacobson radical is 0. So $\mathbf{R}$ is residually discrete local exactly when it is local and the residue field $\mathbf{R}/\text{Rad}(\mathbf{R})$ is a discrete field.

Definition 29. (Monoids and Saturations)

(i) We say that S is a *multiplicative subset* (or a *monoid*) of a ring $\mathbf{R}$ if

$$\begin{cases} 1 \in S \\ \forall\, s,t \in S,\ st \in S. \end{cases}$$

For example, for $a \in \mathbf{R}$, $a^{\mathbb{N}} := \{a^n;\ n \in \mathbb{N}\}$ is a monoid of $\mathbf{R}$.

(ii) A monoid S of a ring $\mathbf{R}$ is said to be *saturated* if we have the implication

$$\forall\, s,t \in \mathbf{R},\ (st \in S \Rightarrow s \in S).$$

(iii) The localization of $\mathbf{R}$ at S will be denoted by $S^{-1}\mathbf{R}$ or $\mathbf{R}_S$. If S is generated by $s \in \mathbf{R}$, we denote $\mathbf{R}_S$ by $\mathbf{R}_s$ or $\mathbf{R}[1/s]$. Note here that $\mathbf{R}_s$ is isomorphic to the ring $\mathbf{R}[T]/(sT - 1)$. Saturating a monoid S (that is, replacing S by its saturation $\bar{S} := \{s \in \mathbf{R},\ \exists\, t \in \mathbf{R} \mid st \in S\}$) does not change the localization $\mathbf{R}_S$. Two monoids are said to be *equivalent* if they have the same saturation.

We keep the same notation for the localization of an $\mathbf{R}$-module.

Definition 30. (Comaximal Elements and Monoids)   Let $\mathbf{R}$ be a ring.

(1) Let $s_1, \ldots, s_k \in \mathbf{R}$. We say that the elements $s_1, \ldots, s_k$ are *comaximal* if $\langle s_1, \ldots, s_k \rangle = \mathbf{R}$.

(2) Let $S, S_1, \ldots, S_n$ be monoids of $\mathbf{R}$.

(i) We say that the monoids $S_1, \ldots, S_n$ are *comaximal* if any ideal of $\mathbf{R}$ meeting all the S_i must contain 1. In other words, if we have:

$$\forall s_1 \in S_1, \ldots,\ \forall s_n \in S_n,\ \exists\, a_1, \ldots, a_n \in \mathbf{R} \quad \mid \quad \sum_{i=1}^{n} a_i s_i = 1,$$

that is, $s_1, \ldots, s_k$ are comaximal elements in $\mathbf{R}$.

For example, if $u_1, \ldots, u_m$ are comaximal elements in $\mathbf{R}$, then the monoids $u_1^{\mathbb{N}}, \ldots, u_m^{\mathbb{N}}$ are comaximal.

(ii) We say that the monoids $S_1, \ldots, S_n$ *cover* the monoid S if S is contained in the S_i and any ideal of $\mathbf{R}$ meeting all the S_i must meet S. In other words, if we have:

$$\forall s_1 \in S_1 \cdots \forall s_n \in S_n\ \exists\, a_1, \ldots, a_n \in \mathbf{R} \quad \mid \quad \sum_{i=1}^{n} a_i s_i \in S.$$

Remark that comaximal monoids remain comaximal when you replace the ring by a bigger one or the multiplicative subsets by smaller ones.

In classical algebra (with the axiom of the prime ideal) this amounts to saying, in the first case, that the Zariski open subsets U_{S_i} cover $\mathrm{Spec}(\mathbf{R})$ and, in the second case, that the Zariski open subsets U_{S_i} cover the open subset U_S. From a constructive point of view, $\mathrm{Spec}(\mathbf{R})$ is a topological space via its open subsets U_S but whose points are often hardly accessible.

We have the following immediate result.

Lemma 31. (Associativity and Transitivity of Coverings)

(1) *(Associativity) If monoids $S_1,\ldots,S_n$ of a ring $\mathbf{R}$ cover a monoid S and each S_ℓ is covered by some monoids $S_{\ell,1},\ldots,S_{\ell,m_\ell}$, then the $S_{\ell,j}$'s cover S.*

(2) *(Transitivity) Let S be a monoid of a ring $\mathbf{R}$ and $S_1,\ldots,S_n$ monoids of the ring $\mathbf{R}_S$. For $\ell=1,\ldots,n$, let V_ℓ be the monoid of $\mathbf{R}$ formed by the denominators of the elements of S_ℓ. Then the monoids $V_1,\ldots,V_n$ cover S.*

Definition and notation 32. Let I and U two subsets of a ring $\mathbf{R}$. We denote by $\mathscr{M}(U)$ the monoid generated by U, $\mathscr{I}_\mathbf{R}(I)$ or $\mathscr{I}(I)$ the ideal generated by I and $\mathscr{S}(I;U)$ the monoid $\mathscr{M}(U)+\mathscr{I}(I)$. If $I=\{a_1,\ldots,a_k\}$ and $U=\{u_1,\ldots,u_\ell\}$, we denote $\mathscr{M}(U)$, $\mathscr{I}(I)$, and $\mathscr{S}(I;U)$ by $\mathscr{M}(u_1,\ldots,u_\ell)$, $\mathscr{I}(a_1,\ldots,a_k)$, and $\mathscr{S}(a_1,\ldots,a_k;u_1,\ldots,u_\ell)$, respectively.

Remark 33. (1) It is clear that if u is equal to a product $u_1\cdots u_\ell$, then the monoids $\mathscr{S}(a_1,\ldots,a_k;u_1,\ldots,u_\ell)$ and $\mathscr{S}(a_1,\ldots,a_k;u)$ are equivalent.

(2) When we localize at $S=\mathscr{S}(I;U)$, the elements of U are forced into being invertible and those of I end up on the radical of $\mathbf{R}_S$.

Accordingly to Henri Lombardi, the "good category" would be that whose objects are couples $(\mathbf{R},I)$, where $\mathbf{R}$ is a commutative ring and I is an ideal contained in the radical of $\mathbf{R}$. Arrows from $(\mathbf{R},I)$ onto $(\mathbf{R}',I')$ are rings homomorphisms $f:\mathbf{R}\to\mathbf{R}'$ such that $f(I)\subseteq I'$. Thus, one can retrieve usual rings by taking $I=0$ and local rings (equipped with the notion of local homomorphism) by taking I equal to the maximal ideal. In order to "localize" an object $(\mathbf{A},I)$ in this category, we use a monoid U and an ideal J in such a way we form the new object $(\mathbf{R}_{\mathscr{S}(J_1;U)},J_1\mathbf{R}_{\mathscr{S}(J_1;U)})$, where $J_1=I+J$.

The following lemma will play a crucial role when we want to reread constructively with an arbitrary ring a proof given in the local case.

Lemma 34. (Lombardi's Trick [98, 101]) *Let U and I be two subsets of a ring $\mathbf{R}$ and consider $a\in\mathbf{R}$. Then the monoids $\mathscr{S}(I;U,a)$ and $\mathscr{S}(I,a;U)$ cover the monoid $\mathscr{S}(I;U)$.*

Proof. For $x\in\mathscr{S}(I;U,a)$ and $y\in\mathscr{S}(I,a;U)$, we have to find a linear combination of the form $x_1x+y_1y\in\mathscr{S}(I;U)$ ($x_1,y_1\in\mathbf{R}$). Write $x=u_1a^k+j_1$, $y=(u_2+j_2)-$

(az) with $u_1, u_2 \in \mathcal{M}(U)$, $j_1, j_2 \in \mathcal{I}(I)$, $z \in \mathbf{R}$. The classical identity $c^k - d^k = (c-d) \times \cdots$ gives a $y_2 \in \mathbf{A}$ such that $y_2 y = (u_2 + j_2)^k - (az)^k = (u_2^k + j_3) - (az)^k$. Just write $z^k x + u_1 y_2 y = u_1 u_2^k + u_1 j_3 + j_1 z^k = u_4 + j_4$. □

It is worth pointing out that, in the lemma above, we have

$$a \in (\mathbf{R}_{\mathcal{S}(I;U,a)})^\times \text{ and } a \in \mathrm{Rad}(\mathbf{R}_{\mathcal{S}(I,a;U)}).$$

Having this lemma in hands, we can state the following general deciphering principle allowing to automatically get a quasi-global version of a theorem from its local version.

General Local-Global Principle 35. (Lombardi [101]) *When rereading an explicit proof given in case $\mathbf{R}$ is local, with an arbitrary ring $\mathbf{R}$, start with $\mathbf{R} = \mathbf{R}_{\mathcal{S}(0;1)}$. Then, at each disjunction (for an element a produced when computing in the local case)*

$$a \in \mathbf{R}^\times \lor a \in \mathrm{Rad}(\mathbf{R}),$$

replace the "current" ring $\mathbf{R}_{\mathcal{S}(I;U)}$ by both $\mathbf{R}_{\mathcal{S}(I;U,a)}$ and $\mathbf{R}_{\mathcal{S}(I,a;U)}$ in which the computations can be pursued. At the end of this rereading, one obtains a finite family of rings $\mathbf{R}_{\mathcal{S}(I_j;U_j)}$ with comaximal monoids $\mathcal{S}(I_j;U_j)$ and finite sets I_j, U_j.

The following examples are frequent and ensue immediately from Lemmas 31 and 34, except the first one which is an easy exercise.

Examples 36. Let $\mathbf{R}$ be a ring, U and I subsets of $\mathbf{R}$, and $S = \mathcal{S}(I;U)$.

(1) Let $s_1, \ldots, s_n \in \mathbf{R}$ be comaximal elements. Then the monoids $S_i = \mathcal{M}(s_i) = s_i^{\mathbb{N}}$ are comaximal.
More generally, if $t_1, \ldots, t_n \in \mathbf{R}$ are comaximal elements in $\mathbf{R}_S$, then the monoids $\mathcal{S}(I;U,t_i)$ cover the monoid S.

(2) Let $s_1, \ldots, s_n \in \mathbf{R}$. The monoids $S_1 = \mathcal{S}(0;s_1)$, $S_2 = \mathcal{S}(s_1;s_2)$, $S_3 = \mathcal{S}(s_1,s_2;s_3)$, ..., $S_n = \mathcal{S}(s_1,\ldots,s_{n-1};s_n)$ and $S_{n+1} = \mathcal{S}(s_1,\ldots,s_n;1)$ are comaximal.
More generally, the monoids

$$V_1 = \mathcal{S}(I;U,s_1), V_2 = \mathcal{S}(I,s_1;U,s_2), V_3 = \mathcal{S}(I,s_1,s_2;U,s_3),\ldots,$$
$$V_n = \mathcal{S}(I,s_1,\ldots,s_{n-1};U,s_n), V_{n+1} = \mathcal{S}(I,s_1,\ldots,s_n;U)$$

cover the monoid S.

(3) If $S, S_1, \ldots, S_n \subseteq \mathbf{R}$ are comaximal monoids and if $b = \frac{a}{s} \in \mathbf{R}_S$, then the monoids

$$\mathcal{S}(I;U,a), \mathcal{S}(I,a;U), S_1, \ldots, S_n$$

are comaximal.

2.1.3.2 From Quasi-Global to Global

Different variant versions of the abstract local-global principle in commutative algebra can be reread constructively: the localization at each prime ideal is replaced by the localization at a finite family of comaximal monoids. In other words, in these "concrete" versions, we affirm that some properties pass from the quasi-global to the global. As an illustration, we cite the following results which often permit to finish our constructive rereading.

Concrete Local-Global Principle 37. *Let* $S_1, \ldots, S_n$ *be comaximal monoids in a ring* $\mathbf{R}$ *and let* $a, b \in \mathbf{R}$. *Then we have the following equivalences:*

(1) *Concrete gluing of equalities:*
$$a = b \text{ in } \mathbf{R} \quad \Longleftrightarrow \quad \forall i \in \{1, \ldots, n\}, \ a/1 = b/1 \text{ in } \mathbf{R}_{S_i}.$$

(2) *Concrete gluing of non zero-divisors:*
$$a \text{ is not a zero-divisor in } \mathbf{R} \quad \Longleftrightarrow$$
$$\forall i \in \{1, \ldots, n\}, \ a/1 \text{ is not a zero-divisor in } \mathbf{R}_{S_i}.$$

(3) *Concrete gluing of units:*
$$a \text{ is a unit in } \mathbf{R} \quad \Longleftrightarrow$$
$$\forall i \in \{1, \ldots, n\}, \ a/1 \text{ is a unit in } \mathbf{R}_{S_i}.$$

(4) *Concrete gluing of solutions of linear systems: let* B *be a matrix* $\in \mathbf{R}^{m \times p}$ *and* C *a column vector* $\in \mathbf{R}^{m \times 1}$.
$$\text{The linear system } BX = C \text{ has a solution in } \mathbf{R}^{p \times 1} \quad \Longleftrightarrow$$
$$\forall i \in \{1, \ldots, n\}, \text{ the linear system } BX = C \text{ has a solution in } \mathbf{R}_{S_i}^{p \times 1}.$$

(5) *Concrete gluing of direct summands: let* M *be a finitely-generated submodule of a finitely-presented module* N.
$$M \text{ is a direct summand of } N \quad \Longleftrightarrow$$
$$\forall i \in \{1, \ldots, n\}, \ M_{S_i} \text{ is a direct summand of } N_{S_i}.$$

Concrete Local-Global Principle 38. (Concrete Gluing of Module Finiteness Properties) *Let* $s_1, \ldots s_n$ *be comaximal elements of a ring* $\mathbf{R}$, *and let* M *be an* $\mathbf{R}$-*module. Then we have the following equivalences:*

(1) M *is finitely-generated if and only if each of the* M_{s_i} *is a finitely-generated* $\mathbf{R}_{s_i}$-*module.*

(2) M *is finitely-presented if and only if each of the* M_{s_i} *is a finitely-presented* $\mathbf{R}_{s_i}$-*module.*

(3) M *is flat if and only if each of the* M_{s_i} *is a flat* $\mathbf{R}_{s_i}$-*module.*

(4) M *is a finitely-generated projective module if and only if each of the* M_{s_i} *is a finitely-generated projective* $\mathbf{R}_{s_i}$-*module.*

(5) *M is projective of rank k if and only if each of the M_{s_i} is a projective $\mathbf{R}_{s_i}$-module of rank k.*

(6) *M is coherent if and only if each of the M_{s_i} is a coherent $\mathbf{R}_{s_i}$-module.*

(7) *M is Noetherian if and only if each of the M_{s_i} is a Noetherian $\mathbf{R}_{s_i}$-module.*

One can rarely find such principles in classical literature. In Quillen's style, the corresponding general principle is in general stated using localizations at all prime ideals, but the proof often brings in a crucial lemma which has exactly the same signification as the corresponding concrete local-global principle. For example, we can state the Concrete Local-Global Principle 38 "à la Quillen" under the following form.

Lemma 39. (Propagation Lemma for Some Module Finiteness Properties) *Let M be an $\mathbf{R}$-module. The following subsets I_k of $\mathbf{R}$ are ideals.*

(1) $I_1 = \{\, s \in \mathbf{R} \,:\, M_s \text{ is a finitely-generated } \mathbf{R}_s\text{-module}\,\}$.

(2) $I_2 = \{\, s \in \mathbf{R} \,:\, M_s \text{ is a finitely-presented } \mathbf{R}_s\text{-module}\,\}$.

(3) $I_3 = \{\, s \in \mathbf{R} \,:\, M_s \text{ is a flat } \mathbf{R}_s\text{-module}\,\}$.

(4) $I_4 = \{\, s \in \mathbf{R} \,:\, M_s \text{ is a finitely-generated projective } \mathbf{R}_s\text{-module}\,\}$.

(5) $I_5 = \{\, s \in \mathbf{R} \,:\, M_s \text{ is a rank } k \text{ projective } \mathbf{R}_s\text{-module}\,\}$.

(6) $I_6 = \{\, s \in \mathbf{R} \,:\, M_s \text{ is a coherent } \mathbf{R}_s\text{-module}\,\}$.

(7) $I_7 = \{\, s \in \mathbf{R} \,:\, M_s \text{ is a Noetherian } \mathbf{R}_s\text{-module}\,\}$.

Remark 40. In general, letting P be a property which is stable under localization, then the following version of the concrete local-global principle:

- for each ring $\mathbf{R}$, if P is true after localizations at comaximal elements of $\mathbf{R}$, then it is true in $\mathbf{R}$,

and its *propagation lemma* version:

- the set $I_P = \{\, s \in \mathbf{R} \,:\, P \text{ is true in } \mathbf{R}_s \,\}$ is an ideal of $\mathbf{R}$,

are equivalent. On the one hand, the propagation lemma version clearly implies the first one. On the other hand, for the converse, if $s, s' \in I_P$ and $t = s + s'$ then $s/1$ and $s'/1$ are comaximal elements of $\mathbf{A}_t$ and P is true in both $(\mathbf{R}_t)_s \simeq (\mathbf{R}_s)_t \simeq \mathbf{R}_{st}$ and $(\mathbf{R}_t)_{s'} \simeq (\mathbf{R}_{s'})_t \simeq \mathbf{R}_{s't}$. Thus, P is true in $\mathbf{A}_t$ by the concrete local-global principle. It is worth pointing out that, in general, for a monoid S, the following implication

- P is true in $\mathbf{R}_S \Rightarrow P$ is true in $\mathbf{R}_s$ for some $s \in S$, is not always true. A property P is said to be *finite-type* if it is stable under localization and satisfies the above-mentioned implication. If P is finite-type then there is an equivalence between the *concrete local-global principle for comaximal elements* and the *concrete local-global principle for comaximal monoids*. This is in general indispensable since the explained rereading system (General Principle 35) naturally produces a local-global version with comaximal monoids rather than with comaximal elements.

For example, for a finitely-presented **R**-module M, the property "*M* is a finitely-generated projective **R**-module" is finite-type. Or, also, for a finitely-presented **R**[X]-module M, the property "*M* is extended from **R**" is finite-type (see Quillen's patching Theorem 45).

It is worth pointing out that the localization at comaximal monoids corresponds in classical literature to the localization in the neighborhood of each prime ideal [148].

2.1.4 The Patchings of Quillen and Vaserstein

We give here a detailed constructive proof by Lombardi and Quitté of the Quillen patching. This comes essentially from [90]. The localization at maximal ideals is replaced by localization at comaximal monoids.

In [112], the constructive Quillen patching (Concrete local-global Principle 4) is given with only a sketch of proof.

Lemma 41. *Let S be a multiplicative subset of a ring* **R** *and consider three matrices* A_1, A_2, A_3 *with entries in* **R**[X] *such that the product* A_1A_2 *is defined and has the same size as* A_3. *If* $A_1A_2 = A_3$ *in* $\mathbf{R}_S[X]$ *and* $A_1(0)A_2(0) = A_3(0)$ *in* **R**, *then there exists* $s \in S$ *such that* $A_1(sX)A_2(sX) = A_3(sX)$ *in* **R**[X].

Proof. All the coefficients of the matrix $A_1A_2 - A_3$ are multiple of X and become zero after localization at S. Thus, there exists $s \in S$ annihilating all of them. Write $A_1A_2 - A_3 = B(X) = XB_1 + X^2B_2 + \cdots + X^kB_k$. We have $sB_1 = sB_2 = \cdots = sB_k = 0$ and, thus, $sB_1 = s^2B_2 = \cdots = s^kB_k = 0$, that is, $B(sX) = A_1(sX)A_2(sX) - A_3(sX) = 0$. □

Lemma 42. *Let S be a multiplicative subset of a ring* **R** *and consider a matrix* $C(X) \in \mathrm{GL}_r(\mathbf{R}_S[X])$. *Then there exists* $s \in S$ *and* $U(X,Y) \in \mathrm{GL}_r(\mathbf{R}[X,Y])$ *such that* $U(X,0) = \mathrm{I}_r$, *and, over* $\mathbf{R}_S[X,Y]$, $U(X,Y) = C(X + sY)C(X)^{-1}$.

Proof. Set $E(X,Y) = C(X+Y)C(X)^{-1}$ and denote $F(X,Y)$ the inverse of $E(X,Y)$. We have $E(X,0) = \mathrm{I}_r$ and, thus, $E(X,Y) = \mathrm{I}_r + E_1(X)Y + \cdots + E_k(X)Y^k$. For some $s_1 \in S$, the $s_1^j E_j$'s can be written without denominators and, thus, we obtain a matrix $E'(X,Y) \in \mathbf{R}[X,Y]^{r \times r}$ such that $E'(X,0) = \mathrm{I}_r$, and, over $\mathbf{R}_S[X,Y]$, $E'(X,Y) = E(X,s_1Y)$. We do the same with F (we can choose the same s_1). Hence we obtain $E'(X,Y)F'(X,Y) = \mathrm{I}_r$ in $\mathbf{R}_S[X,Y]^{r \times r}$ and $E'(X,0)F'(X,0) = \mathrm{I}_r$. Applying Lemma 41 in which we replace X by Y and **R** by **R**[X], we obtain $s_2 \in S$ such that $E'(X,s_2Y)F'(X,s_2Y) = \mathrm{I}_r$. Taking $U = E'(X,s_2Y)$ and $s = s_1s_2$, we obtain the desired result. □

Lemma 43. *Let S be a multiplicative subset of a ring* $\mathbf{R}$ *and* $M \in \mathbf{R}[X]^{p \times q}$. *If* $M(X)$ *and* $M(0)$ *are equivalent over* $\mathbf{R}_S[X]$ *then there exists* $s \in S$ *such that* $M(X + sY)$ *and* $M(X)$ *are equivalent over* $\mathbf{R}[X, Y]$.

Proof. Writing $M(X) = C(X)M(0)D(X)$ with $C(X) \in \mathrm{GL}_q(\mathbf{R}_S[X])$ and $D(X) \in \mathrm{GL}_p(\mathbf{R}_S[X])$, we get

$$M(X + Y) = C(X + Y)C(X)^{-1}M(X)D(X)^{-1}D(X + Y).$$

Applying Lemma 42, we find $s_1 \in S$, $U(X, Y) \in \mathrm{GL}_q(\mathbf{R}[X, Y])$ and $V(X, Y) \in \mathrm{GL}_p(\mathbf{R}[X, Y])$ such that $U(X, 0) = \mathrm{I}_q$, $V(X, 0) = \mathrm{I}_p$, and, over $\mathbf{R}_S[X, Y]$, $U(X, Y) = C(X + s_1 Y)C(X)^{-1}$ and $V(X, Y) = D(X)^{-1}D(X + s_1 Y)$. It follows that $M(X) = U(X, 0)M(X)V(X, 0)$, and over $\mathbf{R}_S[X, Y]$, $M(X + s_1 Y) = U(X, Y)M(X)V(X, Y)$. Applying Lemma 41 (as in Lemma 42), we get $s_2 \in S$ such that $M(X + s_1 s_2 Y) = U(X, s_2 Y)M(X)V(X, s_2 Y)$. The desired result is obtained by taking $s = s_1 s_2$. $\quad\square$

Theorem 44. (Vaserstein) *Let M be a matrix in* $\mathbf{R}[X]$ *and consider* $S_1, \ldots, S_n$ *comaximal multiplicative subsets of* $\mathbf{R}$. *Then* $M(X)$ *and* $M(0)$ *are equivalent over* $\mathbf{R}[X]$ *if and only if, for each* $1 \le i \le n$, *they are equivalent over* $\mathbf{R}_{S_i}[X]$.

Proof. It is easy to see that the set of $s \in \mathbf{R}$ such that $M(X + sY)$ is equivalent to $M(X)$ is an ideal of $\mathbf{R}$. Applying Lemma 43, this ideal meets S_i for each $1 \le i \le n$, and, thus, contains 1. This means that $M(X + Y)$ is equivalent to $M(X)$. To finish, just take $X = 0$. $\quad\square$

Theorem 45. (Quillen's Patching) *Let P be a finitely-presented module over* $\mathbf{R}[X]$ *and consider* $S_1, \ldots, S_n$ *comaximal multiplicative subsets of* $\mathbf{R}$. *Then P is extended from* $\mathbf{R}$ *if and only if for each* $1 \le i \le n$, P_{S_i} *is extended from* $\mathbf{R}_{S_i}$.

Proof. This is a corollary of the previous theorem since, by Definition and Proposition 6, the isomorphism between $P(X)$ and $P(0)$ is nothing but the equivalence of two matrices $A(X)$ and $A(0)$ constructed from a relation matrix $M \in \mathbf{R}^{q \times m}$ of $P \simeq \mathrm{Coker} M$:

$$A(X) = \begin{pmatrix} M(X) & 0_{q,q} & 0_{q,q} & 0_{q,m} \\ 0_{q,m} & \mathrm{I}_q & 0_{q,q} & 0_{q,m} \end{pmatrix}.$$

$\quad\square$

2.1.5 Horrocks' Theorem

Local Horrocks' theorem is the following result.

Theorem 46. (Local Horrocks Extension Theorem)
If $\mathbf{R}$ *is a residually discrete local ring and* P *a finitely-generated projective module over* $\mathbf{R}[X]$ *which is free over* $\mathbf{R}\langle X \rangle$, *then it is free over* $\mathbf{R}[X]$ *(i.e., extended from* $\mathbf{R}$).

Note that is straightforward to see that the hypothesis $M \otimes_{\mathbf{R}[X]} \mathbf{R}\langle X \rangle$ is a free $\mathbf{R}\langle X \rangle$-module is equivalent to the fact that M_f is a free $\mathbf{R}[X]_f$-module for some monic polynomial $f \in \mathbf{R}[X]$. The detailed proof given by Kunz [90] is elementary and constructive, except Lemma 3.13 whose proof is abstract since it uses maximal ideals. In fact this lemma asserts if P is a projective module over $\mathbf{R}[X]$ which becomes free of rank k over $\mathbf{R}\langle X \rangle$, then its kth Fitting ideal equals $\langle 1 \rangle$. This result has the following elementary constructive proof. If $P \oplus Q \simeq \mathbf{R}[X]^m$ then $P \oplus Q_1 = P \oplus (Q \oplus \mathbf{R}[X]^k)$ becomes isomorphic to $\mathbf{R}\langle X \rangle^{m+k}$ over $\mathbf{R}\langle X \rangle$ with Q_1 isomorphic to $\mathbf{R}\langle X \rangle^m$ over $\mathbf{R}\langle X \rangle$. So, we may assume $P \simeq \mathrm{Im}F$, where $G = \mathrm{I}_n - F \in \mathbf{R}[X]^{n \times n}$ is an idempotent matrix, conjugate to a standard projection matrix of rank $n - k$ over $\mathbf{R}\langle X \rangle$. We deduce that $\det(\mathrm{I}_n + TG) = (1 + T)^{n-k}$ over $\mathbf{R}\langle X \rangle$. Since $\mathbf{R}[X]$ is a subring of $\mathbf{R}\langle X \rangle$ this remains true over $\mathbf{R}[X]$. So the sum of all $(n - k)$ principal minors of G is equal to 1 (i.e., the coefficient of T^{n-k} in $\det(\mathrm{I}_n + TG)$). Hence we conclude by noticing that G is a relation matrix for P. For more details see, e.g., [108, 109].

A global version is obtained from a constructive proof of the local one by the Quillen's patching Theorem 45 and applying the General Local-Global Principle 35.

Theorem 47. (Global Horrocks Extension Theorem)
Let S be the multiplicative set of monic polynomials in $\mathbf{R}[X]$, where $\mathbf{R}$ is a ring. If P is a finitely-generated projective module over $\mathbf{R}[X]$ such that P_S is extended from $\mathbf{R}$, then P is extended from $\mathbf{R}$.

Proof. Apply the General Local-Global Principle 35 and conclude with the Concrete Quillen's patching Theorem 45. □

2.1.6 Quillen Induction Theorem

Let $\mathbf{R}$ be a commutative unitary ring. We denote by S the multiplicative subset of $\mathbf{R}[X]$ formed by monic polynomials. Let
$$\mathbf{R}\langle X \rangle := S^{-1}\mathbf{R}[X].$$
The interest in the properties of $\mathbf{R}\langle X \rangle$ branched in many directions and is attested by the abundance of articles on $\mathbf{R}\langle X \rangle$ appearing in the literature (see [69] for a comprehensive list of papers dealing with the ring $\mathbf{R}\langle X \rangle$). The ring $\mathbf{R}\langle X \rangle$ played an important role in Quillen's solution to Serre's problem [145] and its succeeding generalizations to non-Noetherian rings [20, 96, 117] as can be seen in these notes.

Classical Quillen induction is the following one.

Theorem 48. (Quillen Induction)
Suppose that a class of rings $\mathscr{P}$ satisfies the following properties:

(i) *If $\mathbf{R} \in \mathscr{P}$ then $\mathbf{R}\langle X \rangle \in \mathscr{P}$.*

(ii) *If $\mathbf{R} \in \mathscr{P}$ then $\mathbf{R}_\mathfrak{m} \in \mathscr{P}$ for any maximal ideal $\mathfrak{m}$ of $\mathbf{R}$.*

(iii) *If $\mathbf{R} \in \mathscr{P}$ and $\mathbf{R}$ is local, and if M is a finitely-generated projective $\mathbf{R}[X]$-module, then M is extended from $\mathbf{R}$ (that is, free).*

Then, for each $\mathbf{R} \in \mathscr{P}$, if M is a finitely-generated projective $\mathbf{R}[X_1, \ldots, X_n]$-module, then M is extended from $\mathbf{R}$.

Quillen induction needs maximal ideals, it works in classical mathematics but it cannot be fully constructive. One has to replace Quillen's patching with maximal ideals by the constructive form (Theorem 45) with comaximal multiplicative subsets. On the contrary, the "inductive step" in the proof is elementary and is based only on hypotheses (i) and (iii') below (induct on n and use the Global Horrocks extension Theorem 47).

Theorem 49. (Concrete Induction à la Quillen) *Suppose that a class of rings $\mathscr{P}$ satisfies the following properties:*

(i) *If $\mathbf{R} \in \mathscr{P}$ then $\mathbf{R}\langle X \rangle \in \mathscr{P}$.*

(ii') *If $\mathbf{R} \in \mathscr{P}$ then $\mathbf{R}_a \in \mathscr{P}$ for any $a \in \mathbf{R}$.*

(iii') *If $\mathbf{R} \in \mathscr{P}$ and M is a finitely-generated projective $\mathbf{R}[X]$-module, then M is extended from $\mathbf{R}$.*

Then, for each $\mathbf{R} \in \mathscr{P}$, if M is a finitely-generated projective $\mathbf{R}[X_1,\ldots,X_n]$-module, then M is extended from $\mathbf{R}$.

In the case of Serre'problem, $\mathbf{R}$ is a discrete field. So (i) and (iii') are well-known. Remark that (iii') is also given by the Global Horrocks extension Theorem 47. So Quillen's proof is deciphered in a fully constructive way. Moreover, since a zero-dimensional reduced local ring is a discrete field, we obtain the following well-known generalization (see [20]).

Theorem 50. (Quillen-Suslin, Non-Noetherian Version)

(1) *If $\mathbf{R}$ is a zero-dimensional reduced ring then any finitely-generated projective module P over $\mathbf{R}[X_1,\ldots,X_n]$ is extended from $\mathbf{R}$ (i.e., isomorphic to a direct sum of modules $e_i\mathbf{R}[\underline{X}]$ where the e_i's are idempotent elements of $\mathbf{R}$).*

(2) *As a particular case, any finitely-generated projective module of constant rank over $\mathbf{R}[X_1,\ldots,X_n]$ is free.*

(3) *More generally the results work for any zero-dimensional ring.*

Proof. The first point can be obtained from the local case by the constructive Quillen patching Theorem 45. It can also be viewed as a concrete application of the General Local-Global Principle 35.

Let us denote by $\mathbf{R}_{\mathrm{red}}$ the reduced ring associated to a ring $\mathbf{R}$. Recall that $K_0(\mathbf{R})$ is the set isomorphism classes of finitely-generated projective $\mathbf{R}$-modules.

The third point follows from the fact that the canonical map $M \mapsto M_{\mathrm{red}}$, $K_0(\mathbf{R}) \to K_0(\mathbf{R}_{\mathrm{red}})$ is a bijection. Moreover $\mathbf{R}_{\mathrm{red}}[X_1,\ldots,X_n] = \mathbf{R}[X_1,\ldots,X_n]_{\mathrm{red}}$. $\qquad\square$

2.2 Suslin's Proof of Serre's Problem

2.2.1 Making the Use of Maximal Ideals Constructive

The purpose of this subsection is to decipher constructively a lemma of Suslin [165] which played a central role in his second solution of Serre's problem. This lemma says that for a commutative ring $\mathbf{R}$, if $\langle v_1(X),\ldots,v_n(X)\rangle = \mathbf{R}[X]$, where v_1 is monic and $n \geq 3$, then there exist $\gamma_1,\ldots,\gamma_\ell \in E_{n-1}(\mathbf{R}[X])$ (the subgroup of $SL_{n-1}(\mathbf{R}[X])$ generated by elementary matrices) such that

$$\langle \mathrm{Res}(v_1,e_1.\gamma_1{}^t(v_2,\ldots,v_n)),\ldots,\mathrm{Res}(v_1,e_1.\gamma_\ell{}^t(v_2,\ldots,v_n))\rangle = \mathbf{R}.$$

By the constructive proof we give, Suslin's proof of Serre's problem becomes fully constructive. As a matter of fact, the lemma cited above is the only nonconstructive step in Suslin's elementary proof of Serre's problem [165]. Moreover, the new method with which we treat this academic example may be a model for miming constructively abstract proofs in which one works modulo each maximal ideal to prove that a given ideal contains 1. The Concrete local-global principle developed in Sect. 2.1.3 cannot be used here since the proof we want to decipher constructively, instead of passing to the localizations at each maximal ideal, passes to the residue fields modulo each maximal ideal.

In the literature, in order to surmount the obstacle of this lemma which is true for any ring $\mathbf{A}$, constructive mathematicians interested in Suslin's techniques for Suslin's stability theorem and Quillen-Suslin theorem are restricted to a few rings satisfying additional conditions and in which one knows effectively the form of all maximal ideals. For instance, in [62, 64, 93, 94, 97, 135], the authors utilize the facts that for a discrete field $\mathbf{K}$, the ring $\mathbf{K}[X_1,\ldots,X_k]$ is Noetherian and has an effective Nullstellensatz (see the proof of Theorem 4.3 of [135]). For all these reasons, we think that a constructive proof of Suslin's lemma without any restriction on the ring $\mathbf{A}$ will enable the extension of the known algorithms for Suslin's stability (see Theorem 178) and Quillen-Suslin theorems for a wider class of rings. Another feature of our method is that it may be a model for miming constructively abstract proofs passing to all the residue fields (that is, quotients by maximal ideals) in order to prove that an ideal contains 1. Note that we have already treated constructively the other main aspect of utilization of maximal ideals which is the localization at all maximal ideals (see Sect. 2.1.3). It is also worth pointing out that we will also give another constructive proof of the lemma of Suslin in the particular case where $\mathbf{R}$ contains an infinite field using efficient elementary operations.

2.2.2 A Reminder About the Resultant

In this subsection, we content ourselves with a brief outline of *resultant*: an important idea in constructive algebra whose development owes considerably to famous pioneers such as Bezout, Cayley, Euler, Herman, Hurwitz, Kronecker, Macaulay, Noether, and Sylvester, among others.

This subsection will be focused on the few properties of the resultant that we need in our constructive view toward projective modules over polynomial rings.

Definition 51. Let $\mathbf{R}$ be a ring,

$$f = a_0 X^\ell + a_1 X^{\ell-1} + \cdots + a_\ell \in \mathbf{R}[X],\ a_0 \neq 0,\ a_i \in \mathbf{R},$$

and

$$g = b_0 X^m + b_1 X^{m-1} + \cdots + b_m \in \mathbf{R}[X],\ b_0 \neq 0,\ b_i \in \mathbf{R}.$$

The resultant of f and g, denoted by $\mathrm{Res}_X(f,g)$, or simply $\mathrm{Res}(f,g)$ if there is no risk of ambiguity, is the determinant of the $(m+\ell) \times (m+\ell)$ matrix below (called the Sylvester matrix of f and g with respect to X):

$$\mathrm{Syl}(f,\ g,\ X) = \begin{pmatrix} a_0 & & & & b_0 & & & \\ a_1 & a_0 & & & b_1 & b_0 & & \\ a_2 & a_1 & \ddots & & b_2 & b_1 & \ddots & \\ \vdots & & \ddots & a_0 & \vdots & & \ddots & b_0 \\ \vdots & & & a_1 & & & & b_1 \\ a_\ell & & & & b_m & & & \\ & a_\ell & & \vdots & & b_m & & \vdots \\ & & \ddots & & & & \ddots & \\ & & & a_\ell & & & & b_m \end{pmatrix}$$

$$\underbrace{\qquad\qquad}_{m\ \text{columns}}\ \underbrace{\qquad\qquad}_{\ell\ \text{columns}}$$

The resultant is an efficient tool for eliminating variables as can be seen in the following proposition. Applying this proposition in the particular case where $\mathbf{R}[X] = \mathbf{K}[X_1, \ldots, X_n]$, $\mathbf{K}$ a field, $\mathrm{Res}_{X_n}(f,g)$ is in the first elimination ideal $\langle f, g \rangle \cap \mathbf{K}[X_1, \ldots, X_{n-1}]$.

Proposition 52. *Let $\mathbf{R}$ be a ring. Then, for any $f, g \in \mathbf{R}[X]$, there exist $h_1, h_2 \in \mathbf{R}[X]$ such that*

$$h_1 f + h_2 g = \mathrm{Res}_X(f,g) \in \mathbf{R},$$

with $\deg(h_1) \leq m-1$ *and* $\deg(h_2) \leq \ell - 1$.

Proof. First notice that

$$(X^{\ell+m-1}, \ldots, X, 1)\ \mathrm{Syl}(f,\ g,\ X) = (X^{m-1}f, \ldots, f, X^{\ell-1}g, \ldots, g).$$

Thus, by Cramer's rule, considering 1 as the $(\ell + m - 1)$th unknown of the linear system whose matrix is $\mathrm{Syl}(f, g, X)$, $\mathrm{Res}_X(f, g)$ is the determinant of the Sylvester matrix of f and g in which the last row is replaced by $(X^{m-1} f, \ldots, f, X^{\ell-1} g, \ldots, g)$. $\qquad\square$

Corollary 53. *Let* $\mathbf{K}$ *be a discrete field and* $f, g \in \mathbf{K}[X] \setminus \{0\}$. *Then*

(i) $1 \in \langle f, g \rangle \Leftrightarrow \gcd(f, g)$ *is constant* $\Leftrightarrow \mathrm{Res}(f, g) \neq 0$.

(ii) f *and* g *have a common factor* $\Leftrightarrow \gcd(f, g)$ *is nonconstant* $\Leftrightarrow \mathrm{Res}(f, g) = 0$.

Since in these notes we are concerned with the general setting of multivariate polynomials over a ring, we are tempted to say that Corollary 53 remains valid for any ring $\mathbf{R}$, where the condition "$\mathrm{Res}(f, g) \neq 0$" is replaced by "$\mathrm{Res}(f, g) \in \mathbf{R}^\times$". Of course the implication "$\mathrm{Res}(f, g) \in \mathbf{R}^\times \Rightarrow 1 \in \langle f, g \rangle$" is always true by Proposition 52. Unfortunately, the converse does not hold as will be shown by the following example. This is essentially due to the fact that if I is an ideal of a ring $\mathbf{R}$, then modulo I, we have not that $\overline{\mathrm{Res}(f, g)} = \mathrm{Res}(\bar{f}, \bar{g})$ for any $f, g \in \mathbf{R}[X]$.

Example 54. Let $\mathbf{R} = \mathbb{Z}$, $I = 3\mathbb{Z}$, $f = 6X^2 + X$, $g = 3X + 1$.
In $\mathbb{Z}[X]$, we have $1 \in \langle f, g \rangle$ as attested by the identity $3f + (1 - 6X)g = 1$ (this can be found by computing a dynamical Gröbner basis for $\langle f, g \rangle$ as in Sect. 3.3.5. In more details, $S(f, g) = f - 2Xg = -X =: h$, $S(g, h) = g + 3h = 1$). However

$$\mathrm{Res}(f, g) = \begin{vmatrix} 6 & 3 & 0 \\ 1 & 1 & 3 \\ 0 & 0 & 1 \end{vmatrix} = 3 \notin \mathbb{Z}^\times, \mathrm{Res}(\bar{f}, \bar{g}) = \bar{1} \neq \overline{\mathrm{Res}(f, g)} = \bar{0}.$$

As can bee seen in this example, whether $\overline{\mathrm{Res}(f, g)} = \mathrm{Res}(\bar{f}, \bar{g})$ modulo I or not depends mainly on whether the leading coefficients of f and g belong to I or not. We will discuss this fact in the following immediate lemma. The leading coefficient of a polynomial $h \in \mathbf{R}[X]$ will be denoted by $\mathrm{LC}(h)$.

Lemma 55. *Let* I *be an ideal of a ring* $\mathbf{R}$*, and consider two polynomials* $f = a_0 X^\ell + a_1 X^{\ell-1} + \cdots + a_\ell$, $g = b_0 X^m + b_1 X^{m-1} + \cdots + b_m \in \mathbf{R}[X]$ *with* $a_0 \neq 0$ *and* $b_0 \neq 0$, *and such that modulo* I, $\bar{f} \neq \bar{0}$ *and* $\bar{g} \neq \bar{0}$. *Then*

(1) *If* $\overline{\mathrm{LC}(f)} \neq \bar{0}$ *and* $\overline{\mathrm{LC}(g)} \neq \bar{0}$ *then* $\overline{\mathrm{Res}(f, g)} = \mathrm{Res}(\bar{f}, \bar{g})$.

(2) *If* $\overline{\mathrm{LC}(f)} = \bar{0}$ *and* $\overline{\mathrm{LC}(g)} = \bar{0}$ *then* $\overline{\mathrm{Res}(f, g)} = 0$ *(and may be* $\neq \mathrm{Res}(\bar{f}, \bar{g})$*)*.

(3) *If* $\overline{\mathrm{LC}(f)} \neq \bar{0}$ *and* $\overline{\mathrm{LC}(g)} = \bar{0}$ *then* $\overline{\mathrm{Res}(f, g)} = \bar{a}_0^{(\deg g - \deg \bar{g})} \mathrm{Res}(\bar{f}, \bar{g})$.

(4) *If* $\overline{\mathrm{LC}(f)} = \bar{0}$ *and* $\overline{\mathrm{LC}(g)} \neq \bar{0}$ *then* $\overline{\mathrm{Res}(f, g)} = \pm \bar{b}_0^{(\deg f - \deg \bar{f})} \mathrm{Res}(\bar{f}, \bar{g})$.

In fact, for the purpose of generalizing Corollary 53 to rings, we have to suppose that either f or g is monic.

Proposition 56.
Let **R** *be a ring and* $f, g \in \mathbf{R}[X] \setminus \{0\}$ *with f monic. Then*

$$1 \in \langle f, g \rangle \text{ in } \mathbf{R}[X] \iff \text{Res}(f, g) \in \mathbf{R}^\times$$

Proof. A Classical Nonconstructive Proof: we have only to prove the implication "$\Rightarrow$", the implication "$\Leftarrow$" being immediate by virtue of Proposition 52. For this, let $\mathfrak{m}$ be a maximal ideal of **R**. Applying Lemma 55, we have $\overline{\text{Res}(f,g)} = \text{Res}(\bar{f}, \bar{g})$ modulo $\mathfrak{m}$. Moreover, since $\mathbf{R}/\mathfrak{m}$ is a field, then using Corollary 53, we infer that $\overline{\text{Res}(f,g)} \neq \bar{0}$, that is, $\text{Res}(f,g) \notin \mathfrak{m}$. Since this is true for any maximal ideal of **R**, then necessarily $\text{Res}(f,g) \in \mathbf{R}^\times$.

A Constructive Proof: let $h_1, h_2 \in \mathbf{R}[X]$ such that $h_1 f + h_2 g = 1$. Since f is monic, we have $\text{Res}(f, h_2 g) = \text{Res}(f, h_2) \text{Res}(f, g)$ and $\text{Res}(f, h_2 g) = \text{Res}(f, h_1 f + h_2 g) = \text{Res}(f, 1) = 1$. $\square$

2.2.3 A Lemma of Suslin

Recall that for any ring **B** and $n \geq 1$, an $n \times n$ elementary matrix $E_{i,j}(a)$ over **B**, where $i \neq j$ and $a \in \mathbf{B}$, is the matrix with 1s on the diagonal, a on position (i,j) and 0s elsewhere, that is, $E_{i,j}(a)$ is the matrix corresponding to the elementary operation $L_i \to L_i + a L_j$. $E_n(\mathbf{B})$ will denote the subgroup of $\text{SL}_n(\mathbf{B})$ generated by elementary matrices.

Theorem 57. (Suslin's Lemma [165]) *Let* **A** *be a commutative ring. If* $\langle v_1(X), \ldots, v_n(X) \rangle = \mathbf{A}[X]$ *where v_1 is monic and $n \geq 2$, then there exist $\gamma_1, \ldots, \gamma_\ell \in E_{n-1}(\mathbf{A}[X])$ such that, denoting by w_i the first coordinate of $\gamma_i \, {}^t(v_2, \ldots, v_n)$, we have*

$$\langle \text{Res}(v_1, w_1), \ldots, \text{Res}(v_1, w_\ell) \rangle = \mathbf{A}.$$

Proof. For $n = 2$, we have $\text{Res}(f, g) \in \mathbf{A}^\times$ by Proposition 56.

Suppose $n \geq 3$. We can without loss of generality suppose that all the v_i's, for $i \geq 2$, have degrees $< d = \deg v_1$. For the sake of simplicity, we write v_i instead of $\bar{v}_i$. We will use the notation $e_1.x$, where x is a column vector, to denote the first coordinate of x.

Suslin's Proof. It consists in solving the problem modulo an arbitrary maximal ideal $\mathfrak{M}$ using a unique matrix $\gamma^{\mathfrak{M}} \in E_{n-1}(\mathbf{A}/\mathfrak{M})[X]$ which transforms ${}^t(v_2, \ldots, v_n)$ into ${}^t(g, 0 \ldots, 0)$ where g is the gcd of $v_2, \ldots, v_n$ in $(\mathbf{A}/\mathfrak{M})[X]$. This matrix is given by a classical algorithm using elementary operations on ${}^t(v_2, \ldots, v_n)$. One starts by choosing a minimum degree component, say v_2, then the v_i's, $3 \leq i \leq n$, are replaced by their remainders modulo v_2. By iterations, we obtain a column whose all components are zero except the first one. The matrix $\gamma^{\mathfrak{M}}$ lifts as a matrix $\gamma_{\mathfrak{M}} \in E_{n-1}(\mathbf{A}[X])$. It follows that the first component $w_{\mathfrak{M}}$ of $\gamma_{\mathfrak{M}} {}^t(v_2, \ldots, v_n)$ is equal to the gcd of $v_2, \ldots, v_n$ in $(\mathbf{A}/\mathfrak{M})[X]$. Thus, $\text{Res}(v_1, w_{\mathfrak{M}}) \notin \mathfrak{M}$.

Constructive Rereading of Suslin's Proof [180]. Let $u_1(X), \ldots, u_n(X) \in \mathbf{A}[X]$ such that $v_1 u_1 + \cdots + v_n u_n = 1$. Set $w = v_3 u_3 + \cdots + v_n u_n$ and $V = {}^t(v_2, \ldots, v_n)$. We suppose that v_1 has degree d and, for $2 \leq i \leq n$, the formal degree of v_i is $d_i < d$.

This means that v_i has no coefficient of degree $> d_i$ but one does not guarantee that $\deg v_i = d_i$ (it is not necessary to have a zero test inside **A**).

We proceed by induction on $\min_{2 \le i \le n}\{d_i\}$. To simplify, we always suppose that $d_2 = \min_{2 \le i \le n}\{d_i\}$.

For $d_2 = -1$, $v_2 = 0$, and by one elementary operation, we put w in the second coordinate. We have $\mathrm{Res}(v_1, w) = \mathrm{Res}(v_1, v_1 u_1 + w) = \mathrm{Res}(v_1, 1) = 1$ and we are done.

Now, suppose that we can find the desired elementary matrices for $d_2 = m - 1$ and let show that we can do the job for $d_2 = m$.

Let a be the coefficient of degree m of v_2, and consider the ring $\mathbf{B} = \mathbf{A}/\langle a \rangle$. In **B**, all the induction hypotheses are satisfied without changing the v_i nor the u_i. Thus, we can obtain $\Gamma_1, \ldots, \Gamma_k \in E_{n-1}(\mathbf{B}[X])$ such that

$$\langle \mathrm{Res}(v_1, e_1.\Gamma_1 V), \ldots, \mathrm{Res}(v_1, e_1.\Gamma_k V) \rangle = \mathbf{B}.$$

It follows that, denoting by $\Upsilon_1, \ldots, \Upsilon_k$ the matrices in $E_{n-1}(\mathbf{A}[X])$ lifting respectively $\Gamma_1, \ldots, \Gamma_k$, we have

$$\langle \mathrm{Res}(v_1, e_1.\Upsilon_1 V), \ldots, \mathrm{Res}(v_1, e_1.\Upsilon_k V), a \rangle = \mathbf{A}.$$

Let $b \in \mathbf{A}$ such that

$$ab \equiv 1 \bmod \langle \mathrm{Res}(v_1, e_1.\Upsilon_1 V), \ldots, \mathrm{Res}(v_1, e_1.\Upsilon_k V) \rangle = J$$

and consider the ring $\mathbf{C} = \mathbf{A}/J$. Note that in **C**, we have $ab = 1$.

By an elementary operation, we replace v_3 by its remainder modulo v_2, say v_3', and then we exchange v_2 and $-v_3'$. The new column V' obtained has as first coordinate a polynomial with formal degree $m - 1$. The induction hypothesis applies and we obtain $\Delta_1, \ldots, \Delta_r \in E_{n-1}(\mathbf{C}[X])$ such that

$$\langle \mathrm{Res}(v_1, e_1.\Delta_1 V'), \ldots, \mathrm{Res}(v_1, e_1.\Delta_r V') \rangle = \mathbf{C}.$$

Since V' is the image of V by a matrix in $E_{n-1}(\mathbf{C}[X])$, we obtain matrices $\Lambda_1, \ldots, \Lambda_r \in E_{n-1}(\mathbf{C}[X])$ such that

$$\langle \mathrm{Res}(v_1, e_1.\Lambda_1 V), \ldots, \mathrm{Res}(v_1, e_1.\Lambda_r V) \rangle = \mathbf{C}.$$

The matrices Λ_j lift in $E_{n-1}(\mathbf{A}[X])$ as, say $\Psi_1, \ldots, \Psi_r$.
Finally, we obtain

$$\langle \mathrm{Res}(v_1, e_1.\Psi_1 V), \ldots, \mathrm{Res}(v_1, e_1.\Psi_r V) \rangle + J = \mathbf{A},$$

the desired conclusion. $\qquad\square$

Example 58. Take $\mathbf{A} = \mathbb{Z}$ and

$$V = {}^t(v_1, v_2, v_3) = {}^t(x^2 + 2x + 2, 3, 2x^2 + 11x - 3) \in \mathrm{Um}_3(\mathbb{Z}[x]),$$

(taking $u_1 = -2x + 2$, $u_2 = -3x^2 + x - 1$, $u_3 = x$, we have $u_1 v_1 + u_2 v_2 + u_3 v_3 = 1$). It is worth pointing out that the u_i's can be found by constructing a dynamical Gröbner basis for $\langle v_1, v_2, v_3 \rangle$ as in Sect. 3.3.5. Following the algorithm given in the proof of Theorem 57 and keeping the same notation, one has to perform a Euclidean division of v_3 by v_1, so that ${}^t(v_1, v_2, v_3) \xrightarrow{E_{3,1}(-2)} {}^t(v_1, v_2, \tilde{v}_3 = 7x - 7)$, and then passes to the ring $(\mathbb{Z}/3\mathbb{Z})[x]$. This yields to $\ell = 2$, $\gamma_1 = \begin{pmatrix} 1 & 1 \\ 0 & 1 \end{pmatrix}$, $\gamma_2 = I_2 = \begin{pmatrix} 1 & 0 \\ 0 & 1 \end{pmatrix}$, and finally

$$\langle \mathrm{Res}(v_1, e_1 \cdot \gamma_1 \, {}^t(v_2, v_3)), \mathrm{Res}(v_1, e_1 \cdot \gamma_2 \, {}^t(v_2, v_3)) \rangle = \langle 170, 9 \rangle = \mathbb{Z}.$$

This example will be pursued in Sect. 2.2.6, where, as a fruit of the computations above, we will obtain a free basis for the syzygy module

$$\mathrm{Syz}(v_1, v_2, v_3) := \{(w_1, w_2, w_3) \in \mathbb{Z}[x]^3 \mid w_1 v_1 + w_2 v_2 + w_3 v_3 = 0\}.$$

Remark 59. It is easy to see that in Theorem 57, with the hypothesis $\deg v_i \leq d$ for $1 \leq i \leq n$, the number ℓ of matrices γ_j in the group $\mathrm{E}_{n-1}(\mathbf{A}[X])$ is bounded by 2^d. Moreover, each γ_j is the product of at most $2d$ elementary matrices. It is worth pointing out that there is an alternative constructive proof of Suslin's lemma (see Theorem 61) using only $(n-2)d + 1$ matrices γ_j, each of them is the product of $n - 2$ elementary matrices. This is substantially better than the general constructive proof we give above but requires the additional condition that $\mathbf{A}$ has at least $(n-2)d + 1$ elements $y_0, \ldots, y_{(n-2)d}$ such that $y_i - y_j \in \mathbf{A}^\times$ for all $i \neq j$ (for example, if $\mathbf{A}$ contains an infinite field).

2.2.4   A More General Strategy (By "Backtracking") [180]

As already mentioned above, contrary to the local-global principles explained in Sect. 2.1.3, we do not reread a proof in which one localizes at a generic prime ideal $\mathfrak{P}$ or at a generic maximal ideal $\mathfrak{M}$ but a proof in which one passes modulo a generic maximal ideal $\mathfrak{M}$ in order to prove that an ideal $\mathfrak{a}$ of a ring $\mathbf{A}$ contains 1. The classical proof is very often by contradiction: for a generic maximal ideal $\mathfrak{M}$, if $\mathfrak{a} \subseteq \mathfrak{M}$ then $1 \in \mathfrak{M}$. But, in fact, this reasoning hides a concrete fact: $1 = 0$ in the residue field $\mathbf{A}/\mathfrak{M}$ (see [150]). Consequently, this reasoning by contradiction can be converted dynamically into a constructive proof as follows. One has to do the necessary computations as if $\mathbf{A}/\mathfrak{a}$ was a field. Every time one needs to know if an element x_i is null or a unit modulo $\mathfrak{a}$, one has just to force it into being null by adding it to $\mathfrak{a}$. Suppose for example that we have established that $1 \in \mathfrak{a} + \langle x_1, x_2, x_3 \rangle$ (this corresponds in the classical proof to the fact that: $x_1, x_2, x_3 \in \mathfrak{M} \Rightarrow 1 \in \mathfrak{M}$). This means that x_3 is a unit modulo $\mathfrak{a} + \langle x_1, x_2 \rangle$ and, thus, one has to follow the classical proof in case $x_1, x_2 \in \mathfrak{M}$ and x_3 is a unit modulo $\mathfrak{M}$. It is worth pointing out that there is no need of $\mathfrak{M}$ since one has already computed an inverse of x_3 modulo $\mathfrak{a} + \langle x_1, x_2 \rangle$.

For the purpose of illustrating this strategy, let us consider an example of a binary tree corresponding to the computations produced by a "local-global" rereading:

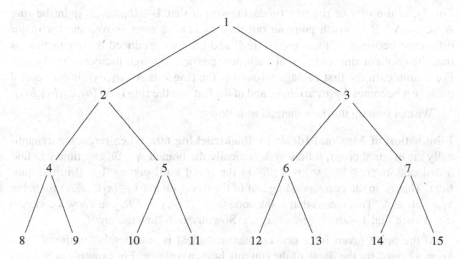

In the tree above, the disjunctions correspond to a test:

$$x \in A_i^{\times} \quad \vee \quad 1 - x \in A_i^{\times},$$

and each node corresponds to a localization A_i of the initial ring A. In order to glue the local solutions (at the terminal nodes, that is, at the leaves), one has to go back from the leaves to the root in a "parallel" way. Now imagine that these disjunctions correspond to a test:

$$x \in A_i^{\times} \quad \vee \quad x = 0 \text{ in } A_i,$$

and each node i corresponds to a quotient A_i of the initial ring A. Following the classical proof which proves that an ideal $\mathfrak{a}$ of A contains 1, one has to start with the leaf which is completely on the right (leaf 15), that is, to follow the path $1 \to 3 \to 7 \to 15$ by considering the successive corresponding quotients $A = A/\langle 0 \rangle, A/\langle a_1 \rangle, A/\langle a_1, a_3 \rangle$, and $A/\langle a_1, a_3, a_7 \rangle$. Using just the information at the leaf 15 where the considered ring is $A/\langle a_1, a_3, a_7 \rangle$ (this information corresponds in the classical proof to the fact that: $a_1, a_3, a_7 \in \mathfrak{M} \Rightarrow 1 \in \mathfrak{M}$), one obtains an element $b_{15} \in A$ such that $1 \in \langle a_1, a_3, a_7, b_{15} \rangle$, or equivalently, a_7 is a unit modulo $\langle a_1, a_3, b_{15} \rangle$. Now, we go back to the node 7 but with a new quotient $A/\langle a_1, a_3, b_{15} \rangle$ (note that at the first passage through 7 the considered quotient ring was $A/\langle a_1, a_3 \rangle$) and we can follow the branch $7 \to 14$ (this corresponds in the classical proof to the fact that: $a_1, a_3 \in \mathfrak{M}$ and a_7 is a unit modulo $\mathfrak{M} \Rightarrow 1 \in \mathfrak{M}$). This will produce an element b_{14} such that $1 \in \langle a_1, a_3, b_{14}, b_{15} \rangle$, or equivalently, a_3 is a unit modulo

$\langle a_1, b_{14}, b_{15} \rangle$. Thus, we can go back to the node 3 through the branch $14 \to 7 \to 3$, and so on. In the end, the entire path followed is

$$1 \to 3 \to 7 \to 15 \to 7 \to 14 \to 7 \to 3 \to 6 \to 13 \to 6 \to 12 \to 6 \to 3 \to 1 \to$$
$$2 \to 5 \to 11 \to 5 \to 10 \to 5 \to 2 \to 4 \to 9 \to 4 \to 8 \to 4 \to 2 \to 1.$$

Finally, at the root of the tree (node 1), we get that $1 \in \langle b_8, \ldots, b_{15} \rangle$ in the ring $\mathbf{A}/\langle 0 \rangle = \mathbf{A}$. It is worth pointing out that, as can be seen above, another major difference between a "local-global tree" and the tree produced by our method is that the quotient ring changes at each new passage through the considered node. For example, in the first passage through 7, the ring was $\mathbf{A}/\langle a_1, a_3 \rangle$, in the second passage it becomes $\mathbf{A}/\langle a_1, a_3, b_{15} \rangle$, and in the last one the ring is $\mathbf{A}/\langle a_1, a_3, b_{14}, b_{15} \rangle$.

We can sum up this new method as follows:

Elimination of Maximal Ideals by Backtracking 60. When rereading dynamically the original proof, follow systematically the branch $x_i \in \mathfrak{M}$ any time you find a disjunction "$x_i \in \mathfrak{M} \ \vee \ x_i \notin \mathfrak{M}$" in the proof until getting $1 = 0$ in the quotient. That is, in the corresponding leaf of the tree, you get $1 \in \langle x_1, \ldots, x_k \rangle$ for some $x_1, \ldots, x_k \in \mathbf{A}$. This means that at the node $\langle x_1, \ldots, x_{k-1} \rangle \subseteq \mathfrak{M}$, you know a concrete $a \in \mathbf{A}$ such that $1 - ax_k \in \langle x_1, \ldots, x_{k-1} \rangle$. So you can follow the proof.

If the proof given for a generic maximal ideal is sufficiently "uniform", you know a bound for the depth of the (infinite branching) tree. For example in Suslin's lemma, the depth is $\deg(v_1)$. So your "finite branching dynamical evaluation" is finite: you get an algorithm.

2.2.5   Suslin's Lemma for Rings Containing an Infinite Field

By the following theorem, we give an elimination process close to that given in [14, Proposition 4.72]. The proof given in [14] wasn't fully constructive as it made use of roots in algebraic closures.

Theorem 61. (Suslin's Lemma, Particular Case, New Formulation)
Let $\mathbf{A}$ *be a commutative ring containing an infinite field* $\mathbf{K}$ *and let us fix a sequence* $(y_i)_{i \in \mathbb{N}}$ *of pairwise distinct elements in* $\mathbf{K}$. *Let* $v_1, \ldots, v_n \in \mathbf{A}[X]$ *such that* v_1 *is monic of degree* d *and* $n \geq 2$. *Then*

$$1 \in \langle v_1, \ldots, v_n \rangle \ \Leftrightarrow \ 1 \in \langle \mathrm{Res}_X(v_1, v_2 + y_i v_3 + \cdots + y_i^{n-2} v_n), 0 \leq i \leq (n-2)d \rangle.$$

Proof. The implication "$\Leftarrow$" is straightforward.
"$\Rightarrow$". Let us denote by $w_i = v_2 + y_i v_3 + \cdots + y_i^{n-2} v_n$ and $r_i = \mathrm{Res}_X(v_1, w_i)$ for $0 \leq i \leq s = (n-2)d$.
A Nonconstructive Proof. To prove that $\langle r_0, \ldots, r_s \rangle = \mathbf{A}$ it suffices to prove that for each maximal ideal $\mathfrak{M}$ of $\mathbf{A}$, there exists $0 \leq i \leq s$ such that $r_i \notin \mathfrak{M}$. For this, let $\mathfrak{M}$ be a maximal ideal of $\mathbf{A}$ and by way of contradiction suppose that $\overline{r_0}, \ldots, \overline{r_s} = 0$ in the

residue field $\mathbf{K} := \mathbf{A}/\mathfrak{M}$. It is worth pointing out that $\overline{\mathrm{Res}_X(v_1, w_i)} = \mathrm{Res}_X(\overline{v_1}, \overline{w_i})$ since v_1 is monic.

This means that for each i there exists $\xi_i \in \overline{\mathbf{K}}$ such that $\overline{v_1}(\xi_i) = \overline{w_i}(\xi_i) = \overline{0}$. But since $\deg_X v_1 = d$, $\overline{v_1}$ has at most d distinct roots and hence there exists at least one root among the ξ_i repeated $n-1$ times. We can suppose that $\xi_1 = \xi_2 = \cdots = \xi_{n-1} := \xi$. Thus, we have:

$$
\begin{pmatrix}
1 & y_1 & \cdots & y_1^{n-2} \\
1 & y_2 & \cdots & y_2^{n-2} \\
\vdots & \vdots & \vdots & \vdots \\
1 & y_{n-1} & \cdots & y_{n-1}^{n-2}
\end{pmatrix}
\begin{pmatrix}
v_2(\xi) \\
v_3(\xi) \\
\vdots \\
v_n(\xi)
\end{pmatrix}
=
\begin{pmatrix}
0 \\
0 \\
\vdots \\
0
\end{pmatrix}.
$$

Since the matrix above is a Vandermonde matrix, its determinant is equal to

$$
\prod_{1 \le i < j \le n-1} (y_j - y_i),
$$

which is invertible in $\mathbf{A}$. Thus, $\overline{v}_1(\xi) = \overline{v}_2(\xi) = \cdots = \overline{v}_n(\xi) = 0$, in contradiction with the fact that $1 \in \langle v_1, \ldots, v_n \rangle$.

A Constructive Proof. Let us denote by $\ell := d + 1$.

Let $Z_0 = \cdots = Z_{n-3} = z_0$,
$\quad Z_{n-2} = \cdots = Z_{2n-5} = z_1$,
$\quad \vdots$

$\quad Z_{(n-2)k} = \cdots = Z_{(n-2)(k+1)-1} = z_k$,

$\quad \vdots$

$\quad Z_{(n-2)(d-1)} = \cdots = Z_{(n-2)d-1} = z_{d-1}$,
$\quad Z_{(n-2)d} = z_d$,

be an enumeration of ℓ indeterminates over $\mathbf{A}$ with $n-2$ repetitions except the last one which is repeated once. Let us denote by

$$
I = \langle v_1(Z_i), w_i(Z_i) \mid 0 \le i \le s \rangle, \quad \mathbf{A}_\ell = \mathbf{A}[Z_0, \ldots, Z_s]/I.
$$

First we prove that $1 = 0$ in $\mathbf{A}_\ell$.

Letting $0 \le i_1 < \cdots < i_{n-1} \le s$, we have:

$$
\begin{pmatrix}
1 & y_{i_1} & \cdots & y_{i_1}^{n-2} \\
1 & y_{i_2} & \cdots & y_{i_2}^{n-2} \\
\vdots & \vdots & \vdots & \vdots \\
1 & y_{i_{n-1}} & \cdots & y_{i_{n-1}}^{n-2}
\end{pmatrix}
\begin{pmatrix}
v_2 \\
v_3 \\
\vdots \\
v_n
\end{pmatrix}
=
\begin{pmatrix}
w_{i_1} \\
w_{i_2} \\
\vdots \\
w_{i_{n-1}}
\end{pmatrix}.
$$

As the matrix above is a Vandermonde matrix, its determinant is equal to

$$
\prod_{1 \le \ell < k \le n-1} (y_{i_k} - y_{i_\ell}),
$$

which is invertible in $\mathbf{A}$. Thus, $v_2,\ldots,v_n \in \langle w_{i_1},\ldots,w_{i_{n-1}} \rangle$ and a fortiori

$$
\begin{aligned}
v_2(Z_{i_1}),\ldots,v_n(Z_{i_1}) \ &\in I + \langle w_{i_2}(Z_{i_1}),\ldots,w_{i_{n-1}}(Z_{i_1}) \rangle \\
&\subseteq I + \langle Z_{i_1} - Z_{i_2},\ldots,Z_{i_1} - Z_{i_{n-1}} \rangle,
\end{aligned}
$$

and hence, using the fact that $1 \in \langle v_1,\ldots,v_n \rangle$, we obtain that

$$
1 \in I + \langle Z_{i_1} - Z_{i_2},\ldots,Z_{i_1} - Z_{i_{n-1}} \rangle.
$$

Thus, for $0 \leq i < j \leq d$,

$$
\begin{aligned}
1 \in I &+ \langle Z_{(n-2)i} - Z_{(n-2)i+1},\ldots,Z_{(n-2)i} - Z_{(n-2)(i+1)-1}, Z_{(n-2)i} - Z_{(n-2)j} \rangle \\
&= I + \langle z_i - z_j \rangle,
\end{aligned}
$$

that is, $z_i - z_j$ is invertible in $\mathbf{A}_\ell$.

On the other hand, by clearing the denominators in the Lagrange interpolation formula, we obtain

$$
v_1(X)\left(\prod_{i \neq j}(z_i - z_j)\right) \in \langle v_1(z_1),\ldots,v_1(z_\ell) \rangle \subseteq \mathbf{A}[z_1,\ldots,z_\ell][X]
$$

(here we need the hypothesis $\ell = \deg v_1 + 1$).

In $\mathbf{A}_\ell$, $\prod_{i \neq j}(z_i - z_j)$ is invertible, $v_1(z_1) = \cdots = v_1(z_\ell) = 0$, thus $v_1(X) = 0$ in $\mathbf{A}_\ell[X]$. Since v_1 is monic, we obtain $1 = 0$ in $\mathbf{A}_\ell$, that is $1 \in I$.

For $0 \leq k \leq s$, denote $I_k = \langle v_1(Z_i),w_i(Z_i) \mid 0 \leq i \leq k \rangle$, $J_k = I_k + \langle r_i \mid k < i \leq s \rangle$ and $\mathbf{A}_k = \mathbf{A}[Z_1,\ldots,Z_k]/I_k$. Note that $I_s = I$, so $1 \in I_s = J_s$. Using Proposition 56, we get by induction on k from s to 0 that $1 \in J_k$: in order to go from $k+1$ to k consider the ring $\mathbf{B}_k = \mathbf{A}[Z_1,\ldots,Z_k]/\langle r_{k+2},\ldots,r_s \rangle$ and apply Proposition 56 with $X = Z_{k+1}$, $a = v_1(Z_{k+1})$, $b = w_{k+1}(Z_{k+1})$. So $1 \in J_0 = \langle r_s,\ldots,r_0 \rangle$. $\square$

Remark 62. Of course, in Theorem 61, it would suffice to suppose that $\mathbf{A}$ contains $(n-2)d + 1$ elements $y_0,\ldots,y_{(n-2)d}$ such that $y_i - y_j \in \mathbf{A}^\times$ for all $i \neq j$.

Example 63. Let $f_1,\ldots,f_n \in \mathbb{Q}[X]$ $(n \geq 2)$ and suppose that $\deg f_1 = \min\{\deg f_i\} = d$. Then

$$
1 \in \langle f_1,\ldots,f_n \rangle \ \Leftrightarrow \ \exists\, 0 \leq i \leq (n-2)d \mid \mathrm{Res}_X(f_1, f_2 + if_3 + \cdots + i^{n-2}f_n) \neq 0.
$$

For example, taking $f_1 = X^5 - X^4 + 3X^2 - 3$, $f_2 = 2X^3 + 3X^2 - X - 4$, $f_3 = 3X^5 + 2X^4 - X^3 - X^2 - 3$ $(n = 3, d = 3)$,

$$
\begin{cases}
\mathrm{Res}_X(f_2,f_1) = 0 \\
\mathrm{Res}_X(f_2,f_1 + f_3) = 0 \\
\mathrm{Res}_X(f_2,f_1 + 2f_3) = 0 \\
\mathrm{Res}_X(f_2,f_1 + 3f_3) = 0
\end{cases}
\Rightarrow 1 \notin \langle f_1,f_2,f_3 \rangle.
$$

2.2.6 Suslin's Algorithm

For any ring $\mathbf{B}$, when we say that a matrix $N \in M_n(\mathbf{B})$ ($n \geq 3$) is in $SL_2(\mathbf{B})$, we mean that it is of the form

$$\begin{pmatrix} N' & 0 & \cdots & 0 \\ 0 & 1 & & \\ \vdots & & \ddots & \\ 0 & & & 1 \end{pmatrix}$$

with $N' \in SL_2(\mathbf{B})$.

Lemma 64. (Translation by the Resultant, [165, Lemma 2.1])
Let $\mathbf{R}$ be a ring, $f_1, f_2 \in \mathbf{R}[X]$, $b, d \in \mathbf{R}$, and $r = \text{Res}(f_1, f_2) \in \mathbf{R}$. Then there exists $B \in SL_2(\mathbf{R}[X])$ such that

$$B\begin{pmatrix} f_1(b) \\ f_2(b) \end{pmatrix} = \begin{pmatrix} f_1(b+rd) \\ f_2(b+rd) \end{pmatrix}.$$

Proof. Take $g_1, g_2 \in \mathbf{R}[X]$ such that $f_1 g_1 + f_2 g_2 = r$, denote by s_1, s_2, t_1, t_2 the polynomials in $\mathbf{R}[X, Y, Z]$ such that

$$\begin{aligned}
f_1(X+YZ) &= f_1(X) + Y s_1(X, Y, Z), \\
f_2(X+YZ) &= f_2(X) + Y s_2(X, Y, Z), \\
g_1(X+YZ) &= g_1(X) + Y t_1(X, Y, Z), \\
g_2(X+YZ) &= g_2(X) + Y t_2(X, Y, Z),
\end{aligned}$$

and set

$$\begin{aligned}
B_{1,1} &= 1 + s_1(b, r, d) g_1(b) + t_2(b, r, d) f_2(b), \\
B_{1,2} &= s_1(b, r, d) g_2(b) - t_2(b, r, d) f_1(b), \\
B_{2,1} &= s_2(b, r, d) g_1(b) - t_1(b, r, d) f_2(b), \\
B_{2,2} &= 1 + s_2(b, r, d) g_2(b) + t_1(b, r, d) f_1(b).
\end{aligned}$$

Then, one can take $B = \begin{pmatrix} B_{1,1} & B_{1,2} \\ B_{2,1} & B_{2,2} \end{pmatrix}$. $\qquad\square$

Algorithm 65. (An algorithm for eliminating variables from unimodular polynomial vectors with coefficients in a ring $\mathbf{A}$ containing infinitely many elements $y_0, y_1, y_2, \ldots$ such that $y_j - y_i \in \mathbf{A}^\times$ for $i \neq j$)

Input: A column $\mathcal{V} = \mathcal{V}(X) = {}^t(v_1(X), \ldots, v_n(X)) \in \text{Um}_n(\mathbf{A}[X])$ such that v_1 is monic.

Output: A matrix $\mathcal{B} \in SL_n(\mathbf{A}[X])$ such that $\mathcal{B}\mathcal{V} = \mathcal{V}(0)$.

Step 1: For $0 \leq i \leq s = (n-2)d$, where $d = \deg_X v_1$, set $w_i = v_2 + y_i v_3 + \cdots + y_i^{n-2} v_n$, compute $r_i := \text{Res}_X(v_1, w_i)$ and find $\alpha_0, \ldots, \alpha_s \in \mathbf{A}$ such that $\alpha_0 r_0 + \cdots + \alpha_s r_s = 1$ (here we use Theorem 61).
For $0 \leq i \leq s$, compute $f_i, g_i \in \mathbf{A}[X]$ such that $f_i v_1 + g_i w_i = r_i$ (use Proposition 52).

Step 2: Set

$$b_{s+1} := 0,$$
$$b_s := \alpha_s r_s X,$$
$$b_{s-1} := b_s + \alpha_{s-1} r_{s-1} X,$$
$$\vdots$$

$b_0 := b_1 + \alpha_0 r_0 X = X$ (this follows from the fact that $X = \sum_{i=0}^{s} \alpha_i r_i X$).

Step 3: For $1 \leq i \leq s+1$, find $\mathcal{B}_i \in \mathrm{SL}_n(\mathbf{A}[X])$ such that $\mathcal{B}_i \mathcal{V}(b_{i-1}) = \mathcal{V}(b_i)$.

In more details, let γ_i be the matrix corresponding to the elementary operation $L_2 \to L_2 + \sum_{j=3}^{n} y_i^{j-2} L_j$, that is,

$$\gamma_i := E_{2,n}(y_i^{n-2}) \cdots E_{2,3}(y_i).$$

For $3 \leq j \leq n$, set $F_{i,j} := \frac{v_j(b_{i-1}) - v_j(b_i)}{b_{i-1} - b_i} = \frac{v_j(b_{i-1}) - v_j(b_i)}{\alpha_i r_i X} \in \mathbf{A}[X]$, so that one obtains

$$
\begin{aligned}
v_j(b_{i-1}) - v_j(b_i) &= \alpha_i r_i X F_{i,j} = \alpha_i X F_{i,j} f_i(b_{i-1}) v_1(b_{i-1}) \\
&\quad + \alpha_i X F_{i,j} g_i(b_{i-1}) w_i(b_{i-1}) \\
&= \sigma_{i,j} v_1(b_{i-1}) + \tau_{i,j} w_i(b_{i-1}),
\end{aligned}
$$

with

$$\sigma_{i,j} := \alpha_i X F_{i,j} f_i(b_{i-1}), \tau_{i,j} := \alpha_i X F_{i,j} g_i(b_{i-1}) \in \mathbf{A}[X].$$

Let $\Gamma_i \in \mathrm{E}_n(\mathbf{A}[X])$ be the matrix corresponding to the elementary operations: $L_j \to L_j - \sigma_{i,j} L_1 - \tau_{i,j} L_2, 3 \leq j \leq n$, that is

$$\Gamma_i := \prod_{j=3}^{n} E_{j,1}(-\sigma_{i,j}) E_{j,2}(-\tau_{i,j}).$$

Set

$$B_{i,2} := \Gamma_i \gamma_i \in \mathrm{E}_n(\mathbf{A}[X]),$$

so that we have

$$
B_{i,2} \mathcal{V}(b_{i-1}) = \begin{pmatrix} v_1(b_{i-1}) \\ w_i(b_{i-1}) \\ v_3(b_i) \\ \vdots \\ v_n(b_i) \end{pmatrix}.
$$

Following Lemma 64, set

$$s_{i,1}(X,Y,Z) := \tfrac{v_1(X+YZ) - v_1(X)}{Y} \in \mathbf{A}[X,Y,Z],$$
$$s_{i,2}(X,Y,Z) := \tfrac{w_i(X+YZ) - w_i(X)}{Y} \in \mathbf{A}[X,Y,Z],$$
$$t_{i,1}(X,Y,Z) := \tfrac{f_i(X+YZ) - f_i(X)}{Y} \in \mathbf{A}[X,Y,Z],$$
$$t_{i,2}(X,Y,Z) := \tfrac{g_i(X+YZ) - g_i(X)}{Y} \in \mathbf{A}[X,Y,Z],$$
$$C_{i,1,1} := 1 + s_{i,1}(b_{i-1}, r_i, -\alpha_i X) f_i(b_{i-1}) + t_{i,2}(b_{i-1}, r_i, -\alpha_i X) w_i(b_{i-1}) \in \mathbf{A}[X],$$
$$C_{i,1,2} = s_{i,1}(b_{i-1}, r_i, -\alpha_i X) g_i(b_{i-1}) - t_{i,2}(b_{i-1}, r_i, -\alpha_i X) v_1(b_{i-1}) \in \mathbf{A}[X],$$
$$C_{i,2,1} = s_{i,2}(b_{i-1}, r_i, -\alpha_i X) f_i(b_{i-1}) - t_{i,1}(b_{i-1}, r_i, -\alpha_i X) w_i(b_{i-1}) \in \mathbf{A}[X],$$
$$C_{i,2,2} = 1 + s_{i,2}(b_{i-1}, r_i, -\alpha_i X) g_i(b_{i-1}) + t_{i,1}(b_{i-1}, r_i, -\alpha_i X) v_1(b_{i-1}) \in \mathbf{A}[X],$$

$$C_i := \begin{pmatrix} C_{i,1,1} & C_{i,1,2} \\ C_{i,2,1} & C_{i,2,2} \end{pmatrix} \in SL_2(\mathbf{A}[X]).$$

Note that

$$C_i \begin{pmatrix} v_1(b_{i-1}) \\ w_i(b_{i-1}) \end{pmatrix} = \begin{pmatrix} v_1(b_i) \\ w_i(b_i) \end{pmatrix}.$$

Set

$$B_{i,1} := \gamma_i^{-1} \begin{pmatrix} C_i & 0 \\ 0 & I_{n-2} \end{pmatrix},$$

with

$$\gamma_i^{-1} = E_{2,3}(-y_i) \cdots E_{2,n}(-y_i^{n-2}).$$

Set

$$\mathscr{B}_i := B_{i,1} B_{i,2} \in SL_n(\mathbf{A}[X]),$$

so that $\mathscr{B}_i \mathscr{V}(b_{i-1}) = \mathscr{V}(b_i)$.

Step 4: $\mathscr{B} := \mathscr{B}_{s+1} \cdots \mathscr{B}_1$.

Example 66. Let $V = \begin{pmatrix} x+y^2-1 \\ -x+y^2-2xy \\ x-y^3+2 \end{pmatrix} \in Um_3(\mathbb{Q}[x,y])$.

Algorithm 65 has been implemented using the Computer Algebra System MAPLE. The code of our algorithm (UnimodElimination) gives a matrix $B \in SL_3(\mathbb{Q}[x,y])$ eliminating one variable. In this example, $BV = V(0,y)$.

```
> V := matrix([[x+y² - 1],[-x+y² - 2*x*y],[x - y³ +2]]);
> B := UnimodElimination(V,x);
B := matrix([[1+27/151*x-56/151*x*y-24/151*x*y²-8/151*y³*
x,-35/151*x-4/151*x*y²-14/151*x*y,-62/151*x-8/151*x*y²-
28/151*x*y],[2/151*x*y+56/151*y³*x+16/151*y⁴*x+136/151*x*
y²-27/151*x,1+84/151*x*y+8/151*y³*x+32/151*x*y²+35/151*
x,152/151*x*y+16/151*y³*x+64/151*x*y²+62/151*x],[-56/151*
x*y-8/151*y³*x-24/151*x*y²+27/151*x,-35/151*x-4/151*x*
y²-14/151*x*y,1-62/151*x-8/151*x*y²-28/151*x*y]])
> VV := expandvector(multiply(B,V));
VV := matrix([[-1+y²],[y²],[2-y³]])
```

One can read that

$$V = \begin{pmatrix} x+y^2-1 \\ -x+y^2-2xy \\ x-y^3+2 \end{pmatrix},$$

$151B =$

$$\begin{pmatrix} 151+27x-56xy-24xy^2-8y^3x & -35x-4xy^2-14xy & -62x-8xy^2-28xy \\ 2xy+56y^3x+16y^4x+136xy^2-27x & 151+84xy+8y^3x+32xy^2+35x & 152xy+16y^3x+64xy^2+62x \\ -56xy-8y^3x-24xy^2+27x & -35x-4xy^2-14xy & 151-62x-8xy^2-28xy \end{pmatrix},$$

$$BV = V(0,y) = \begin{pmatrix} y^2-1 \\ y^2 \\ -y^3+2 \end{pmatrix}.$$

Let us fix an infinite sequence of pairwise distinct elements (y_i) in **K** and use the notation $\underline{X} = (X_1, \ldots, X_k)$.

Algorithm 67. (An algorithm for the Quillen-Suslin theorem: case of $\mathbf{K}[X_1, \ldots, X_k]$, where **K** is an infinite field

Input: One column $\mathscr{V} = \mathscr{V}(\underline{X}) = {}^t(v_1(\underline{X}), \ldots, v_n(\underline{X})) \in \mathrm{Um}_n(\mathbf{K}[\underline{X}])$ such $\max_{1 \leq i \leq n} \{\deg v_i\} = d$ (here by degree we mean total degree), where $d \geq 2$.
Output: A matrix G in $\mathrm{SL}_n(\mathbf{K}[\underline{X}])$ such that $G\mathscr{V} = {}^t(1, 0, \ldots, 0)$.

For j from k to 1 perform steps 1 and 2:

Step 1: Make a linear change of variables so that v_1 becomes monic at X_j.

Step 2: Perform Algorithm 65 with $\mathbf{A} = \mathbf{K}[X_1, \ldots, X_{j-1}]$ and $X = X_j$. Output the new $\mathscr{V}$.

Example 68. (Example 66 Continued)
$$\text{Let } V = \begin{pmatrix} v_1 \\ v_2 \\ v_3 \end{pmatrix} = \begin{pmatrix} x + y^2 - 1 \\ -x + y^2 - 2xy \\ x - y^3 + 2 \end{pmatrix} \in \mathrm{Um}_3(\mathbb{Q}[x,y]).$$
Recall that the syzygy module of (v_1, v_2, v_3) is

$$\mathrm{Syz}(v_1, v_2, v_3) := \{{}^t(w_1, w_2, w_3) \in \mathbb{Q}[x,y]^3 \text{ such that } w_1 v_1 + w_2 v_2 + w_3 v_3 = 0\}.$$

Recall also that since ${}^t(v_1, v_2, v_3) \in \mathrm{Um}_3(\mathbb{Q}[x,y])$, $\mathrm{Syz}(v_1, v_2, v_3)$ is a projective $\mathbb{Q}[x,y]$-module which is free of rank 2 by the Quillen-Suslin Theorem 50. A generating set for $\mathrm{Syz}(v_1, v_2, v_3)$ can be obtained using Gröbner bases techniques (see for example [43, 73]). For this, let us open a SINGULAR Session (for more details see [73]):

```
> ringB = 0, (x, y), dp;
> idealI = x + y2 − 1, −x + y2 − 2xy, x − y3 + 2;
> moduleN = syz(I);
> N;
N[1] = 2y3 * gen(1) + 2xy * gen(1) + 2y2 * gen(3) + y2 * gen(2) − y2 * gen(1) +
2x * gen(3) + x * gen(2) − x * gen(1) − 2 * gen(3) − gen(2) − 4 * gen(1)
N[2] = 4xy2 * gen(1) − 14y3 * gen(1) + 4xy * gen(3) + 2xy * gen(2) − 12xy *
gen(1) − 14y2 * gen(3) − 7y2 * gen(2) + 7y2 * gen(1) − 10x * gen(3) − 5x * gen(2) +
5x * gen(1) − 2y * gen(2) + 12 * gen(3) + 11 * gen(2) + 24 * gen(1)
N[3] = 8x2y * gen(1) − 98y3 * gen(1) + 8x2 * gen(3) + 4x2 * gen(2) − 4x2 *
gen(1) − 98xy * gen(1) − 98y2 * gen(3) − 49y2 * gen(2) + 53y2 * gen(1) − 98x *
gen(3) − 53x * gen(2) + 25x * gen(1) + 4y * gen(3) − 12y * gen(2) + 8y * gen(1) +
94 * gen(3) + 61 * gen(2) + 188 * gen(1)
```

One can read that $\mathrm{Syz}(v_1, v_2, v_3) = \langle u_1, u_2, u_3 \rangle$ with

$$u_1 = {}^t(2y^3 + 2xy - y^2 - x - 4, y^2 + x - 1, 2y^2 + 2x - 2),$$

$$u_2 = {}^t(4xy^2 - 14y^3 - 12xy + 7y^2 + 5x + 24, 2xy - 7y^2 - 5x - 2y + 11,$$
$$4xy - 14y^2 - 10x + 12),$$
$$u_3 = {}^t(8x^2y - 98y^3 - 4x^2 - 98xy + 53y^3 + 25x + 8y + 188,$$
$$4x^2 - 49y^2 - 53x - 12y + 61, 8x^2 - 98y^2 + 4y + 94).$$

But this is not a **minimal** set of generators for $\mathrm{Syz}(v_1, v_2, v_3)$!
In order to obtain such a minimal generating set one has to compute a free basis for
$\mathrm{Syz}(v_1, v_2, v_3)$. We have implemented Algorithm 67 using the Computer Algebra
System MAPLE. It computes a matrix $G \in \mathrm{SL}_3(\mathbb{Q}[x,y])$ such that $GV = {}^t(1,0,0)$.

$G := matrix([[-1 + 60/151 * x * y^3 + 540/151 * x * y^2 + 62/151 * x * y - 108/151 *$
$x + 2 * y^2 - 128/151 * x * y^5 - 272/151 * x * y^4 - 32/151 * x * y^6, -40/151 * x *$
$y^2 + 266/151 * x * y + 140/151 * x - 72/151 * x * y^4 - 172/151 * x * y^3 + 3 - 2 *$
$y^2 - 16/151 * x * y^5, 248/151 * x - 48/151 * x * y^2 + 484/151 * x * y - 144/151 *$
$x * y^4 - 312/151 * x * y^3 - 32/151 * x * y^5], [-y^2 + 64/151 * x * y^5 + 144/151 * x *$
$y^4 + 2/151 * x * y^3 - 190/151 * x * y^2 + 27/151 * x - 2/151 * x * y + 16/151 * x *$
$y^6, 36/151 * x * y^4 + 90/151 * x * y^3 + 38/151 * x * y^2 - 1 - 35/151 * x - 84/151 *$
$x * y + y^2 + 8/151 * x * y^5, 60/151 * x * y^2 + 72/151 * x * y^4 + 164/151 * x * y^3 -$
$152/151 * x * y - 62/151 * x + 16/151 * x * y^5], [2 - 190/151 * x * y^3 - 344/151 *$
$x * y^2 - 172/151 * x * y + 135/151 * x - y^3 + 64/151 * x * y^6 + 160/151 * x * y^5 +$
$26/151 * x * y^4 + 16/151 * x * y^7, -76/151 * x * y^2 - 210/151 * x * y - 175/151 *$
$x + 36/151 * x * y^5 + 98/151 * x * y^4 + 54/151 * x * y^3 - 2 + y^3 + 8/151 * x *$
$y^6, -310/151 * x - 152/151 * x * y^2 - 388/151 * x * y + 92/151 * x * y^3 + 72/151 *$
$x * y^5 + 180/151 * x * y^4 + 16/151 * x * y^6 + 1]])$

One can read $G = \begin{pmatrix} {}^t\varepsilon_0 \\ {}^t\varepsilon_1 \\ {}^t\varepsilon_2 \end{pmatrix}$, where

$\varepsilon_0 =$

$$\frac{1}{151} \begin{pmatrix} -151 + 60xy^3 + 540xy^2 + 62xy - 108x + 302y^2 - 128xy^5 - 272xy^4 - 32xy^6 \\ -40xy^2 + 266xy + 140x - 72xy^4 - 172xy^3 + 453 - 302y^2 - 16xy^5 \\ 248x - 48xy^2 + 484xy - 144xy^4 - 312xy^3 - 32xy^5 \end{pmatrix},$$

$\varepsilon_1 =$

$$\frac{1}{151} \begin{pmatrix} -151y^2 + 64xy^5 + 144xy^4 + 2xy^3 - 190xy^2 + 27x - 2xy + 16xy^6 \\ 36xy^4 + 90xy^3 + 38xy^2 - 151 - 35x - 84xy + 151y^2 + 8xy^5 \\ 60xy^2 + 72xy^4 + 164xy^3 - 152xy - 62x + 16xy^5 \end{pmatrix}, \text{ and}$$

$\varepsilon_2 =$

$$\frac{1}{151} \begin{pmatrix} 302 - 190xy^3 - 344xy^2 - 172xy + 135x - 151y^3 + 64xy^6 + 160xy^5 + 26xy^4 + 16xy^7 \\ -76xy^2 - 210xy - 175x + 36xy^5 + 98xy^4 + 54xy^3 - 302 + 151y^3 + 8xy^6 \\ -310x - 152xy^2 - 388xy + 92xy^3 + 72xy^5 + 180xy^4 + 16xy^6 + 151 \end{pmatrix}.$$

So, $(\varepsilon_1, \varepsilon_2)$ is a free basis for $\mathrm{Syz}(v_1, v_2, v_3)$. A minimal parametrization of the set
$\mathscr{E}$ of all inverses of V is
$$\mathscr{E} := \{U = (u_1, u_2, u_3) \in \mathbb{Q}[x,y]^{1 \times 3} \mid UV = 1\} = \{\varepsilon_0 + \alpha\varepsilon_1 + \beta\varepsilon_2, \alpha, \beta \in \mathbb{Q}[x,y]\}.$$

The following algorithm is due to Suslin [165]. We skip some details as they were already given in Algorithm 65.

Algorithm 69. (An algorithm for eliminating variables from unimodular polynomial vectors with coefficients in a ring, general case

Input: A column $\mathscr{V} = \mathscr{V}(X) = {}^t(v_1(X),\ldots,v_n(X)) \in \mathrm{Um}_n(\mathbf{A}[X])$ such that v_1 is monic.

Output: A matrix $\mathscr{B} \in \mathrm{SL}_n(\mathbf{A}[X])$ such that $\mathscr{B}\mathscr{V} = \mathscr{V}(0)$.

Step 1: Find $\gamma_0,\ldots,\gamma_s \in \mathrm{E}_{n-1}(\mathbf{A}[X])$ such that, denoting $w_i = e_1.\gamma_i{}^t(v_2,\ldots,v_n)$ and $r_i = \mathrm{Res}(v_1, w_i)$, we can find $\alpha_0,\ldots,\alpha_s \in \mathbf{A}$ such that $\alpha_0 r_0 + \cdots + \alpha_s r_s = 1$ (here we use the algorithm given in the proof of Theorem 57).
For $0 \le i \le s$, compute $f_i, g_i \in \mathbf{A}[X]$ such that $f_i v_1 + g_i w_i = r_i$ (use Proposition 52).

Step 2: Perform steps 2–4 of Algorithm 65 doing the necessary small changes.

Example 70. Take $\mathbf{A} = \mathbb{Z}$ and $V = {}^t(x^2 + 2x + 2, 3, 2x^2 + 11x - 3) \in \mathrm{Um}_3(\mathbb{Z}[x])$.

A generating set for $\mathrm{Syz}(v_1, v_2, v_3)$ can be obtained by computing a dynamical Gröbner basis for the ideal $\langle v_1, v_2, v_3 \rangle$ (see Sect. 3.3.5). A dynamical computation gives

$$\mathrm{Syz}(v_1,v_2,v_3) = \langle \begin{pmatrix} 3 \\ -X^2 - 2X - 2 \\ 0 \end{pmatrix}, \begin{pmatrix} 0 \\ -2X^2 - 11X + 3 \\ 3 \end{pmatrix},$$

$$\begin{pmatrix} -2X^3 - 11X^2 - 18X \\ 7X^3 + 14X^2 + 14X \\ X^3 + 2X^2 + 2X \end{pmatrix}, \begin{pmatrix} -21 - 6X \\ 14 + 21X \\ 3X \end{pmatrix}, \begin{pmatrix} -4X^3 - 36X^2 - 71X + 21 \\ 14X^3 + 77X^2 - 21X \\ 2X^3 + 11X^2 - 3X + 14 \end{pmatrix} \rangle.$$

But of course as mentioned above this is not a minimal generating set for Syz (v_1, v_2, v_3) as it is a rank 2 free $\mathbb{Z}[x]$-module (by the Lequain-Simis-Vasconcelos Theorem, see Corollary 142). Following Algorithm 69 and doing the computations by hand (assisted by the computer algebra system MAPLE), we get a matrix $G \in \mathrm{SL}_3(\mathbb{Z}[x])$ such that

$$GV = \begin{pmatrix} 1 \\ 0 \\ 0 \end{pmatrix}.$$

```
> V := matrix(3, 1, [x^2 + 2*x + 2, 3, 2*x^2 + 11*x - 3]);
> G := matrix([[2 + 29142*x^2 + 340*x + 4788*x^3, -25686*x^2 - 2394*x^3 -
272*x - 1, -6192*x^2 - 2394*x^3 - 44*x], [-3 - 43713*x^2 - 510*x - 7182*
x^3, 38529*x^2 + 3591*x^3 + 408*x + 2, 9288*x^2 + 3591*x^3 + 66*x], [12 +
204092*x^2 + 2975*x + 33516*x^3, -179851*x^2 - 16758*x^3 - 2429*x -
7, -43393*x^2 - 16758*x^3 - 434*x + 1]])
> det(G);
1
> F := expandvector(multiply(G, V));
F := matrix([[1], [0], [0]])
```

Thus,

$$\left(\begin{array}{c} -3 - 43713x^2 - 510x - 7182x^3 \\ 38529x^2 + 3591x^3 + 408x + 2 \\ 9288x^2 + 3591x^3 + 66x \end{array} \right), \left(\begin{array}{c} 12 + 204092x^2 + 2975x + 33516x^3 \\ -179851x^2 - 16758x^3 - 2429x - 7 \\ -43393x^2 - 16758x^3 - 434x + 1 \end{array} \right)$$

is a free basis for $\mathrm{Syz}(v_1, v_2, v_3)$.

> $inverse(G);$
$matrix([[x^2 + 2 * x + 2, 5586 * x^3 + 14465 * x^2 + 146 * x + 1, 1197 * x^3 + 3096 * x^2 + 22 * x], [3,2,0], [2 * x^2 + 11 * x - 3, 11172 * x^3 + 68032 * x^2 + 999 * x + 2, 2394 * x^3 + 14571 * x^2 + 170 * x + 1]])$

The matrix G^{-1} is a completion of V into an invertible matrix as V is the first column of G^{-1}.

2.2.7 Suslin's Solution to Serre's Problem

Theorem 71. (Unimodular Completion Theorem) *Let* $\mathbf{K}$ *be a field,* $\mathbf{R} = \mathbf{K}[X_1, \ldots, X_r]$ *and consider a unimodular vector*

$$f = {}^t(f_1(X_1, \ldots, X_r), \ldots, f_n(X_1, \ldots, X_r)),$$

in $\mathbf{R}^{n \times 1}$. *Then, there exists a matrix* $H \in \mathrm{SL}_n(\mathbf{R})$ *such that* $Hf = {}^t(1, 0, \ldots, 0)$. *In other words,* f *is the first column of a matrix in* $\mathrm{SL}_n(\mathbf{R})$.

Proof. If $n = 1$ or 2, the result is straightforward. If $n > 2$ and $r = 1$, the result comes from the fact that $\mathbf{R}$ is a PID. It is explicitly given by a Smith reduction of the column matrix f. For $r \geq 2$, we make an induction on r. If the field $\mathbf{K}$ has enough elements (for example, if it is infinite), we can make a linear change of variables so that one of the f_i becomes monic. Else, we make a change of variables " à la Nagata": $Y_r = X_r$, and for $1 \leq j < r$, $Y_j = X_j + X_r^{d^j}$, with a sufficiently large integer d. It suffices now to use Algorithm 69. $\qquad \square$

Theorem 72. (Suslin's Solution to Serre's Problem) *Let* $\mathbf{K}$ *be a field,* $\mathbf{R} = \mathbf{K}[X_1, \ldots, X_r]$ *and* M *a finitely-generated projective* $\mathbf{R}$-*module. Then* M *is free.*

Proof. By virtue of Definition and Proposition 27, we know that M is stably free, i.e., we have an isomorphism

$$\varphi : \mathbf{R}^k \oplus M \longrightarrow \mathbf{R}^{\ell + k}$$

for some integers k and ℓ. If $k = 0$, there is nothing to prove. Suppose that $k > 0$. The vector $f = \varphi((e_{k,1}, 0_M))$ (where $e_{k,1}$ is the first vector in the canonical basis of $\mathbf{R}^k$) is unimodular. To see this, just consider the linear form λ over $\mathbf{R}^{\ell+k}$ mapping y

$(y \in \mathbf{R}^{\ell+k})$ to the first coordinate of $\varphi^{-1}(y)$. We have $\lambda(y_1,\ldots,y_{k+\ell}) = u_1 y_1 + \cdots + u_{k+\ell} y_{k+\ell}$ and $\lambda(f) = 1$.

Consider f as a column vector. Taking the composition of φ with the isomorphism given in Theorem 71, we obtain an isomorphism ψ mapping $(e_{k,1}, 0_M)$ to $e_{k+\ell,1}$. By passing modulo $\mathbf{A}(e_{k,1}, 0_M)$ and modulo $\mathbf{A}e_{k+\ell,1}$, we get an isomorphism

$$\theta : \mathbf{R}^{k-1} \oplus M \longrightarrow \mathbf{R}^{\ell+k-1}.$$

$\square$

2.2.8   A Simple Result About Coherent Rings

Let $\mathbf{A}$ be a ring with a test "$x = 0$".

For a polynomial $g = \sum_j a_j X^j \in \mathbf{A}[X]$, we set $\mathrm{coeff}_X(f,k) := a_k$. Since the degree of g is known, we denote by $\mathrm{LT}(g)$, $\mathrm{LM}(g)$, $\mathrm{LC}(g)$ respectively the leading term of g, its leading monomial and its leading coefficient.

We denote by $\mathbf{A}[X]_k$ the free submodule of rank $k+1$ of $\mathbf{A}[X]$ generated by $1, X, \ldots, X^k$. If I is an ideal of $\mathbf{A}[X]$, we denote by I_k the submodule $I \cap \mathbf{A}[X]_k$. If $\mathbf{A}$ is discrete, we denote by $\mathrm{LT}(I)$ the ideal $\langle \mathrm{LT}(f) : f \in I \rangle$.

If the ring is not known to be discrete (i.e., with a test "$x = 0$"), for $f \in \mathbf{A}[X]$, $\langle \mathrm{LT}(f) \rangle$ denotes the ideal generated by the terms $a_k X^k$ of f for all k s.t. $\mathrm{coeff}(f,\ell) = 0$ for $\ell > k$. And, for a subset $E \subseteq \mathbf{A}[X]$, $\mathrm{LT}(E)$ denotes the ideal $\sum_{f \in E} \langle \mathrm{LT}(f) \rangle$.

In this subsection we don't assume $\mathbf{A}$ to be a discrete ring.

Proposition 73. *Let $I = \langle f, f_1, \ldots, f_s \rangle$ be a finitely-generated ideal of $\mathbf{A}[X]$, with f monic of degree n. Then*

(1) *I_{n-1} is a finitely-generated module,*

(2) *$I = \langle I_{n-1} \rangle + \langle f \rangle = I_{n-1} \oplus \langle f \rangle$,*

(3) *$\mathrm{LT}(I) = \mathrm{LT}(I_{n-1}) + \langle X^n \rangle$.*

Proof. Let $\mathbf{B} = \mathbf{A}[X]/\langle f \rangle$ be the quotient algebra, which is a free $\mathbf{A}$-module with basis $1, x, \ldots, x^{n-1}$ ($x = \overline{X}$ is the class of X modulo f), let $\psi : \mathbf{B}^s \to \mathbf{B}$ be the *generalized Sylvester map*

$$(\overline{g_1}, \ldots, \overline{g_s}) \mapsto \sum_{i=1}^s \overline{g_i} \, \overline{f_i}.$$

Then clearly $I_{n-1} \mathbf{B}$ is generated by the image of ψ, which is the module generated by all the $x^k \overline{f_i}$ with $0 \le k < n, 1 \le i \le s$.

In matrix form, we get the *generalized Sylvester matrix* associated to the polynomials $f, f_1, \ldots, f_s$ denoted by $\mathrm{Syl}_X(f, f_1, \ldots, f_s)$ which is the matrix with the following columns:

$$\mathrm{Rem}(f_1, f), \ldots, \mathrm{Rem}(f_s, f), \mathrm{Rem}(X f_1, f), \ldots, \mathrm{Rem}(X f_s, f), \ldots, \mathrm{Rem}$$
$$(X^{n-1} f_1, f),$$

$$\ldots, \mathrm{Rem}(X^{n-1} f_s, f)$$

(where $\mathrm{Rem}(g,f)$ denotes the remainder of the division of g by f) in the basis $(X^{n-1},\ldots,X,1)$. And I_{n-1} is the module generated by the columns of Syl_X $(f,f_1,\ldots,f_s)$. $\qquad\square$

Example 74. *If*

$$f(X) = X^3 + 3X^2 + 4, \ f_1(X) = 4X^2 + 5X + 3,$$
$$f_2(X) = -3X^2 + 2X + 3, \ f_3(X) = 2X^2 - X + 7,$$

then

$$\mathrm{Syl}_X(f,f_1,f_2,f_3) = \begin{pmatrix} 4 & -3 & 2 & -7 & 11 & -7 & 24 & -30 & 28 \\ 5 & 2 & -1 & 3 & 3 & 7 & -16 & 12 & -8 \\ 3 & 3 & 7 & -16 & 12 & -8 & 28 & -44 & 28 \end{pmatrix}.$$

Theorem 75. *Let* $\mathbf{A}$ *be a coherent ring and* $I = \langle f,f_1,\ldots,f_s \rangle$ *a finitely-generated ideal of* $\mathbf{A}[X]$, *with* f *monic. Then*

(1) *the elimination ideal* $I_0 = I \cap \mathbf{A}$,

(2) *the elimination modules* $I_k = I \cap \mathbf{A}[X]_k$, *and*

(3) *the leading ideal* $\mathrm{LT}(I)$

are finitely-generated.

Proof. Let $\pi_k : \mathbf{A}[X]_k \to \mathbf{A}$ be the coordinate form $f \mapsto \mathrm{coeff}(f,k)$. We know that I_{n-1} is a finitely-generated module. For $k \geq n$, the module $I_k = I_{n-1} \oplus f(\mathbf{A} + X\mathbf{A} + \ldots + X^{k-n}\mathbf{A})$ is finitely-generated. For $k < n-1$, the module I_k is finitely-generated because $I_k = I_{n-1} \cap \mathbf{A}[X]_k$, and these two modules are finitely-generated submodules of the module $\mathbf{A}[X]_{n-1}$, which is isomorphic to $\mathbf{A}^n$, hence coherent. So I_k and $\pi_k(I_k)$ are finitely-generated $\mathbf{A}$-modules. Thus, the leading ideal

$$\mathrm{LT}(I) = \pi_0(I_0) + \pi_1(I_1)\langle X \rangle + \cdots + \pi_{n-1}(I_{n-1})\langle X^{n-1} \rangle + \langle X^n \rangle$$

is finitely-generated. $\qquad\square$

Let us describe with more details a computation corresponding to the above proof. We assume that $\deg(f) = 5$ and we want to know I_2 and the ideal generated by the terms of degree 2 for polynomials in I_2, that is $\pi_2(I_2) \cdot \langle X^2 \rangle$, where $\pi_2 : I_2 \to \mathbf{A}$ is the coordinate form $f \mapsto \mathrm{coeff}(f,2)$. Suppose further that the generalized Sylvester matrix has the following pattern

$$\begin{array}{c} X^4 \\ X^3 \\ X^2 \\ X \\ 1 \end{array} \begin{pmatrix} c_1 & c_2 & c_3 & c_4 & \cdots\cdots \\ b_1 & b_2 & b_3 & b_4 & \cdots\cdots \\ a_1 & a_2 & a_3 & a_4 & \cdots\cdots \\ v_1 & v_2 & v_3 & v_4 & \cdots\cdots \\ u_1 & u_2 & u_3 & u_4 & \cdots\cdots \end{pmatrix}$$

with ℓ columns. We have

$$\pi_2(I_2) = \left\{ \sum_{i=1}^{\ell} \alpha_i a_i \right\}$$

for all $(\alpha_1, , \ldots, \alpha_\ell)$ that are linear dependence relations for the family

$$U = \left(\begin{pmatrix} c_1 \\ b_1 \end{pmatrix}, \ldots, \begin{pmatrix} c_\ell \\ b_\ell \end{pmatrix} \right) \in (\mathbf{A}^2)^\ell.$$

Similarly

$$I_2 = \left\{ \sum_{i=1}^{\ell} \alpha_i (u_i + v_i X + a_i X^2) \right\}$$

for the same $(\alpha_1, , \ldots, \alpha_\ell)$'s.

Since $\mathbf{A}$ is a coherent ring, $\mathbf{A}^2$ is a coherent $\mathbf{A}$-module and the module of relations for U is finitely-generated.

2.3   Constructive Definitions of Krull Dimension

This section is taken from the papers [34, 39, 42, 100].

The constructive theory of Krull dimension presented here was mainly developed by Lombardi. The first constructive definition of Krull dimension was given by Joyal and his student Español. It was essentially formulated in terms of Zariski lattice of the ring [18, 56, 57, 58, 59, 60, 83, 84] and was difficult to use. A more usable equivalent constructive definition of Krull dimension via singular sequences appeared independently in [100]. The notions of Krull boundary ideal and Krull boundary monoid appeared for the first time in [39].

2.3.1   Ideals and Filters

Let S be a monoid (a multiplicative subset) of a ring $\mathbf{R}$. If M is an $\mathbf{R}$-module, then the $\mathbf{R}_S$-module M_S is obtained by extension of the scalars from $\mathbf{R}$ to $\mathbf{R}_S$. In particular, if M is finitely-generated, finitely-presented or projective, then so is M_S.

Recall that S is said to be *saturated* if

$$\forall s, t \in \mathbf{R}, \ st \in S \ \Rightarrow \ s \in S.$$

A saturated monoid is also called a *filter*. Note that denoting by

$$\overline{S} = \{ s \in \mathbf{R}, \exists t \in \mathbf{R} \text{ such that } st \in S \},$$

$\overline{S}$ is a saturated monoid of $\mathbf{R}$ called the *saturation* of S, and we have

$$\mathbf{R}_{\overline{S}} = \mathbf{R}_S.$$

Note that there is a duality between ideals and filters. On the one hand, ideals are used to pass to the quotient, that is, to force the elements of the considered ideal $\mathfrak{a}$ of $\mathbf{R}$ into being zero in $\mathbf{R}/\mathfrak{a}$. On the other hand, filters are used to localize, that is, to force the elements of the considered monoid into being invertible.

An ideal is prime if and only if its complementary if a filter. A filter whose complementary is an ideal is called a *prime filter*.

The duality between ideals and filters is also a duality between addition and multiplication as can be seen by the axioms defining ideals (resp., prime ideals) and filters (resp., prime filters):

$$\text{Ideal } \mathscr{I} \qquad \text{Filter } \mathscr{F}$$

$$\vdash 0 \in \mathscr{I} \qquad\qquad \vdash 1 \in \mathscr{F}$$

$$x \in \mathscr{I}, y \in \mathscr{I} \vdash x+y \in \mathscr{I} \qquad x \in \mathscr{F}, y \in \mathscr{F} \vdash xy \in \mathscr{F}$$

$$x \in \mathscr{I} \vdash xy \in \mathscr{I} \qquad\qquad xy \in \mathscr{F} \vdash x \in \mathscr{F}$$

$$\text{prime} \qquad\qquad \text{prime}$$

$$xy \in \mathscr{I} \vdash x \in \mathscr{I} \vee y \in \mathscr{I} \qquad x+y \in \mathscr{F} \vdash x \in \mathscr{F} \vee y \in \mathscr{F}$$

$$1 \in \mathscr{I} \vdash \text{False} \qquad\qquad 0 \in \mathscr{F} \vdash \text{False}$$

2.3.2 Zariski Lattice

Notation 76. If $\mathfrak{a}$ is an ideal of $\mathbf{R}$, we denote $D_{\mathbf{R}}(\mathfrak{a}) = \sqrt{\mathfrak{a}}$ the radical of $\mathfrak{a}$, that is, the set of all $x \in \mathbf{R}$ such that $x^k \in \mathfrak{a}$ for some $k \in \mathbb{N}$.
If $\mathfrak{a} = \langle x_1, \ldots, x_n \rangle$, we often denote $D_{\mathbf{R}}(x_1, \ldots, x_n)$ instead of $D_{\mathbf{R}}(\mathfrak{a})$.
Note that $D_{\mathbf{R}}(0) = \sqrt{(0)} = \{x \in \mathbf{R} \mid x \text{ nilpotent}\}$ and $D_{\mathbf{R}}(x_1, \ldots, x_n) = D_{\mathbf{R}}(1)$ if and only if $1 \in \langle x_1, \ldots, x_n \rangle$.

Definition 77. We denote $\operatorname{Zar}\mathbf{R}$ the set of all the $D_{\mathbf{R}}(x_1, \ldots, x_n)$, where $n \in \mathbb{N}$ and $x_1, \ldots, x_n \in \mathbf{R}$. This set is ordered by inclusion.

Fact 78. $\operatorname{Zar}\mathbf{R}$ is a distributive lattice equipped with

$$D_{\mathbf{R}}(\mathfrak{a}_1) \vee D_{\mathbf{R}}(\mathfrak{a}_2) = D_{\mathbf{R}}(\mathfrak{a}_1 + \mathfrak{a}_2) \quad \& \quad D_{\mathbf{R}}(\mathfrak{a}_1) \wedge D_{\mathbf{R}}(\mathfrak{a}_2) = D_{\mathbf{R}}(\mathfrak{a}_1 \mathfrak{a}_2).$$

$\operatorname{Zar}\mathbf{R}$ is called the *Zariski lattice* of the ring $\mathbf{R}$.

2.3.3 Krull Boundary

Let us recall the classical definition of the Krull dimension of a ring $\mathbf{R}$. A finite chain $\mathfrak{p}_0 \subsetneq \mathfrak{p}_1 \subsetneq \cdots \subsetneq \mathfrak{p}_n$ of $n+1$ proper prime ideals of $\mathbf{R}$ is said to have length n. If $\mathbf{R}$ has no proper prime ideal (that is, $\mathbf{R}$ is trivial), we say that $\mathbf{R}$ has Krull dimension -1. If there is a nonnegative integer d such that $\mathbf{R}$ contains a chain of proper prime ideals of length d, but no such chain of length $d+1$, we say that $\mathbf{R}$ has Krull dimension d, and we write $\operatorname{Kdim}\mathbf{R} = d$ or, simply, $\dim\mathbf{R} = d$. Otherwise, we say that $\mathbf{R}$ is infinite dimensional. For example, a field or a finite product of fields has Krull dimension 0; $\mathbb{Z}$ or more generally a PID which is not a field has Krull dimension 1.

Definition 79. *Let* **R** *be a ring and* $x \in$ **R**.

(1) *The upper Krull boundary of* x *in* **R** *is the quotient ring* $\mathbf{R}^{\{x\}} := \mathbf{R}/\mathrm{K}_{\mathbf{R}}(x)$, *where* $\mathrm{K}_{\mathbf{R}}(x) := \langle x \rangle + (\mathrm{D}_{\mathbf{R}}(0) : x) = \langle x \rangle + \{ b \in \mathbf{R}, \ bx \ \text{is nilpotent} \}$.

   We will say that $\mathrm{K}_{\mathbf{R}}(x)$ *is the Krull boundary ideal of* x.

(2) *The lower Krull boundary of* x *in* **R** *is the localized ring* $\mathbf{R}_{\{x\}} := \mathbf{R}_{S_{\{x\}}}$, *where* $S_{\{x\}} := x^{\mathbb{N}}(1 + x\mathbf{R}) = \{ x^k(1 + xy), k \in \mathbb{N}, y \in \mathbf{R} \}$.

   We will say that $S_{\{x\}}$ *is the Krull boundary monoid of* x.

The terminology above is legitimated by the following geometric case: if $\mathbf{R} = \mathbf{K}[V]$ is the ring of rational functions over an affine variety V, an element $f \in \mathbf{R}$ represents a function over V whose zeroes form an affine subvariety W. Hence, $\mathbf{R}/\mathrm{D}_{\mathbf{R}}(\mathrm{K}_{\mathbf{R}}(f))$, which is the reduced ring associated to $\mathbf{R}^{\{f\}}$, is the ring $\mathbf{K}[W']$, where W' is the boundary of W in V.

The following theorem gives an inductive elementary characterization of the Krull dimension starting from dimension -1 which means that the ring is trivial $(1 = 0)$. This inductive characterization corresponds to the geometrical intuition that a variety is of dimension $\leq k$ if and only if any subvariety has a boundary of dimension $< k$.

Theorem 80. *For any ring* **R** *and* $\ell \in \mathbb{N}$, *the following assertions are equivalent:*

(i) *Kdim*$\mathbf{R} \leq \ell$.

(ii) *For any* $x \in \mathbf{R}$, *Kdim*$\mathbf{R}^{\{x\}} \leq \ell - 1$.

(iii) *For any* $x \in \mathbf{R}$, *Kdim*$\mathbf{R}_{\{x\}} \leq \ell - 1$.

(iv) *For any* $x_0, \dots, x_\ell \in \mathbf{R}$, *there exist* $a_0, \dots, a_\ell \in \mathbf{R}$ *and* $m_0, \dots, m_\ell \in \mathbb{N}$ *such that*

$$x_0^{m_0}(x_1^{m_1} \cdots (x_\ell^{m_\ell}(1 + a_\ell x_\ell) + \cdots + a_1 x_1) + a_0 x_0) = 0.$$

Proof. Let us first prove the equivalence between assertions (i) and (ii). Recall that for any monoid S of **R**, the prime ideals of R_S are of the form $S^{-1}\mathfrak{p} := \{ \frac{t}{s}, t \in \mathfrak{p}, s \in S \}$, where $\mathfrak{p}$ is a prime ideal of **R** not meeting S. The desired equivalence results from the following two immediate affirmations:

(a) For any $x \in \mathbf{R}$ and any maximal ideal $\mathfrak{m}$ of **R**, $S_{\{x\}} \cap \mathfrak{m} \neq \emptyset$.

(b) If $\mathfrak{m}$ is a maximal ideal of **R**, and if $x \in \mathfrak{m} \setminus \mathfrak{p}$, where $\mathfrak{p}$ is a prime ideal contained in $\mathfrak{m}$, then $S_{\{x\}} \cap \mathfrak{p} = \emptyset$.

Thus, if $\mathfrak{p}_0 \subsetneq \mathfrak{p}_1 \subsetneq \cdots \subsetneq \mathfrak{p}_\ell$ is a chain of proper prime ideals of **R** with $\mathfrak{p}_\ell$ maximal, then for any $x \in \mathbf{R}$, when localizing at $S_{\{x\}}$, it will be shortened to at least $S_{\{x\}}^{-1}\mathfrak{p}_0 \subsetneq S_{\{x\}}^{-1}\mathfrak{p}_1 \subsetneq \cdots \subsetneq S_{\{x\}}^{-1}\mathfrak{p}_{\ell-1}$, and to exactly $S_{\{x\}}^{-1}\mathfrak{p}_0 \subsetneq S_{\{x\}}^{-1}\mathfrak{p}_1 \subsetneq \cdots \subsetneq S_{\{x\}}^{-1}\mathfrak{p}_{\ell-1}$ if $x \in \mathfrak{p}_\ell \setminus \mathfrak{p}_{\ell-1}$.

The equivalence between assertions (i) and (iii) can be proved in a dual way, just replace prime ideals by prime filters. Recall that for any ideal $\mathfrak{J}$ of $\mathbf{R}$, the prime filters of $\mathbf{R}/\mathfrak{J}$ are of the form $(S+\mathfrak{J})/\mathfrak{J}$, where S is a prime filter of $\mathbf{R}$ not meeting $\mathfrak{J}$. Affirmations (a) and (b) are, thus, replaced by the following dual affirmations (a') and (b'):

(a') For any $x \in \mathbf{R}$ and any maximal filter S of $\mathbf{R}$, $S \cap K_{\mathbf{R}}(x) \neq \emptyset$.

(b') If S is a maximal filter of $\mathbf{R}$, and if $x \in S \setminus S'$, where $S' \subsetneq S$ is a prime filter, then $S' \cap K_{\mathbf{R}}(x) = \emptyset$.

Let us prove by induction on ℓ that the assertions (iii) and (iv) (for example) are equivalent. If $\ell = 0$, this is trivial. Suppose that the result is true for ℓ. If S is a monoid of $\mathbf{R}$, then $\mathrm{Kdim}\mathbf{R}_S \leq \ell$ if and only if for any $x_0, \ldots, x_\ell \in \mathbf{R}$, there exist $a_0, \ldots, a_\ell \in \mathbf{R}$, $m_0, \ldots, m_\ell \in \mathbb{N}$, and $s \in S$ such that $x_0^{m_0}(x_1^{m_1} \cdots (x_\ell^{m_\ell}(s + a_\ell x_\ell) + \cdots + a_1 x_1) + a_0 x_0) = 0$. Just replace s by an arbitrary element of the form $x_{\ell+1}^{m_{\ell+1}}(1 + a_{\ell+1} x_{\ell+1})$. $\qquad\square$

Remark 81. It is easy to see that if S is a monoid of a ring $\mathbf{R}$ and $\ell \in \mathbb{N}$, then $\mathrm{Kdim}(S^{-1}\mathbf{R}) \leq \ell$ if and only if for any $x_0, \ldots, x_\ell \in \mathbf{R}$, there exist $a_0, \ldots, a_\ell \in \mathbf{R}$, $s \in S$, and $m_0, \ldots, m_\ell \in \mathbb{N}$ such that

$$x_0^{m_0}(x_1^{m_1} \cdots (x_\ell^{m_\ell}(s + a_\ell x_\ell) + \cdots + a_1 x_1) + a_0 x_0) = 0.$$

2.3.4 Pseudo-Regular Sequences and Krull Dimension

Definition 82. Let $(x_1, \ldots, x_\ell)$ be a sequence of length ℓ in a ring $\mathbf{R}$.

- We say that the sequence $(x_1, \ldots, x_\ell)$ is singular (or collapses) if there exist $a_1, \ldots, a_\ell \in \mathbf{R}$ and $m_1, \ldots, m_\ell \in \mathbb{N}$ such that

$$x_1^{m_1}(x_2^{m_2} \cdots (x_\ell^{m_\ell}(1 + a_\ell x_\ell) + \cdots + a_2 x_2) + a_1 x_1) = 0.$$

- We say that the sequence $(x_1, \ldots, x_\ell)$ is pseudo-regular if it does not collapse.

Definition 83. Let $(x_1, \ldots, x_\ell)$ be a sequence of length ℓ in a ring $\mathbf{R}$.

- We say that an element x of an $\mathbf{R}$-module M is *regular* if its annihilator $\mathrm{Ann}(x)$ is null. If $M = \mathbf{R}$, we say also that x is not a zero-divisor.

- We say that the sequence $(x_1, \ldots, x_\ell)$ is *regular* if each x_i is regular in the ring $\mathbf{R}/\langle x_j; \ j < i \rangle$. Note that we adopt Bourbaki's definition of regular sequences as we do not suppose that $1 \notin \langle x_1, \ldots, x_\ell \rangle$.

The connection between singular and regular sequences is given by the following straightforward proposition.

Proposition 84. *If a sequence $(x_1, \ldots, x_\ell)$ is both singular and regular then $1 \in \langle x_1, \ldots, x_\ell \rangle$.*

Using Theorem 82 and the notion of pseudo-regular sequence, we can now formulate a constructive definition of Krull dimension.

Definition 85. (Constructive Definition of Krull Dimension via Singular Sequences)

(i) We say that a ring **R** has dimension -1 (in short, Kdim**R** $= -1$) if it is trivial $(1 = 0)$. Otherwise, we say that **R** has dimension ≥ 0.

(ii) We say that a ring **R** has dimension $\leq \ell - 1$ (in short, Kdim**R** $\leq \ell - 1$) if each sequence of length ℓ is singular (or collapses).

(iii) We say that a ring **R** has dimension $\geq \ell$ (in short, Kdim**R** $\geq \ell$) if there exists a pseudo-regular sequence of length ℓ.

(iv) We say that a ring **R** has dimension ℓ (in short, Kdim**R** $= \ell$) if its dimension is $\geq \ell$ and $\leq \ell$ at the same time.

(v) We say that a ring **R** has *finite Krull dimension* or is *finite-dimensional* if Kdim**R** $\leq \ell$ for some $\ell \in \mathbb{N}$.

Examples 86.   (1) A ring **R** has dimension ≤ 0 if and only if

$$\forall x \in \mathbf{R}, \; \exists n \in \mathbb{N}, \; \exists a \in \mathbf{R} \; \mid \; x^n = ax^{n+1}. \tag{2.3}$$

Here, it is worth pointing out the inherent difficulty of the constructive point of view: in order to be zero-dimensional constructively, a field must be *discrete*, i.e., it must have a zero test (the field of real numbers is not discrete. In numerical applications, the reals are known only via their rational approximations).

Any finite ring **R** ($\mathbb{Z}/m\mathbb{Z}$ for example) has Krull dimension ≤ 0. To see this, denoting by $k = \sharp(\mathbf{R})$ ($k \geq 2$; we suppose that **R** is not trivial) and considering $x \in \mathbf{R}$, necessarily, there exist $0 \leq r < r' \leq k$ such that $x^{r'} = x^r$.

(2) A ring **R** is local zero-dimensional if and only if

$$\forall x \in \mathbf{R}, \; x \text{ is invertible or nilpotent}. \tag{2.4}$$

For example, if p is a prime number and $k \in \mathbb{N}$, then the ring $\mathbb{Z}/p^k\mathbb{Z}$ is local and has Krull dimension ≤ 0.

(3) A ring **R** has dimension ≤ 1 if and only if

$$\forall a,b \in \mathbf{R}, \; \exists n \in \mathbb{N}, \; \exists x,y \in \mathbf{R} \; \mid \; a^n(b^n(1+xb)+ya) = 0. \tag{2.5}$$

For example, if p is a prime number then the ring

$$\mathbb{Z}_{p\mathbb{Z}} := \{\frac{a}{b} \in \mathbb{Q} \mid a \in \mathbb{Z} \text{ and } b \in \mathbb{Z} \setminus p\mathbb{Z}\}$$

is local and has Krull dimension ≤ 1. To see that $\mathrm{Kdim}(\mathbb{Z}_{p\mathbb{Z}}) \leq 1$, for $c \in \mathbb{Z} \setminus \{0\}$, denoting by $v_p(c) := \max\{k \in \mathbb{N} \mid p^k$ divides $c\}$ (the valuation of c at p), then, for any $x = \frac{a}{b}, x' = \frac{a'}{b'} \in \mathbb{Z}_{p\mathbb{Z}} \setminus \{0\}$ with $a, a' \in p\mathbb{Z} \setminus \{0\}$ and $b, b' \in \mathbb{Z} \setminus p\mathbb{Z}$, x' divides $x^{v_p(a')}$.

Another example is the ring $\mathbb{Z}$. To see this, let $a, b \in \mathbb{Z} \setminus \{-1, 0, 1\}$. Computing successively $d_1 = \gcd(a, b)$, $d_2 = \gcd(\frac{a}{d_1}, b), \ldots, d_n = \gcd(\frac{a}{d_1 \cdots d_{n-1}}, b)$, we eventually factorize a as

$$a = d_1 \cdots d_n a' \text{ with } d_i, a' \in \mathbb{Z}, \ d_i \mid b, \text{ and } \gcd(a', b) = 1.$$

Writing a Bezout identity $ca' + db = 1$ for some $c, d \in \mathbb{Z}$, we have

$$b^n(1 - db) \in \langle a \rangle.$$

For example, take $a = 700$ and $b = 6$. We have $d_1 = \gcd(700, 6) = 2$, $d_2 = \gcd(350, 6) = 2$, with $\gcd(175, 6) = 1$, and $175 - 29 \times 6 = 1$ as a Bezout identity. We infer that

$$6^2(1 + 29 \times 6) \in \langle 700 \rangle.$$

2.3.5 Krull Dimension of a Polynomial Ring Over a Discrete Field

We first need the following intermediary result.

Proposition 87. *Let* $\mathbf{K}$ *be a discrete field,* $\mathbf{R}$ *a* $\mathbf{K}$*-algebra, and* $x_1, \ldots, x_\ell \in \mathbf{R}$ *algebraically dependent over* $\mathbf{K}$*. Then the sequence* $(x_1, \ldots, x_\ell)$ *is singular.*

Proof. Let $Q(x_1, \ldots, x_\ell) = 0$ be an algebraic relation over $\mathbf{K}$ testifying the dependence between the x_i's. Let us order the monomials of Q with nonzero coefficients by the lexicographic order. We can without loss of generality suppose that the first nonzero coefficient of Q is 1. Denoting this monomial by $x_1^{m_1} \cdots x_\ell^{m_\ell}$, it is clear that Q can be written in the form

$$Q = x_1^{m_1} \cdots x_\ell^{m_\ell} + x_1^{m_1} \cdots x_\ell^{1+m_\ell} R_\ell + x_1^{m_1} \cdots x_{\ell-1}^{1+m_{\ell-1}} R_{\ell-1} + \cdots$$
$$+ x_1^{m_1} x_2^{1+m_2} R_2 + x_1^{1+m_1} R_1,$$

the desired collapse. $\qquad\qquad\qquad\qquad\qquad\qquad\qquad\qquad\qquad\qquad\qquad\qquad\qquad \square$

Theorem 88. *If* $\mathbf{K}$ *is a discrete field, then the Krull dimension of* $\mathbf{K}[X_1, \ldots, X_\ell]$ *is equal to* ℓ*.*

Proof. Just use Proposition 87 and the fact that the sequence $(X_1, \ldots, X_\ell)$ is pseudo-regular since it is regular (see Proposition 84). $\qquad\qquad\qquad\qquad\qquad\qquad \square$

Note that we have painlessly obtained this fundamental result quashing the common opinion that constructive proofs are necessarily more complicated than classical proofs.

2.3.6   Application to the Stable Range Theorem

In this subsection, we present a simple and elegant constructive proof of the stable range theorem due to Coquand [31].

Lemma 89. $\sqrt{\langle y, b \rangle} = \sqrt{\langle y + b, by \rangle}$.

Proof. It is clear that $\sqrt{\langle y + b, by \rangle} \subseteq \sqrt{\langle y, b \rangle}$. The converse follows from the identity $y^2 = (y + b)y - yb$. □

Lemma 90. *If by is nilpotent then* $1 \in \sqrt{\langle y, b \rangle} \Leftrightarrow 1 \in \sqrt{\langle y + b \rangle}$.

Proof. By virtue of Lemma 89, $1 \in \sqrt{\langle y, b \rangle} \Leftrightarrow 1 \in \sqrt{\langle y + b, by \rangle} \Leftrightarrow 1 \in \sqrt{\langle y + b \rangle}$ (*by* being nilpotent). □

Theorem 91. (Kronecker's Theorem) *If the Krull dimension of a ring $\mathbf{R}$ is $< s$ then for any $a, b_1, \ldots, b_s \in \mathbf{R}$ such that $1 \in \langle a, b_1, \ldots, b_s \rangle$, there exist $x_1, \ldots, x_s \in \mathbf{R}$ such that $1 \in \langle b_1 + ax_1, \ldots, b_s + ax_s \rangle$.*

Proof. We proceed by induction on s. If $s = 0$ the result is clear as the ring $\mathbf{R}$ is trivial. If $s > 0$, let I be the ideal boundary of b_s. We have $b_s \in I$ and the dimension of $\mathbf{R}/I < s - 1$. By induction, we can find $x_1, \ldots, x_{s-1}$ such that

$$1 \in \langle b_1 + ax_1, \ldots, b_{s-1} + ax_{s-1} \rangle$$

in $\mathbf{R}/I$. This means that there exists $x_s \in \mathbf{R}$ such that $b_s x_s$ is nilpotent and

$$1 \in \langle b_1 + ax_1, \ldots, b_{s-1} + ax_{s-1}, b_s, x_s \rangle.$$

Now, to obtain the desired result, one has only to reason modulo

$$\langle b_1 + ax_1, \ldots, b_{s-1} + ax_{s-1} \rangle$$

and to use Lemmas 89 and 90. □

As an immediate consequence, we get the following so-called stable range theorem.

Theorem 92. (Stable Range Theorem) *Let $\mathbf{R}$ be a ring of dimension $\leq d$, $n \geq d + 1$, and let $v = (v_0, \ldots, v_n) \in \mathrm{Um}_{n+1}(\mathbf{R})$. Then there exists $E \in \mathrm{E}_{n+1}(R)$ such that $Ev = (1, 0, \ldots, 0)$.*

Corollary 93. (Stable Range Theorem, bis) *For any ring $\mathbf{R}$ with Krull dimension $\leq d$, all finitely-generated stably free $\mathbf{R}$-modules of rank $> d$ are free.*

Proof. Use Proposition 22 and Theorem 92. □

2.3.7 Serre's Splitting Theorem and Forster-Swan Theorem

The constructive proofs we will give for Serre's Splitting Theorem and Forster-Swan Theorem are due to Coquand, Lombardi and Quitté. This constructive approach reveals the purely matrix aspect of these important theorems.

We adopt the following definition for the rank of a matrix.

Definition 94. *(Rank and Determinantial Ideals of a Matrix)*

(1) Recall that for a matrix G with entries in $\mathbf{R}$ and $k \in \mathbb{N}^*$, $\mathscr{D}_k(G)$ denotes the determinantial ideal of G of order k, that is, the ideal generated by all the minors of G of size k, with the convention $\mathscr{D}_0(G) = \langle 1 \rangle$.

(2) A linear application φ between two finite-rank free $\mathbf{R}$-modules (or, similarly, a matrix φ with entries in $\mathbf{R}$) is said to have

  – rank $\leq k$ if $\mathscr{D}_{k+1}(\varphi) = 0$,

  – rank $\geq k$ if $\mathscr{D}_k(\varphi) = \langle 1 \rangle$,

  – rank k if it has rank $\leq k$ and $\geq k$.

Definition 95. Let $\mathbf{R}$ be a ring and $n \in \mathbb{N}^*$.

(1) We say that $\mathrm{Sdim}\mathbf{R} < n$ if for every matrix F of rank $\geq n$ with entries in $\mathbf{R}$, there is a linear combination of its columns which is unimodular, i.e., $\mathscr{D}_n(F) = \langle 1 \rangle \Rightarrow \exists X \mid \mathscr{D}_1(FX) = \langle 1 \rangle$.

  We say that $\mathrm{Sdim}\mathbf{R} < 0$ if $\mathbf{R}$ is trivial.

(2) We say that $\mathrm{Gdim}\mathbf{R} < n$ if for every matrix $F = [C_0, C_1, \dots, C_p]$ (the C_i's stand for the columns of F) such that $\mathscr{D}_1(C_0) + \mathscr{D}_n([C_1, \dots, C_p]) = \langle 1 \rangle$, there exist $\lambda_1, \dots, \lambda_p \in \mathbf{R}$ such that $C_0 + \lambda_1 C_1 + \cdots + \lambda_p C_p$ is unimodular.

(3) It is clear that:

  $\mathrm{Sdim}\mathbf{R} = \mathrm{Sdim}(\mathbf{R}/\mathrm{Rad}(\mathbf{R}))$ & $\mathrm{Gdim}\mathbf{R} = \mathrm{Gdim}(\mathbf{R}/\mathrm{Rad}(\mathbf{R}))$.

  $\mathrm{Gdim}\mathbf{R} < n \Rightarrow \mathrm{Sdim}\mathbf{R} < n$.

  We will see in Corollary 115 that if $\mathbf{R}$ is n-stable (see Definition 105) then $\mathrm{Gdim}\mathbf{R} < n$.

Theorem 96. (Serre's Splitting Theorem) *Let $\mathbf{R}$ be a ring and M a finitely-generated projective $\mathbf{R}$-module of rank $r \geq k$ (or, more generally, an $\mathbf{R}$-module which is isomorphic to the image of a matrix of rank $r \geq k$). If $\mathrm{Sdim}\mathbf{R} < k$ (for example if $\mathrm{Kdim}\mathbf{R} < k$, see Corollary 116) then $M \cong N \oplus \mathbf{R}^{r-k+1}$, where N is isomorphic to the image of a matrix of rank $k - 1$.*

Proof. By induction, it suffices to prove that $M \cong N_1 \oplus \mathbf{R}$ where N_1 is isomorphic to the image of a matrix of rank $\geq k-1$. For this, let $F \in \mathbf{R}^{n \times m}$ be a matrix with $\mathcal{D}_k(F) = \langle 1 \rangle$ and $\mathrm{Im}(F) = M$. As $\mathrm{Sdim}\mathbf{R} < k$, there exists a unimodular vector $u \in \mathrm{Im}(F)$. It follows that $\mathbf{R}u \cong \mathbf{R}$, $\mathbf{R}u$ is a direct summand in $\mathbf{R}^n$ (see Proposition 2), and, thus, in M also. To see that N_1 has rank $\geq k-1$, it suffices to see it locally. $\qquad \square$

Theorem 97. (Forster-Swan Theorem) *Let $k \in \mathbb{N}$ and $\mathbf{R}$ be a ring with $\mathrm{Gdim}\mathbf{R} \leq k$ (for example if $\mathrm{Kdim}\mathbf{R} \leq k$, see Corollary 116). If a finitely-presented $\mathbf{R}$-module M is locally generated by r elements then it is generated by $k + r$ elements.*

Proof. Let $\{v_0, \ldots, v_p\}$ be a generating set for M with $p \geq k + r$ and denote by F a presentation matrix corresponding to this generating set. By assumption, we have $\langle 1 \rangle = \mathcal{F}_r(M) = \mathcal{D}_{p+1-r}(F)$ (see Remark 9 for the definition of the Fitting ideals $\mathcal{F}_i(M)$). As $p + 1 - r \geq k + 1$, we have $\langle 1 \rangle = \mathcal{D}_{k+1}(F)$.

Let us denote by $F = \begin{bmatrix} L_0 \\ L_1 \\ \vdots \\ L_p \end{bmatrix}$, where $L_0, \ldots, L_p$ are the rows of F. As $\mathrm{Gdim}\mathbf{R} \leq k$

and ${}^t F$ has rank $\geq k + 1$, there exist $\lambda_1, \ldots, \lambda_p \in \mathbf{R}$ such that $L'_0 := L_0 + \lambda_1 L_1 + \cdots + \lambda_p L_p$ is unimodular. Denote by $L'_0 = (\alpha_0, \ldots, \alpha_m)$ with $\alpha_0 \beta_0 + \cdots + \alpha_m \beta_m = 1$

for some $\beta_0, \ldots, \beta_m \in \mathbf{R}$. The matrix $F' = \begin{bmatrix} L'_0 \\ L_1 \\ \vdots \\ L_p \end{bmatrix} = [C'_0 \ C'_1 \ \cdots \ C'_m]$, where the

C'_i's denote the columns of F', is a new presentation matrix of M corresponding to the generating set $\{v'_0 = v_0, v'_1 = v_1 - \lambda_1 v_0, \ldots, v'_p = v_p - \lambda_p v_0\}$. As $\beta_0 C'_0 + \cdots + \beta_m C'_m = {}^t(1, *, \ldots, *)$, we infer that $v'_0 \in \langle v'_1, \ldots, v'_p \rangle$ and, thus, $M = \langle v'_1, \ldots, v'_p \rangle$. $\qquad \square$

2.3.8   Support on a Ring and n-Stability

Definition 98.

(i) A *support* on a ring $\mathbf{R}$ is a map $D : \mathbf{R} \to \mathbf{T}$ from $\mathbf{R}$ to a distributive lattice $\mathbf{T}$ satisfying the following properties:

  – $D(0_{\mathbf{R}}) = 0_{\mathbf{T}}, D(1_{\mathbf{R}}) = 1_{\mathbf{T}}$,

  – $D(ab) = D(a) \wedge D(b)$,

  – $D(a + b) \leq D(a) \vee D(b)$.

We will denote by $D(x_1, \ldots, x_n) := D(x_1) \vee \cdots D(x_n)$.

(ii) Let D be a support on a ring $\mathbf{R}$. We say that two sequences $(x_0,\ldots,x_n)$ and $(y_0,\ldots,y_n)$ of elements in $\mathbf{R}$ are D-*complementary* if the following inequalities hold:

$$\begin{cases} D(y_0 x_0) = D(0) \\ D(y_1 x_1) \le D(y_0, x_0) \\ \vdots \quad \vdots \quad \vdots \\ D(y_n x_n) \le D(y_{n-1}, x_{n-1}) \\ D(1) = D(y_n, x_n) \end{cases} \qquad (2.6)$$

(iii) A support D on a ring $\mathbf{R}$ is said to of *Krull dimension* $\le n$ (in short, Kdim$D \le n$), if every sequence $(x_0,\ldots,x_n)$ in $\mathbf{R}$ has a D-complementary sequence.

Example 99. (Zariski Lattice) Let $\mathbf{R}$ be a ring. Recall that for $a_1,\ldots,a_n \in \mathbf{R}$, we denote by $\mathfrak{a} = \langle a_1,\ldots,a_n \rangle$,

$$D_{\mathbf{R}}(a_1,\ldots,a_n) = D_{\mathbf{R}}(\mathfrak{a}) = \sqrt{\mathfrak{a}} \ \& \ \mathrm{Zar}(\mathbf{R}) = \{ D_{\mathbf{R}}(a_1,\ldots,a_n) \mid n \in \mathbb{N}, a_i \in \mathbf{R} \}.$$

$\mathrm{Zar}(\mathbf{R})$ ordered under inclusion is a distributive lattice called *Zariski lattice* with

$$D_{\mathbf{R}}(0) = 0_{\mathrm{Zar}(\mathbf{R})}, \quad D_{\mathbf{R}}(\mathfrak{a}_1) \vee D_{\mathbf{R}}(\mathfrak{a}_2) = D_{\mathbf{R}}(\mathfrak{a}_1 + \mathfrak{a}_2),$$
$$D_{\mathbf{R}}(1) = 1_{\mathrm{Zar}(\mathbf{R})}, \quad D_{\mathbf{R}}(\mathfrak{a}_1) \wedge D_{\mathbf{R}}(\mathfrak{a}_2) = D_{\mathbf{R}}(\mathfrak{a}_1 \, \mathfrak{a}_2).$$

Within classical mathematics, $D_{\mathbf{R}}(a_1,\ldots,a_n)$ can be seen as a quasi-compact open subset of the prime spectrum $\mathrm{Spec}(\mathbf{R})$ of $\mathbf{R}$. It corresponds to $\{\mathfrak{p} \in \mathrm{Spec}(\mathbf{R}) \mid \langle a_1,\ldots,a_n \rangle \nsubseteq \mathfrak{p}\}$.
The map $D_{\mathbf{R}} : \mathbf{R} \to \mathrm{Zar}(\mathbf{R})$ is a support called *Zariski support*.

By the following, we see that a ring has the same Krull dimension as his Zariski support.

Proposition 100. *Let $\mathbf{R}$ be a ring and $n \in \mathbb{N}$. Then $\mathbf{R}$ has Krul dimension $\le n$ if and only if its Zariski support has Krul dimension $\le n$, i.e., for every sequence $(x_0,\ldots,x_n)$ of elements in $\mathbf{R}$ there exists a sequence $(y_0,\ldots,y_n)$ of elements in $\mathbf{R}$ such that:*

$$\begin{cases} D_{\mathbf{R}}(y_0 x_0) = D_{\mathbf{R}}(0) \\ D_{\mathbf{R}}(y_1 x_1) \le D_{\mathbf{R}}(y_0, x_0) \\ \vdots \quad \vdots \quad \vdots \\ D_{\mathbf{R}}(y_n x_n) \le D_{\mathbf{R}}(y_{n-1}, x_{n-1}) \\ D_{\mathbf{R}}(1) = D_{\mathbf{R}}(y_n, x_n) \end{cases} \qquad (2.7)$$

Proof. "$\Rightarrow$" Let $x_0,\ldots,x_n \in \mathbf{R}$. As the Krull dimension of $\mathbf{R}$ is $\le n$, the sequence $(x_0,\ldots,x_n)$ collapses, that is, there exist $a_0,\ldots,a_n \in \mathbf{R}$ and $m_0,\ldots,m_n \in \mathbb{N}$ such that

$$x_0^{m_0}(x_1^{m_1} \cdots (x_n^{m_n}(1 + a_n x_n) + \cdots + a_1 x_1) + a_0 x_0) = 0.$$

Take $y_n = 1 + a_n x_n$ and then $y_{r-1} = x_r^{m_r} y_r + a_{r-1} x_{r-1}$ successively for $r = n,\ldots,1$.
"$\Leftarrow$" We proceed by induction on n.

For $n=0$, let $x_0 \in \mathbf{R}$ and consider $y_0 \in \mathbf{R}$ such that $D_\mathbf{R}(y_0 x_0) = D_\mathbf{R}(0)$ and $D_\mathbf{R}(1) = D_\mathbf{R}(y_0, x_0)$. It follows that there exist $\ell \in \mathbb{N}$ and $c, d \in \mathbf{R}$ such that $y_0^\ell x_0^\ell = 0$ and $cx_0 + dy_0 = 1$. Thus, $(1 - cx_0)^\ell x_0^\ell = 0$ with $(1 - cx_0)^\ell \in 1 + x_0 \mathbf{R}$, as desired (see Example 86-(1)).

Now, let us denote $\mathbf{T} := \mathbf{R}^{\{x_0\}} = \mathbf{R}/(\langle x \rangle + (D_\mathbf{R}(0) : x_0))$. From Equalities (2.7), we infer that:

$$\begin{cases} D_\mathbf{T}(\bar{y}_1 \bar{x}_1) \leq D_\mathbf{T}(\bar{y}_0, \bar{x}_0) \\ D_\mathbf{T}(\bar{y}_2 \bar{x}_2) \leq D_\mathbf{T}(\bar{y}_1, \bar{x}_1) \\ \quad \vdots \qquad \vdots \qquad \vdots \\ D_\mathbf{T}(\bar{y}_n \bar{x}_n) \leq D_\mathbf{T}(\bar{y}_{n-1}, \bar{x}_{n-1}) \\ D_\mathbf{T}(\bar{1}) = D_\mathbf{T}(\bar{y}_n, \bar{x}_n) \end{cases} \qquad (2.8)$$

Now, as $y_0 x_0$ is nilpotent in $\mathbf{R}$, we have $\langle y_0, x_0 \rangle \subseteq D_\mathbf{R}(0) : x_0$. Since $(y_1 x_1)^m \in \langle y_0, x_0 \rangle$ for some $m \in \mathbb{N}$, we deduce that $\bar{y}_1 \bar{x}_1$ is nilpotent in $\mathbf{T}$, or also that $D_\mathbf{T}(\bar{y}_1 \bar{x}_1) = D_\mathbf{T}(\bar{0})$. By the induction hypothesis, we have $\mathrm{Kdim}\mathbf{T} \leq n-1$ and, thus, $\mathrm{Kdim}\mathbf{R} \leq n$ by virtue of Theorem 80. $\qquad \square$

Our goal now is to generalize Kronecker's Theorem 91 to supports. For this, we need the following two lemmas.

Lemma 101. *Let D be a support on a ring $\mathbf{R}$. Then, for every $u, v \in \mathbf{R}$, we have*

$$D(u,v) = D(u+v, uv) = D(u+v) \vee D(uv).$$

In particular, if $D(uv) = D(0)$, we have $D(u,v) = D(u+v)$.

Proof. As $D(uv) = D(u) \wedge D(v) \leq D(u) \vee D(v)$ and $D(u+v) \leq D(u) \vee D(v)$, we have $D(u+v, uv) = D(u+v) \vee D(uv) \leq D(u) \vee D(v)$.
Conversely, as $u^2 = (u+v)u - uv$, we have $D(u^2) = D(u) \leq D((u+v)u) \vee D(uv) \leq D(u+v) \vee D(uv)$. $\qquad \square$

Lemma 102. *Let D be a support on a ring $\mathbf{R}$. If $(b_1, \ldots, b_n)$ and $(x_1, \ldots, x_n)$ are two complementary sequences in $\mathbf{R}$ ($n \geq 1$) then, for any $a \in \mathbf{R}$, we have:*

$$D(a, b_1, \ldots, b_n) = D(b_1 + ax_1, \ldots, b_n + ax_n),$$

or also:

$$D(a) \leq D(b_1 + ax_1, \ldots, b_n + ax_n).$$

Proof. As

$$\begin{cases} D(b_1 x_1) = D(0) \\ D(b_2 x_2) \leq D(b_1, x_1) \\ \quad \vdots \qquad \vdots \qquad \vdots \\ D(b_n x_n) \leq D(b_{n-1}, x_{n-1}) \\ D(1) = D(b_n, x_n) \end{cases}$$

we infer that

$$\begin{cases} D(ab_1x_1) = D(0) \\ D(ab_2x_2) \le D(b_1, ax_1) \\ \vdots \qquad \vdots \qquad \vdots \\ D(ab_nx_n) \le D(b_{n-1}, ax_{n-1}) \\ D(a) = D(b_n, ax_n) \end{cases}$$

Thus, by virtue of Lemma 101, we have

$$\begin{cases} D(a) \le D(b_n + ax_n) \vee D(b_n ax_n) \\ D(ab_nx_n) \le D(b_{n-1} + ax_{n-1}) \vee D(b_{n-1}ax_{n-1}) \\ \vdots \qquad \vdots \qquad \vdots \\ D(ab_3x_3) \le D(b_2 + ax_2) \vee D(b_2 ax_2) \\ D(ab_2x_2) \le D(b_1 + ax_1) \vee D(b_1 ax_1) = D(b_1 + ax_1) \end{cases}$$

Combining the inequalities above, one obtains
$$D(a) \le D(b_1 + ax_1) \vee D(b_2 + ax_2) \cdots D(b_n + ax_n) = D(b_1 + ax_1, \ldots, b_n + ax_n).$$

$\square$

As an immediate consequence of Lemma 102, one obtains:

Theorem 103. (Kronecker's Theorem for Supports) *Let D be a support of Krull dimension $\le n$ on a ring $\mathbf{R}$. Then for any finitely-generated ideal $\mathfrak{a}$ of $\mathbf{R}$, there exists an ideal $\mathfrak{b}$ of $\mathbf{R}$ generated by $n+1$ elements such that $D(\mathfrak{a}) = D(\mathfrak{b})$.*
More precisely, for any $b_1, \ldots, b_{n+r} \in \mathbf{R}$ ($r \ge 2$) there exist $c_j \in \langle b_{n+2}, \ldots, b_{n+r} \rangle$, $1 \le j \le n+1$, such that $D(b_1, \ldots, b_{n+r}) = D(b_1 + c_1, \ldots, b_{n+1} + c_{n+1})$.

In particular, considering the Zariski support, one obtains:

Corollary 104. (Kronecker's Theorem, bis) *In a ring of Krull dimension $\le n$, every finitely-generated ideal has the same radical as an ideal generated by $n+1$ elements.*

Definition 105. (*n*-Stable Support)

(1) Let $n \ge 1$. A support D on a ring $\mathbf{R}$ is said to be *n-stable* if for any $a \in \mathbf{R}$ and $L \in \mathbf{R}^n$, there exists $X \in \mathbf{R}^n$ such that $D(L, a) = D(L + aX)$, i.e., $D(a) \le D(L + aX)$.

(2) Let $n \ge 1$. A ring $\mathbf{R}$ is said to be *n*-stable if its Zariski support $D_{\mathbf{R}}$ is *n*-stable.

(3) A ring $\mathbf{R}$ is said to be 0-stable if it is trivial.

Remark 106. If a support D on a ring $\mathbf{R}$ is *n*-stable then for any $a \in \mathbf{R}$ and $L \in \mathbf{R}^n$, there exists $X \in \mathbf{R}^n$ such that $D(L, a) = D(L + a^2X)$, i.e., $D(a) \le D(L + a^2X)$. This ensues from the fact that $D(a) = D(a^2)$ and $D(L, a) = D(L, a^2)$.

Example 107.

(1) Let D be a support on a ring $\mathbf{R}$ and $n \in \mathbb{N}$. If $\mathrm{Kdim}(D) < n$ then, by virtue of Lemma 102, D is n-stable. In particular, if $\mathrm{Kdim}\mathbf{R} < n$ then $\mathbf{R}$ is n-stable.

(2) A valuation ring $\mathbf{V}$ is 1-stable as for any a, $b \in \mathbf{V}$, if a/b then $\langle a, b \rangle = \langle a \rangle$, and if b/a then $\langle a, b \rangle = \langle b \rangle$.

Definition 108. (Localization and Quotient of a Support) Let $\mathbf{R}$ be a ring and $D : \mathbf{R} \to \mathbf{T}$ a support on $\mathbf{R}$.
(1) If $\mathfrak{a}$ is a finitely-generated ideal of $\mathbf{R}$, then, denoting by π the projection $\mathbf{T} \to \mathbf{T}/(D(\mathfrak{a}) = 0)$, one obtains a support $D/\mathfrak{a} = \pi \circ D : \mathbf{R} \to \mathbf{T}/(D(\mathfrak{a}) = 0)$.

It is clear that if D is n-stable then so is $D/\mathfrak{a}$.

(2) If $u \in \mathbf{R}$, then, denoting by p the projection $\mathbf{T} \to \mathbf{T}/(D(u) = 1)$, one obtains a support $D[\frac{1}{u}] = p \circ D : \mathbf{R} \to \mathbf{T}[\frac{1}{u}] := \mathbf{T}/(D(u) = 1)$.

It is clear that if D is n-stable then so is $D[\frac{1}{u}]$.

The following lemma is particularly important. Let us first take an example which speaks for itself: consider the ring $\mathbb{Z}_{1+p\mathbb{Z}}[X]$ where p is a prime number. Since $\mathrm{Kdim}\mathbb{Z}_{1+p\mathbb{Z}}[X] \leq 2$, we know by Lemma 102 that $\mathbb{Z}_{1+p\mathbb{Z}}[X]$ is 3-stable. But, using Proposition 109 below, one can do better: as

$$\mathbb{Z}_{1+p\mathbb{Z}}[X][\frac{1}{p}] = \mathbb{Q}[X] \ \& \ (\mathbb{Z}_{1+p\mathbb{Z}}[X])/\langle p \rangle = \mathbb{F}_p[X] \ \& \ \mathrm{Kdim}\mathbb{Q}[X] = \mathrm{Kdim}\mathbb{F}_p[X] = 1,$$

we infer that $\mathbb{Z}_{1+p\mathbb{Z}}[X]$ is 2-stable. More generally, consider a ring $\mathbf{A}$ of Krull dimension ≤ 1 and suppose that there exists a regular element r in $\mathrm{Rad}(\mathbf{A})$. Since $\mathrm{Kdim}\mathbf{A}[\frac{1}{r}] \leq 0$ and $\mathrm{Kdim}(\mathbf{A}/\langle r \rangle) \leq 0$, we have $\mathrm{Kdim}\mathbf{A}[\frac{1}{r}][X] \leq 1$ and $\mathrm{Kdim}(\mathbf{A}/\langle r \rangle)[X] \leq 1$, and thus, $\mathbf{A}[X]$ is 2-stable.

Lemma 109. (Coquand's Lemma) *Let D be a support on a ring $\mathbf{R}$ and pick $a \in \mathbf{R}$. Then, D is n-stable if and only if both $D/\langle a \rangle$ and $D[\frac{1}{a}]$ are n-stable. In particular, $\mathbf{R}$ is n-stable if and only if both $\mathbf{R}/\langle a \rangle$ and $\mathbf{R}[\frac{1}{a}]$ are n-stable.*

Proof. The implication "$\Rightarrow$" is immediate.
"$\Leftarrow$" Let $b \in \mathbf{R}$ and $L \in \mathbf{R}^n$. Since $D/\langle a \rangle$ is n-stable, there exists $Y \in \mathbf{R}^n$ such that $D(b) \leq D(L + bY)$ in $\mathbf{T}/(D(a) = 0)$, i.e., in $\mathbf{T}$ we have:

$$D(b) \leq D(a) \vee D(L + bY) \tag{2.9}$$

Now, as $D[\frac{1}{a}]$ is n-stable, considering the element ab and the vector $L + bY$, there exists $Z \in \mathbf{R}^n$ such that $D(ab) \leq D(L + bY + abZ)$ in $\mathbf{T}/(D(a) = 1)$, i.e., in $\mathbf{T}$ we have:

$$D(ab) \wedge D(a) \leq D(L + bX),$$

and, thus, since $D(ab) = D(a) \wedge D(b)$, we have

$$D(ab) \leq D(L + bX) \tag{2.10}$$

But, as $\langle a, L + bX \rangle = \langle a, L + bY \rangle$, we have $D(a, L + bX) = D(a, L + bY)$. Inequalities (2.9) and (2.10) become

$$D(b) \leq D(a) \vee D(L + bX) \quad \& \quad D(a) \wedge D(b) \leq D(L + bX).$$

It follows that $D(b) \leq D(L + bX)$ (see Exercise 383). □

Theorem 110. (n-Stable Induction) *Let $\mathscr{F}$ be a class of rings containing the class of zero-dimensional rings, contained in the class of finite-dimensional rings, and satisfying the following induction rule:*

$$\forall \mathbf{R} \in \mathscr{F}, \; \exists a \in \mathbf{R} \mid \forall \mathbf{A} \in \{\mathbf{R}/\langle a \rangle, \; \mathbf{R}[\tfrac{1}{a}]\}, \; \mathbf{A} \in \mathscr{F} \; \& \; \mathrm{Kdim}\mathbf{A} < \sup(1, \mathrm{Kdim}\mathbf{R}).$$

Then, for each $\mathbf{R} \in \mathscr{F}$ and $n \in \mathbb{N}$, $\mathbf{R}[X_1, \ldots, X_n]$ is $(n+1)$-stable.

Proof. Let $\mathbf{R} \in \mathscr{F}$ and $n \in \mathbb{N}$. To prove that $\mathbf{R}[X_1, \ldots, X_n]$ is $(n+1)$-stable, we proceed by induction on $\mathrm{Kdim}\mathbf{R}$. If $\mathrm{Kdim}\mathbf{R} \leq 0$, then as $\mathrm{Kdim}\mathbf{R}[X_1, \ldots, X_n] \leq n$, the result follows from Lemma 102. Else, there exists $a \in \mathbf{R}$ such that for $\mathbf{A} \in \{\mathbf{R}/\langle a \rangle, \; \mathbf{R}[\tfrac{1}{a}]\}$, either $\mathrm{Kdim}\mathbf{A} \leq 0$ or ($\mathbf{A} \in \mathscr{F} \; \& \; \mathrm{Kdim}\mathbf{A} < \mathrm{Kdim}\mathbf{R}$). Using the induction hypothesis, the fact "$\mathrm{Kdim}\mathbf{A} \leq 0 \Rightarrow \mathbf{A}[X_1, \ldots, X_n]$ is $(n+1)$-stable", and Lemma 109, we obtain the desired result. □

Corollary 111. *If $\mathbf{R}$ is a domain of Krull dimension 1 such that $\mathrm{Rad}(\mathbf{R}) \neq \{0\}$ (for example, a local domain of Krull dimension 1), or a valuation ring of finite Krull dimension, or a ring with finite Zariski lattice (finite prime spectrum), then, for any $n \in \mathbb{N}$, $\mathbf{R}[X_1, \ldots, X_n]$ is $(n+1)$-stable.*

Proof. First case: $\mathbf{R}$ is a domain of Krull dimension 1 such that $\mathrm{Rad}(\mathbf{R}) \neq \{0\}$. It is clear that the class $\mathscr{F}_1$ of domains of Krull dimension 1 such that $\mathrm{Rad}(\mathbf{R}) \neq \{0\}$ and rings of Krull dimension ≤ 0 satisfies the induction hypotheses of Theorem 110. As a matter of fact, if $\mathbf{R}$ is a one-dimensional domain then for any $a \in \mathrm{Rad}(\mathbf{R}) \setminus \{0\}$, we have $\mathrm{Kdim}\mathbf{R}[\tfrac{1}{a}] \leq 0 \; \& \; \mathrm{Kdim}(\mathbf{R}/\langle a \rangle) \leq 0$.

Second case: $\mathbf{R}$ is a valuation ring of finite Krull dimension. It is clear that the class $\mathscr{F}_2$ of valuation rings with finite Krull dimension and rings of Krull dimension ≤ 0 satisfies the induction hypotheses of Theorem 110. As a matter of fact, if $\mathbf{R}$ is a valuation ring with $1 \leq \mathrm{Kdim}\mathbf{R} \leq \ell$ for some $\ell \in \mathbb{N}^*$, picking $a \in \mathbf{R}$ which is neither invertible nor nilpotent, then both of the rings $\mathbf{R}[\tfrac{1}{a}]$ and $\mathbf{R}/\langle a \rangle$ are in $\mathscr{F}_2$ and $\max(\mathrm{Kdim}(\mathbf{R}[\tfrac{1}{a}]), \mathrm{Kdim}(\mathbf{R}/\langle a \rangle)) < \mathrm{Kdim}\mathbf{R}$.

Third case: $\mathbf{R}$ has a finite prime spectrum. Denote by $\mathscr{F}_3$ the class of rings with finite prime spectrum and rings of Krull dimension ≤ 0. It is clear that for $\mathbf{A} \in \mathscr{F}_3$ and $a \in \mathbf{A}$, both of the rings $\mathbf{A}[\tfrac{1}{a}]$ and $\mathbf{A}/\langle a \rangle$ are in $\mathscr{F}_3$. Now, suppose that $\mathbf{R}$ is a ring with finite number of prime ideals and with Krull dimension ≥ 1. Let us denote by $\{\mathfrak{m}_1, \ldots, \mathfrak{m}_s\}$ the set of maximal ideals of $\mathbf{R}$ which are not minimal, and $\{\mathfrak{p}_1, \ldots, \mathfrak{p}_r\}$

the set of minimal prime ideals of $\mathbf{R}$. As $\cap_{i=1}^{s} \mathfrak{m}_i \not\subseteq \cup_{j=1}^{r} \mathfrak{p}_j$, one can pick an element $a \in \cap_{i=1}^{s} \mathfrak{m}_i \setminus \cup_{j=1}^{r} \mathfrak{p}_j$, and thus, we have $\max(\mathrm{Kdim}(\mathbf{R}[\frac{1}{a}]), \mathrm{Kdim}(\mathbf{R}/\langle a \rangle)) < \mathrm{Kdim}\mathbf{R}$. $\qquad \square$

It is worth pointing out that Corollary 111 can be extended to supports.

Corollary 112. *If* $\mathbf{V}$ *is a valuation ring, then for any* $n \in \mathbb{N}$, $\mathbf{V}[X_1, \ldots, X_n]$ *is* $(n+1)$-*stable.*

Proof. Let $a \in \mathbf{V}$ and $L = (\ell_1, \ldots, \ell_{n+1}) \in \mathbf{V}^{n+1}$. Our goal is to find $X = (x_1, \ldots, x_{n+1}) \in \mathbf{V}^{n+1}$ such that $D_{\mathbf{V}}(L, a) = D_{\mathbf{V}}(L + aX)$. Let us denote by $\mathbf{V}_1$ the finite-dimensional subring of $\mathbf{V}$ generated by the coefficients of $a, \ell_1, \ldots, \ell_{n+1}$. Then $\mathbf{V}_2 := \{c/b \mid c, b \in \mathbf{V}_1 \ \& \ b \text{ is regular and divides } c \text{ in} \mathbf{V}\}$ is a valuation subring of $\mathbf{V}$ with finite Krull dimension. As $\mathbf{V}_2[X_1, \ldots, X_n]$ is $(n+1)$-stable by Corollary 111, we can find $X \in \mathbf{V}_2^{n+1} \subseteq \mathbf{V}^{n+1}$ satisfying the required equality. $\qquad \square$

Our goal now is to prove that if $\mathbf{R}$ is a n-stable ring then $\mathrm{Gdim}\mathbf{R} < n$. The proof of this important fact relies on the following lemma.

Lemma 113. (Coquand's Lemma [108]) *Let* D *be a support on a ring* $\mathbf{R}$. *If* D *is* n-*stable then for any matrix* $F = [C_0, C_1, \ldots, C_n] \in \mathbf{R}^{n \times (n+1)}$ *(the* C_i's *stand for the columns of* F*), denoting by* $G = [C_1, \ldots, C_n]$ *and* $\delta = \det G$, *there exist* $\lambda_1, \ldots, \lambda_n \in \mathbf{R}$ *such that*

$$D(C_0, \delta) \le D(C_0 + \delta(\lambda_1 C_1 + \cdots + \lambda_n C_n)).$$

Proof. We want to find $\Lambda = (\lambda_1, \ldots, \lambda_n) \in \mathbf{R}^n$ such that

$$D(\delta) \le D(C_0 + \delta G \Lambda).$$

Let us denote by $\tilde{G}$ the adjugate matrix of G (the transpose of the cofactor matrix of G) and $L = \tilde{G} C_0$. For any $\Lambda \in \mathbf{R}^n$, we have $\tilde{G}(C_0 + \delta G \Lambda) = L + \delta^2 \Lambda$, and, thus, $D(L + \delta^2 \Lambda) \le D(C_0 + \delta G \Lambda)$. As D is n-stable then, by virtue of Remark 106, there exists $\Lambda \in \mathbf{R}^n$ such that $D(\delta) \le D(L + \delta^2 \Lambda)$, and, thus, $D(\delta) \le D(C_0 + \delta G \Lambda)$, as desired. $\qquad \square$

Theorem 114. (Coquand's Theorem on n-Stability and G-Dimension [33]) *Let* D *be a support on a ring* $\mathbf{R}$ *and* $n \in \mathbb{N}$. *If* D *is* n-*stable then for any matrix* $F = [C_0, C_1, \ldots, C_p] \in \mathbf{R}^{m \times (p+1)}$ *(the* C_i's *stand for the columns of* F*), denoting by* $G = [C_1, \ldots, C_p]$, *there exist* $\lambda_1, \ldots, \lambda_p \in \mathscr{D}_n(G)$ *such that*

$$D(C_0, \mathscr{D}_n(G)) \le D(C_0 + \lambda_1 C_1 + \cdots + \lambda_p C_p).$$

In particular, if $D(C_0, \mathscr{D}_n(G)) = 1$ *then there exist* $\lambda_1, \ldots, \lambda_p \in \mathscr{D}_n(G)$ *such that* $D(C_0 + \lambda_1 C_1 + \cdots + \lambda_p C_p) = 1$.

Proof. Let δ_1 be a minor of size n of G. Write $\delta_1 = \det(\Gamma_1)$ where Γ_1 is the matrix extracted from G corresponding to the minor δ_1. Let us denote by $\Gamma_{1,0}$ the vector extracted from C_0 by keeping only the rows of Γ_1. By Lemma 113, we know that there exists $Y_1 \in \mathbf{R}^n$ such that

$$D(\delta_1) \leq D(\Gamma_{1,0} + \delta_1 \Gamma_1 Y_1) \leq D(C_0 + \delta_1 G \Lambda_1) = D(C_0 + \lambda_{1,1} C_1 + \cdots + \lambda_{1,p} C_p),$$

where $\Lambda_1 = (\lambda_{1,1}, \ldots, \lambda_{1,p}) \in \mathbf{R}^p$ $(\lambda_{1,j} \in \mathscr{D}_n(G))$ is obtained by completing Y_1 with zeroes. Thus,

$$D(C_0, \delta_1) \leq D(C_0 + \lambda_{1,1} C_1 + \cdots + \lambda_{1,p} C_p).$$

Repeating this process with a new minor δ_2, one obtains $\lambda_{2,1}, \ldots, \lambda_{2,p} \in \mathscr{D}_n(G)$ such that

$$D(C_0, \delta_1, \delta_2) \leq D(C_0 + \lambda_{2,1} C_1 + \cdots + \lambda_{2,p} C_p),$$

and so on.

□

Corollary 115. *Let* $\mathbf{R}$ *be a ring and* $n \in \mathbb{N}$. *If* $\mathbf{R}$ *is n-stable then* $\mathrm{Gdim}\mathbf{R} < n$.

Corollary 116. *Let* $\mathbf{R}$ *be a ring and* $n \in \mathbb{N}$. *We have the following implications:*

$$\mathrm{Kdim}\mathbf{R} < n \Rightarrow \mathbf{R} \text{ is } n-\text{stable} \Rightarrow \mathrm{Gdim}\mathbf{R} < n \Rightarrow \mathrm{Sdim}\mathbf{R} < n.$$

Proof. Use Example 107.(1) and Corollary 115.

□

Corollary 117. *Let* $\mathbf{R}$ *be a ring of Krull dimension* ≤ 0, *or a domain of Krull dimension 1 such that* $\mathrm{Rad}(\mathbf{R}) \neq \{0\}$ *(for example, a local domain of Krull dimension 1), or a valuation ring, or a ring with finite Zariski lattice. Then for* $n \in \mathbb{N}^*$, *if* M *is a finitely-generated projective* $\mathbf{R}[X_1, \ldots, X_n]$-*module of rank* $r \geq n+1$ *then* $M \cong N \oplus \mathbf{R}^{r-n}$ *where* N *is isomorphic to the image of a matrix of rank* n.

Proof. Use Corollaries 111, 112 and 115, and Serre's Splitting Theorem 96.

□

Before giving the next two consequences of Corollary 117, let us first recall the definition of a seminormal ring together with the Traverso-Swan-Coquand Theorem (for a constructive proof, see Exercise 381).

Definition 118. A ring $\mathbf{R}$ will be called *seminormal* if for every $b, c \in \mathbf{R}$ satisfying $b^2 = c^3$ there exists $a \in \mathbf{R}$ such that $a^3 = b$ and $a^2 = c$.
Note that an integral ring $\mathbf{R}$ is seminormal if and only if for every $b, c \in \mathbf{R}$ satisfying $b^2 = c^3$, we have $\frac{b}{c} \in \mathbf{R}$.

Theorem 119. (Traverso-Swan-Coquand) *For a ring R. The following assertions are equivalent:*

(i) *All finitely-generated projective* $\mathbf{R}[X]$-*modules of rank 1 are extended from* $\mathbf{R}$.

(ii) *For all $n \geq 1$, all finitely-generated projective $\mathbf{R}[X_1, \ldots, X_n]$-modules of rank 1 are extended from $\mathbf{R}$.*

(iii) $\mathbf{R}_{\text{red}} := \mathbf{R}/D_{\mathbf{R}}(0)$ *is seminormal.*

Proof. See Exercise 381. □

In the particular case where $n = 1$, we obtain the following two consequences of Corollary 117:

Corollary 120. *Let $\mathbf{R}$ be a ring of Krull dimension ≤ 1, or more generally, a ring R which is locally of finite prime spectrum (i.e., for any maximal ideal $\mathfrak{m}$ of $\mathbf{R}$, the set of prime ideals of $\mathbf{R}$ contained in $\mathfrak{m}$ is finite; of course, this last hypothesis is not constructive). Then finitely-generated projective $\mathbf{R}[X]$-modules are extended from $\mathbf{R}$ if and only if $\mathbf{R}_{\text{red}}$ is seminormal.*

Theorem 121. *(Bass-Simis-Vasconcelos [163]) If $\mathbf{R}$ is a valuation ring (resp., an arithmetical ring) then every finitely-generated projective $\mathbf{R}[X]$-module is free (resp., is extended from $\mathbf{R}$).*

Proof. By the Quillen's patching Theorem 45, it suffices to prove the result for valuation rings. By Corollary 117, it suffices to deal with the case of a rank 1 projective $\mathbf{R}[X]$-module. It is well-known that a reduced valuation ring $\mathbf{R}$ is without zero-divisors (for $a, b, c \in \mathbf{R}$, $ab = 0 \ \& \ b = ac \Rightarrow a^2 c = 0 \Rightarrow (ac)^2 = 0 \Rightarrow ac = 0 \Rightarrow b = 0$). Thus, $\mathbf{V} := \mathbf{R}_{\text{red}}$ is a valuation ring without zero-divisors, and a fortiori, it is seminormal. To see this, let $b, c \in \mathbf{V}$ such that $b^2 = c^3$, and let us try to find $z \in \mathbf{V}$ such that $b = z^3$ and $c = z^2$.

If $b = xc$ for some $x \in \mathbf{V}$ then $c^2(x^2 - c) = 0$, and thus, as $\mathbf{V}$ is without zero-divisors, either $c = b = 0$ ($z = 0$ suits) or $c = x^2$ and $b = xc = x^3$ ($z = x$ suits).

If $c = yb$ for some $y \in \mathbf{V}$ then $b^2(y^3 b - 1) = 0$, and thus, either $b = c = 0$ ($z = 0$ suits) or $b, c \in \mathbf{V}^{\times}$ ($z = bc^{-1}$ suits).

The desired result follows from Theorem 119.

□

2.4 Projective Modules Over $\mathbf{R}[X_1, \ldots, X_n]$, $\mathbf{R}$ an Arithmetical Ring

2.4.1 A Constructive Proof of Brewer-Costa-Maroscia Theorem

The aim of this subsection is to prove constructively the following theorem [20, 117] due to Maroscia and Brewer & Costa which is a remarkable generalization of the Quillen-Suslin Theorem 72 since it is free of any Noetherian hypothesis.

Theorem 122. *If $\mathbf{R}$ is a Prüfer domain of Krull dimension ≤ 1, then each finitely-generated projective module over the ring $\mathbf{R}[X_1, \ldots, X_n]$ is extended. In particular, if $\mathbf{R}$ is a Bezout domain of Krull dimension ≤ 1, then each finitely-generated projective module over $\mathbf{R}[X_1, \ldots, X_n]$ is free.*

We will also propose in this Sect. 2.4 an alternative simpler constructive proof of Theorem 122 (see Remark 137).

2.4.1.1   Krull Dimension ≤ 1

In order to use constructively the hypothesis that **R** has Krull dimension ≤ 1, we recall the following constructive meaning of Krull dimension ≤ 1:

A ring **R** *has Krull dimension* ≤ 1 *if and only if*

$$\forall a,b \in \mathbf{R}, \ \exists n \in \mathbb{N}, \ \exists x,y \in \mathbf{R} \ | \ a^n(b^n(1+xb)+ya)=0 \qquad (2.11)$$

or equivalently

$$\forall a,b \in \mathbf{R}, \ \exists n \in \mathbb{N} \ | \ a^n b^n \in a^n b^{n+1}\mathbf{R} + a^{n+1}\mathbf{R}. \qquad (2.12)$$

In the sequel, we will consider the family of identities in (2.11) as the constructive meaning of the hypothesis that **R** has Krull dimension ≤ 1.
To simplify the computation of collapses related to Krull dimension ≤ 1, we introduce the following ideal $I_{\mathbf{R}}(a,b)$.

Notation 123. If a,b are two elements of a ring **R**, we denote by $I_{\mathbf{R}}(a,b)$ the set of all $z \in \mathbf{R}$ such that there exist $x,y \in \mathbf{R}$ and $n \in \mathbb{N}$ satisfying $a^n(b^n(z+xb)+ya)=0$. In other words,
$$I_{\mathbf{R}}(a,b) = \cup_{n \in \mathbb{N}}(a^n b^{n+1}\mathbf{R} + a^{n+1}\mathbf{R} : a^n b^n \mathbf{R}).$$

So, the sequence (a,b) collapses if and only if $1 \in I_{\mathbf{R}}(a,b)$.

Lemma 124.

- $I_{\mathbf{R}}(a,b)$ *is an ideal of* **R**,

- $z \in I_{\mathbf{R}}(a,b) \ \Rightarrow \ uvz \in I_{\mathbf{R}}(ua,vb)$,

- *if* $\varphi : \mathbf{R} \to \mathbf{T}$ *is an homomorphism, then* $\varphi(I_{\mathbf{R}}(a,b)) \subset I_{\mathbf{T}}(\varphi(a),\varphi(b))$,

- *the Krull dimension of* **R** *is* $\leq 1 \iff \forall a,b \in \mathbf{R}, I_{\mathbf{R}}(a,b) = \langle 1 \rangle$.

2.4.1.2   A Crucial Result

Definition 125.

- A ring is *Noetherian* if every nondecreasing sequence $(I_n)_{n \in \mathbb{N}}$ of finitely-generated ideals pauses (i.e., there exists $n \in \mathbb{N}$ such that $I_n = I_{n+1}$) [138, 139, 140, 141, 149, 159].

- A ring **R** is said to be *equipped with a divisibility test* if given $a, b \in \mathbf{R}$, one can answer the question $a \in? \langle b \rangle$, and in case of positive answer, one can explicitly find $c \in \mathbf{R}$ such that $a = bc$.

- A ring is said to be *strongly discrete* if it is equipped with a membership test to finitely-generated ideals, i.e., given $a, b_1, \ldots, b_n \in \mathbf{R}$, one can answer the question $a \in? \langle b_1, \ldots, b_n \rangle$, and in case of positive answer, one can explicitly find $c_1, \ldots, c_n \in \mathbf{R}$ such that $a = b_1 c_1 + \cdots + b_n c_n$.

- A ring $\mathbf{R}$ is a *valuation ring* if it is equipped with a divisibility test and every two elements are comparable under division, i.e., given $a, b \in \mathbf{R}$, either $a \mid b$ or $b \mid a$. In particular, a valuation ring is local, strongly discrete, and residually discrete.

- A ring $\mathbf{R}$ is *Bezout* if each finitely-generated ideal is principal.

- A ring $\mathbf{R}$ is *arithmetical* if each finitely-generated ideal is locally principal.

  A constructive characterization of arithmetical rings is that they are strongly discrete and satisfy the following property:

$$\forall x, y \in \mathbf{R} \quad \exists s, t, a, b \in \mathbf{R} \left\{ \begin{array}{rcl} sx & = & ay \\ bx & = & ty \\ s + t & = & 1 \end{array} \right. \tag{2.13}$$

  See [49] or [101] for detailed explanations about this characterization. Property (2.13) amounts to saying that each finitely-generated ideal becomes principal after localization at a finite family of comaximal monoids.

- A ring is called a *Prüfer ring* if it is arithmetical and reduced.

- An integral domain is called a *Prüfer domain* if it is arithmetical.

- Recall that a *coherent ring* is a ring in which finitely-generated ideals are finitely-presented. We say that a ring $\mathbf{A}$ is *stably coherent* if $\mathbf{A}[X_1, \ldots, X_n]$ is coherent for every n.

- A *pp-ring* is a ring in which principal ideals are projective, which means that the annihilator of each element is idempotent.

- A coherent Prüfer ring is often called a *semi-hereditary ring*. Since a finitely-presented module is flat if and only if it is projective, coherent Prüfer rings are characterized by the fact that finitely-generated ideals are projective. And an arithmetical ring is a coherent Prüfer ring if and only if it is a *pp-ring*.

- A *Dedekind ring* is an arithmetical coherent Noetherian ring.

Let us recall some well-known results concerning Bezout rings. A Bezout ring is reduced and coherent if and only if it is a pp-ring. Over a Bezout pp-ring, each constant rank projective module is free. Over a Bezout domain each finitely-generated projective module is free. For a constructive approach of all previously cited facts see [49, 101].

The following result of Brewer & Costa is an important intermediate result for Quillen Induction.

Theorem 126. *If* $\mathbf{R}$ *is a Prüfer domain with Krull dimension* ≤ 1 *then so is* $\mathbf{R}\langle X \rangle$.

Next, we will give a constructive proof of a slightly more general version of the result above.

Theorem 127. ([112]) *If* $\mathbf{R}$ *is a coherent Prüfer ring with Krull dimension* ≤ 1 *then so is* $\mathbf{R}\langle X \rangle$.

2.4.1.3 A Local Theorem

In the sequel, the letters a, b, c will denote elements of $\mathbf{R}$ and f, g, h elements of $\mathbf{R}[X]$. We will prove a local version of Theorem 127.

A local Prüfer ring is nothing but a reduced valuation ring. From a constructive point of view, a local coherent Prüfer ring is a ring $\mathbf{R}$ satisfying the following hypotheses:

$$\begin{cases} \forall x \in \mathbf{R} & x^2 = 0 \;\Rightarrow\; x = 0 \\ \forall x, y \in \mathbf{R} & \exists z\, x = zy \quad \text{or} \quad \exists z\, y = zx \\ \forall x \in \mathbf{R} & x \in \mathbf{R}^\times \quad \text{or} \quad x \in \mathrm{Rad}(\mathbf{R}) \\ \forall x \in \mathbf{R} & \mathrm{Ann}(x) = 0 \quad \text{or} \quad \mathrm{Ann}(x) = 1 \end{cases} \tag{2.14}$$

For example, the constructive meaning of the third item is that for each element $x \in \mathbf{R}$, we are able either to find an y such that $xy = 1$ or to find for each z an y such that $(1 + xz)y = 1$. The properties (2.14) amounts to saying that $\mathbf{R}$ is a valuation domain.

The first two properties imply that the ring has no zero-divisors ($xy = 0, x = zy \Rightarrow zy^2 = 0 \Rightarrow (zy)^2 = 0 \Rightarrow zy = 0 \Rightarrow x = 0$), thus, in classical mathematics, the last two properties are automatically satisfied.

Denoting $\mathrm{Rad}(\mathbf{R})$ by $\mathscr{R}$, we easily infer that

$$\begin{cases} \forall x, y \in \mathbf{R} & \exists z \in \mathscr{R}\; x = zy \;\text{ or }\; \exists z \in \mathscr{R}\; y = zx \;\text{ or }\; \exists u \in \mathbf{R}^\times\; y = ux \\ \forall x, y \in \mathbf{R} & xy = 0 \;\Rightarrow\; \qquad\qquad (x = 0 \;\text{ or }\; y = 0) \end{cases} \tag{2.15}$$

The following easy lemmas are useful for the proof of our Theorem 132.

Lemma 128. *If the ring* $\mathbf{R}$ *satisfies* (2.15), *then each* $F \in \mathbf{R}[X]$ *can be written as* $F = af$ *with* $f = bf_1 + f_2$ *where* $b \in \mathrm{Rad}(\mathbf{R})$ *and* f_2 *is monic.*

Proof. By the first property in (2.15), there is one coefficient of F, say a, dividing all the others. Thus, we can write $F = af$ for some $f \in \mathbf{R}[X]$ with at least one coefficient equal to 1. Now, write $f = f_2 + f_3$ with f_2 monic and all the coefficients of f_3 are in $\mathrm{Rad}(\mathbf{R})$. Again, there is one coefficient in f_3, say b, dividing all the others. Thus, $f_3 = bf_1$ for some $f_1 \in \mathbf{R}[X]$. $\qquad\square$

Lemma 129. *If* $\mathbf{R}$ *has Krull dimension* ≤ 1, $c \in \mathbf{R}$ *is regular and* $b \in \mathrm{Rad}(\mathbf{R})$, *then* c *divides a power of* b.

Proof. Just use the equality (2.11) and the fact that $1 + b\mathbf{R} \subseteq \mathbf{R}^\times$. $\qquad\square$

Corollary 130. *If* $\mathbf{R}$ *has Krull dimension* ≤ 1 *and* $f = b f_1 + f_2 \in \mathbf{R}[X]$ *with* $b \in$ Rad$(\mathbf{R})$ *and* f_2 *monic, then for every regular* $c \in \mathbf{R}$, $\langle f, c \rangle$ *contains a monic.*

Proof. Using Lemma 129, we know that there exists $n \in \mathbb{N}$ such that c divides b^n. Thus, the monic polynomial $f_2^n \in \langle f, b^n \rangle \subseteq \langle f, c \rangle$.

□

Remark 131. Let $\mathbf{R}$ be a ring and $x, y \in \mathbf{R}$. If $\langle x, y \rangle = \langle g \rangle$ for some $g \in \mathbf{R}$ then g is a gcd of x and y in the following sense: an element in $\mathbf{R}$ divides both x and y if and only if it divides g.

A local version of Theorem 127 is Theorem 132.

Theorem 132. *If* $\mathbf{R}$ *is a local coherent Prüfer ring (that is, it satisfies (2.14)) and has Krull dimension* ≤ 1, *then* $\mathbf{R}\langle X \rangle$ *is a Bezout domain with Krull dimension* ≤ 1.

Proof. We first prove that $\mathbf{R}\langle X \rangle$ is a Bezout domain. It is a domain (each element is zero or regular) since $\mathbf{R}$ is a domain. Since $\mathbf{R}$ is a discrete gcd domain (that is, each pair of nonzero elements has a greatest common divisor) so is $\mathbf{R}[X]$ (see for example Theorem IV.4.7 of [120]) and $\mathbf{R}\langle X \rangle$ as well. Recall that a gcd ring $\mathbf{B}$ is Bezout if and only if

$$\forall x, y \in \mathbf{B}, \quad (\gcd(x,y) = 1 \implies \langle x, y \rangle = \langle 1 \rangle).$$

To prove that $\mathbf{R}\langle X \rangle$ is Bezout, consider $F, G \in \mathbf{R}\langle X \rangle$ such that $\gcd(F, G) = 1$ and let us show that $1 \in \langle F, G \rangle$. We may assume w.l.o.g. that $F \neq 0$ and $G \neq 0$. Since monic polynomials are invertible in $\mathbf{R}\langle X \rangle$, we may also assume that $F, G \in \mathbf{R}[X]$. We need to show that $\langle F, G \rangle_{\mathbf{R}[X]}$ contains a monic polynomial. Letting $H = \gcd(F, G)_{\mathbf{R}[X]}$, H divides $\gcd(F, G)_{\mathbf{R}\langle X \rangle} = 1$ (in $\mathbf{R}\langle X \rangle$) and so the leading coefficient of H is invertible in $\mathbf{R}$. Using the equality $\langle F, G \rangle_{\mathbf{R}[X]} = H \langle F/H, G/H \rangle_{\mathbf{R}[X]}$, we see that we may suppose $H = 1$. Following Lemma 128, we have $F = a f = a (b f_1 + f_2)$, $G = a' g = a' (b' g_1 + g_2)$, with $b, b' \in$ Rad$(\mathbf{R})$ and f_2, g_2 monic. In $\mathbf{R}\langle X \rangle$ we have:

$$\gcd(F, G) = \gcd(a f, a' g) = 1 \Rightarrow \gcd(a, a') = 1.$$

Thus, $\gcd(F, G) = 1$ in $\mathbf{R}\langle X \rangle$ implies that either a or a' is invertible in $\mathbf{R}$. Suppose for example that $a = 1$. The fact that $\gcd(F, G)_{\mathbf{R}[X]} = 1$ yields that the gcd in $\mathbf{K}[X]$ (where $\mathbf{K}$ is the quotient field of $\mathbf{R}$) is equal to 1, that is, there is a regular element c in $\mathbf{R} \cap \langle F, G \rangle_{\mathbf{R}[X]}$. By Corollary 130, we get a monic polynomial in $\langle c, F \rangle_{\mathbf{R}[X]} \subseteq \langle F, G \rangle_{\mathbf{R}[X]}$, as desired.

Now, let us check that the Krull dimension of $\mathbf{R}\langle X \rangle$ is ≤ 1. The Krull dimension of $\mathbf{K}[X]$ is ≤ 1, and more precisely, for all $F, G \in \mathbf{R}[X]$ (keeping the same notations as above), we have an explicit collapse in $\mathbf{K}[X]$ (see Theorem 88) which can be rewritten in $\mathbf{R}[X]$ (by clearing the denominators) as follows:

$$\exists n \in \mathbb{N}, \ \exists h_1, h_2 \in \mathbf{R}[X], \ \exists w \in \mathbf{R} \setminus \{0\} \quad F^n (G^n (w + h_1 G) + h_2 F) = 0.$$

This means that $\exists w \in \mathbf{R} \setminus \{0\}$, such that $w \in I_{\mathbf{R}\langle X \rangle}(F, G)$. Moreover, we have $1 \in I_{\mathbf{R}}(a, a')$ and a fortiori $1 \in I_{\mathbf{R}\langle X \rangle}(a, a')$, implying that $f g \in I_{\mathbf{R}\langle X \rangle}(a f, a' g) =$

$I_{\mathbf{R}\langle X\rangle}(F,G)$. Finally, since the gcd in $\mathbf{R}\langle X\rangle$ of w and fg is equal to 1 (this is due to the fact that fg is primitive), the ideal $I_{\mathbf{R}\langle X\rangle}(F,G)$, which contains w and fg, contains 1.

Finally the fact that $\mathbf{R}\langle X\rangle$ is a pp-ring can be easily checked under the only hypothesis that $\mathbf{R}$ is a pp-ring. □

2.4.1.4 A Quasi-Global Version

Applying the General Local-Global Principle 35 to the proof of Theorem 132 above, we get an algorithmic proof for the following quasi-global proposition.

Proposition 133. *Let* $\mathbf{R}$ *be a coherent Prüfer ring with Krull dimension* ≤ 1. *Considering* $F,G \in \mathbf{R}[X]$:

- *There exists a family* (S_i) *of comaximal monoids of* $\mathbf{R}$ *such that in each* $\mathbf{R}_{S_i}\langle X\rangle$ *the ideal* $\langle F,G\rangle$ *is finitely-generated and projective.*

- *There exists a family* (S_i) *of comaximal monoids of* $\mathbf{R}$ *such that in each* $\mathbf{B}_i = \mathbf{R}_{S_i}\langle X\rangle$ *we have a collapse* $I_{\mathbf{B}_i}(F,G) = \langle 1\rangle$.

An immediate corollary of Proposition 133 is Theorem 127. This is due to the fact that finitely-generated ideals are projective and that two elements producing a collapse are local properties, i.e., it suffices to check them after localizations at a family of comaximal monoids [34, 49, 101].

Let $\mathscr{F}$ be the class of coherent Prüfer rings of Krull dimension ≤ 1. This class clearly satisfies the localization property (ıı') in Theorem 49 (Concrete Induction à la Quillen). It satisfies (i) (in the same theorem) by Theorem 127.
Theorem 132 above asserts that if $\mathbf{R} \in \mathscr{F}$ is local, then $\mathbf{R}\langle X\rangle$ is a Bezout domain. In particular, every projective module over $\mathbf{R}\langle X\rangle$ is free. Combined with Theorem 47 (the Global Horrocks extension theorem), we obtain condition (iii') in Theorem 49. Finally we constructively get:

Theorem 134. (Brewer-Costa-Maroscia Theorem [20, 117]) *If* $\mathbf{R}$ *is a coherent Prüfer ring with Krull dimension* ≤ 1, *then every finitely-generated projective module over* $\mathbf{R}[X_1,\ldots,X_n]$ *is extended. In particular, if* $\mathbf{R}$ *is a Bezout pp-ring with Krull dimension* ≤ 1, *then every constant rank projective module over* $\mathbf{R}[X_1,\ldots,X_n]$ *is free.*

2.4.2 The Theorem of Lequain, Simis and Vasconcelos

Let $\mathbf{R}$ be a ring. We denote by $\mathbf{R}(X)$ the localization of $\mathbf{R}[X]$ at primitive polynomials, i.e., polynomials whose coefficients generate the whole ring $\mathbf{R}$. Of course, the ring $\mathbf{R}(X)$ is also a localization of $\mathbf{R}\langle X\rangle$ and we have $\mathbf{R}[X] \subseteq \mathbf{R}\langle X\rangle \subseteq \mathbf{R}(X)$. The containment $\mathbf{R}\langle X\rangle \subseteq \mathbf{R}(X)$ becomes an equality if and only if $\mathbf{R}$ has Krull dimension ≤ 0 (in short, Kdim $\mathbf{R} \leq 0$) [80] (see Remark 138).

The construction $\mathbf{R}(X)$ turned out to be an efficient tool for proving results on $\mathbf{R}$ via passage to $\mathbf{R}(X)$.

As seen in the previous subsection, the restriction in Brewer-Costa-Maroscia theorem to Prüfer domains with Krull dimension ≤ 1 is due to the fact that $\mathbf{R}\langle X \rangle$ is a Prüfer domain if and only if $\mathbf{R}$ is a Prüfer domain with Krull dimension ≤ 1. Subsequently, in order to generalize the Quillen-Suslin theorem to Prüfer domains and seeing that the class of Prüfer domains is not stable under the formation $\mathbf{R}\langle X \rangle$, Lequain and Simis [96] found a clever way to bypass this difficulty by proving the following new induction theorem.

Theorem 135. (Lequain-Simis Induction) *Suppose that a class of rings $\mathscr{F}$ satisfies the following properties:*

(i) *If $\mathbf{R} \in \mathscr{F}$ then every nonmaximal prime ideal of $\mathbf{R}$ has finite height.*

(ii) *If $\mathbf{R} \in \mathscr{F}$ then $\mathbf{R}[X]_{\mathfrak{p}[X]} \in \mathscr{F}$ for any prime ideal $\mathfrak{p}$ of $\mathbf{R}$.*

(iii) *If $\mathbf{R} \in \mathscr{F}$ then $\mathbf{R}_{\mathfrak{p}} \in \mathscr{F}$ for any prime ideal $\mathfrak{p}$ of $\mathbf{R}$.*

(iv) *If $\mathbf{R} \in \mathscr{F}$ and $\mathbf{R}$ is local then any finitely-generated projective module over $\mathbf{R}[X]$ is free.*

Then, for each $\mathbf{R} \in \mathscr{F}$, all finitely-generated projective $\mathbf{R}[X_1, \ldots, X_n]$-modules are extended from $\mathbf{R}$.

Note here that if $\mathbf{R}$ is local with maximal ideal $\mathfrak{m}$, then $\mathbf{R}(X) = \mathbf{R}[X]_{\mathfrak{m}[X]}$. When coupled with Bass-Simis-Vasconcelos Theorem 121 asserting that over a valuation ring $\mathbf{V}$, all finitely-generated projective $\mathbf{V}[X]$-modules are free, the Lequain-Simis Induction Theorem yields to the elegant Lequain-Simis Vasconcelos Theorem 142 (see Corollary 142).

In this subsection, we will prove that for any ring $\mathbf{R}$ with Krull dimension $\leq d$, the ring $\mathbf{R}\langle X \rangle$ "dynamically behaves like the ring $\mathbf{R}(X)$ or a localization of a polynomial ring of type $(S^{-1}\mathbf{R})[X]$ with S a multiplicative subset of $\mathbf{R}$ and the Krull dimension of $S^{-1}\mathbf{R}$ is $\leq d - 1$".

As application of this dynamical comparison between the rings $\mathbf{R}(X)$ and $\mathbf{R}\langle X \rangle$, we will give a constructive variation of Lequain-Simis Induction Theorem—using a simple proof (see Theorem 140). Note that Lequain and Simis put considerable effort for proving this marvellous theorem and they used some quite complicated technical steps.

2.4.2.1   A Dynamical Comparison Between the Rings $\mathbf{R}(X)$ and $\mathbf{R}\langle X \rangle$ [55]

By the following theorem, we prove that for any ring $\mathbf{R}$ with Krull dimension $\leq d$, the ring $\mathbf{R}\langle X \rangle$ "dynamically behaves like the ring $\mathbf{R}(X)$ or a localization of a polynomial ring of type $(S^{-1}\mathbf{R})[X]$ with S a multiplicative subset of $\mathbf{R}$ and the Krull dimension of $S^{-1}\mathbf{R}$ is $\leq d - 1$". This can be schematized as follows:

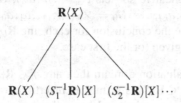

with $\mathrm{Kdim}(S_i^{-1}\mathbf{R}) < \mathrm{Kdim}\,\mathbf{R}$.

Theorem 136. *Let $d \in \mathbb{N}$ and $\mathbf{R}$ be a ring with Krull dimension $\leq d$. Then for any primitive polynomial $f \in \mathbf{R}[X]$, there exist comaximal monoids $V_1,\ldots,V_s$ of $\mathbf{R}\langle X\rangle$ such that for each $1 \leq i \leq s$, either f is invertible in $\mathbf{R}\langle X\rangle_{V_i}$ or $\mathbf{R}\langle X\rangle_{V_i}$ is a localization of $(S_{\mathbf{R},a_i}^{-1}\mathbf{R})[X]$, where $S_{\mathbf{R},a_i} = a_i^{\mathbb{N}}(1 + a_i\mathbf{R})$, for some coefficient a_i of f (note that $\mathrm{Kdim}\,S_{\mathbf{R},a_i}^{-1}\mathbf{R} \leq d - 1$).*

Proof.
First Case: $\mathbf{R}$ is Residually Discrete Local. Observe that any primitive polynomial $f \in \mathbf{R}[X]$ can be written in the form $f = g + u$ where $g, u \in \mathbf{R}[X]$, all the coefficients of g are in the Jacobson radical $\mathrm{Rad}(\mathbf{R})$ of $\mathbf{R}$ and u is quasi monic (that is, the leading coefficient of u is invertible). If the degree of u is k, then $g = \sum_{j>k} a_j X^j$. Now we open two branches: we localize $\mathbf{R}\langle X\rangle$ at the comaximal multiplicative subsets generated by f and g.

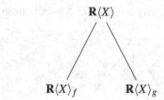

In $\mathbf{R}\langle X\rangle_f$, f is clearly invertible.

In $\mathbf{R}\langle X\rangle_g$, write $g = \sum_{j=k+1}^{m} a_j X^j$, where the $a_j \in \mathrm{Rad}(\mathbf{R})$. It follows that the multiplicative subsets $\mathscr{M}(a_{k+1}),\ldots,\mathscr{M}(a_s)$ are comaximal in $\mathbf{R}\langle X\rangle_g$. Note that for any $k+1 \leq i \leq m$, $\mathscr{M}(a_i)^{-1}(\mathbf{R}\langle X\rangle_g)$ is a localization of the polynomial ring $\mathbf{R}_{a_i}[X]$ and $\dim \mathbf{R}_{a_i} < \dim \mathbf{R}$.

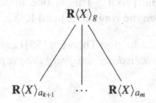

General Case: **R** *Arbitrary.* Apply the General Local-Global Principle 35. Precisely this gives the following computation. First we remark that since f is primitive, say $f = \sum_{j=0}^{m} a_j X^j$, the multiplicative subsets $U_m = \mathcal{M}(a_m)$, $U_{m-1} = \mathcal{S}_\mathbf{R}(a_m; a_{m-1})$, $\ldots, U_k = \mathcal{S}_\mathbf{R}(a_m, \ldots, a_{k+1}; a_k), \ldots, U_0 = \mathcal{S}_\mathbf{R}(a_m, \ldots, a_1; a_0)$ are comaximal in **R**. It is now sufficient to prove the conclusion for each ring $\mathbf{R}_{U_i}$. And this conclusion is obtained from the proof given for the first case.   □

Remark 137. If **R** is a valuation domain then any $f \in \mathbf{R}[X]$ is easily written as $f = ag$ where $a \in \mathbf{R}$ and $g \in \mathbf{R}[X]$ is primitive, invertible in $\mathbf{R}(X)$. From this fact, it follows easily that $\mathbf{R}(X)$ is again a valuation domain, and if $\mathrm{Kdim}\,\mathbf{R} \leq d$ then $\mathrm{Kdim}\,\mathbf{R}(X) \leq d$. So by Theorem 136, we painlessly get constructively that:

(i) If **R** is a valuation domain with $\mathrm{Kdim}\,\mathbf{R} \leq 1$ then $\mathbf{R}\langle X \rangle$ is a Prüfer domain with $\mathrm{Kdim}\,\mathbf{R} \leq 1$ (we retrieve a very simple constructive proof of the Brewer-Costa-Maroscia Theorem 134). As a matter of fact, it is clear that in this case, in one of the $\mathbf{R}\langle X \rangle_{U_i}$, the computations are done like in $\mathbf{R}(X)$, while the other $\mathbf{R}\langle X \rangle_{U_i}$ are localizations of the polynomial ring $\mathbf{K}[X]$ where **K** is the quotient field of **R**.

(ii) If **R** is a Prüfer domain with $\mathrm{Kdim}\,\mathbf{R} \leq 1$ then so is $\mathbf{R}\langle X \rangle$.
This is obtained from (i) by application of the General Constructive Rereading Principle.

Remark 138. If $\mathrm{Kdim}\,\mathbf{R} = 0$ then clearly $\mathbf{R}\langle X \rangle = \mathbf{R}(X)$ (the rings $S_{a_i}^{-1}\mathbf{R}$ in Theorem 136 being trivial). This constructive proof is simpler than the one given in [80].

Example 139. $\mathbb{Z}\langle X \rangle$ behaves locally as $\mathbb{Z}(X)$ or a localization $S_i^{-1}\mathbb{Q}[X]$ of $\mathbb{Q}[X]$.

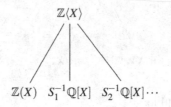

$$\mathbb{Z}(X) \quad S_1^{-1}\mathbb{Q}[X] \quad S_2^{-1}\mathbb{Q}[X] \cdots$$

2.4.2.2   The Lequain-Simis Induction Theorem

We propose here a constructive variation of Lequain-Simis Induction Theorem (Theorem 135) using a simple proof. This is one important application of our dynamical comparison between the rings $\mathbf{R}(X)$ and $\mathbf{R}\langle X \rangle$.

Theorem 140. (Constructive Induction Theorem [55]) *Let $\mathscr{F}$ be a class of commutative rings with finite Krull dimensions satisfying the properties below:*

(ii') *If* $\mathbf{R} \in \mathscr{F}$ *then* $\mathbf{R}(X) \in \mathscr{F}$.

(iii) *If* $\mathbf{R} \in \mathscr{F}$ *then* $\mathbf{R}_S \in \mathscr{F}$ *for each multiplicative subset* S *in* $\mathbf{R}$.

(iv') *If* $\mathbf{R} \in \mathscr{F}$ *then any finitely-generated projective module over* $\mathbf{R}[X]$ *is extended from* $\mathbf{R}$.

Then, for each $\mathbf{R} \in \mathscr{F}$, *all finitely-generated projective* $\mathbf{R}[X_1,\ldots,X_n]$-*modules are extended from* $\mathbf{R}$.

Proof. We reason by double induction on the number n of variables and the Krull dimension of the base ring $\mathbf{R}$.

For the initialization of the induction there is no problem since if $n = 1$ there is nothing to prove and for polynomial rings over zero-dimensional rings (see Theorem 50) the result is true constructively.

We assume that the construction is given with n variables for rings in $\mathscr{F}$. Then we consider the case of $n + 1$ variables and we give the proof by induction on the dimension of the ring $\mathbf{R} \in \mathscr{F}$. We assume that the dimension is $\leq d + 1$ with $d \geq 0$ and the construction has been done for rings of dimension $\leq d$.

Let P be a finitely-generated projective $\mathbf{R}[X_1,\ldots,X_n,Y]$-module. Let us denote X for $X_1,\ldots,X_n$. The module P can be seen as the cokernel of a presentation matrix $M = M(X,Y)$ with entries in $\mathbf{R}[X,Y]$. Let $A(X,Y)$ be the associated enlarged matrix (as in the proof of Theorem 45).

Using the induction hypothesis over n and (ii') we know that $A(X,Y)$ and $A(0,Y)$ are equivalent over the ring $\mathbf{R}(Y)[X]$. This means that there exist matrices Q_1, R_1 with entries in $\mathbf{R}[X,Y]$ such that

$$Q_1 A(X,Y) = A(0,Y)R_1 \qquad (2.16)$$

$$\text{where} \quad \det(Q_1) \text{ and} \det(R_1) \quad \text{are primitive polynomials in } \mathbf{R}[Y].$$

$$(2.17)$$

We first want to show that $A(X,Y)$ and $A(0,Y)$ are equivalent over $\mathbf{R}\langle Y\rangle[X]$. Using the Vaserstein's patching, for doing this job it is sufficient to show that A and $A(0,Y)$ are equivalent over $\mathbf{R}\langle Y\rangle[X]_{\mathscr{M}_i}$ for comaximal multiplicative subsets $\mathscr{M}_i$.

We consider the primitive polynomial $f = \det(Q_1)\det(R_1) \in \mathbf{R}[Y]$ and we apply Theorem 136. We get comaximal subsets $V_1,\ldots,V_s$ of $\mathbf{R}\langle Y\rangle$ such that for each $1 \leq i \leq s$, either f is invertible in $\mathbf{R}\langle Y\rangle_{V_i}$ or $\mathbf{R}\langle Y\rangle_{V_i}$ is a localization of $\mathbf{R}_{a_i}[Y]$ for some $a_i \in \mathbf{R}$ such that $\mathbf{R}_{a_i}$ has Krull dimension $\leq d$.

In the first case $\det(Q_1)$ and $\det(R_1)$ are invertible in $\mathbf{R}\langle Y\rangle_{V_i}$. This implies that $A(X,Y)$ and $A(0,Y)$ are equivalent over $\mathbf{R}\langle Y\rangle[X]_{V_i}$.

In the second case, by induction hypothesis on the dimension, $A(X,Y)$ and $A(0,0)$ are equivalent over $\mathbf{R}_{a_i}[Y][X]$. An immediate consequence is that $A(X,Y)$ and $A(0,Y)$ are equivalent over $\mathbf{R}_{a_i}[Y][X]$. Finally they are also equivalent over $\mathbf{R}\langle Y\rangle[X]_{V_i}$ which is a localization of the previous ring.

Now we know that there exist invertible matrices Q, R over the ring $\mathbf{R}\langle Y\rangle[X] \subseteq (\mathbf{R}[X])\langle Y\rangle$ such that

$$QA(X,Y) = A(0,Y)R.$$

We know also that $A(0,0)$ and $A(0,Y)$ are equivalent over $\mathbf{R}[Y] \subseteq (\mathbf{R}[X])\langle Y \rangle$ (case $n = 1$) and $A(0,0)$ and $A(X,0)$ are equivalent over $\mathbf{R}[X] \subseteq (\mathbf{R}[X])\langle Y \rangle$. So $A(X,0)$ and $A(X,Y)$ are equivalent over $(\mathbf{R}[X])\langle Y \rangle$, and by virtue of Horrocks Theorem 47, P is extended from $\mathbf{R}[X]$, i.e., $A(X,0)$ and $A(X,Y)$ are equivalent over $\mathbf{R}[X,Y]$. By induction hypothesis, P is extended from $\mathbf{R}$. $\qquad\square$

Remark 141. In fact, the proof does not use any multiplicative subset of rings $\mathbf{R}$ in $\mathscr{F}$, but only multiplicative subsets obtained by iterating localizations at some $\mathscr{S}(a_1, \dots, a_k; u)$.

Corollary 142. (Lequain-Simis-Vasconcelos Theorem) *For any arithmetical ring* $\mathbf{R}$, *all finitely-generated projective* $\mathbf{R}[X_1, \dots, X_n]$*-modules are extended from* $\mathbf{R}$.

Proof. By the Quillen's patching Theorem 45, it suffices to prove the result for valuation rings. As in the proof of Corollary 112 from Corollary 111, we can suppose that $\mathbf{R}$ is a finite-dimensional valuation ring.

Let us denote by $\mathscr{F}$ the class of finite-dimensional valuation rings. It is immediate that the hypotheses (ii') and (iii) in the Constructive Induction Theorem 140 are satisfied by $\mathscr{F}$. The hypothesis (iv') is nothing but the Bass-Simis-Vasconcelos Theorem 121. $\qquad\square$

Corollary 143. *For any Bezout domain* $\mathbf{R}$, *all finitely-generated projective* $\mathbf{R}[X_1, \dots, X_n]$*-modules are free.*

As always constructive proofs work in classical mathematics and Theorem 140 applies. Moreover, in classical mathematics, we get the following variation:

Theorem 144. (New Classical Induction Theorem) *Let* $\mathscr{F}$ *be a class of commutative rings with finite Krull dimensions satisfying the properties below:*

(ii) *If* $\mathbf{R} \in \mathscr{F}$ *and* $\mathbf{R}$ *is local then* $\mathbf{R}(X) \in \mathscr{F}$.

(iii') *If* $\mathbf{R} \in \mathscr{F}$ *then* $\mathbf{R}_S \in \mathscr{F}$ *for each multiplicative set* S *in* $\mathbf{R}$.

(iv) *If* $\mathbf{R} \in \mathscr{F}$ *and* $\mathbf{R}$ *is local then any finitely-generated projective module over* $\mathbf{R}[X]$ *is extended from* $\mathbf{R}$.

Then, for each $\mathbf{R} \in \mathscr{F}$, *all finitely-generated projective* $\mathbf{R}[X_1, \dots, X_n]$*-modules are extended from* $\mathbf{R}$.

Proof. From (ii) and (iv) we deduce (ii') and (iv') in Theorem 140 by using the abstract Quillen's patching that uses maximal ideals. $\qquad\square$

2.5 Suslin's Stability Theorem

2.5.1 "Obvious" Syzygies

Recall that for a ring $\mathbf{R}$ and $u = (u_1, \dots, u_m) \in \mathbf{R}^m$ ($m \geq 1$), the syzygy module of u is

$$\mathrm{Syz}(u) = \mathrm{Syz}(u_1, \dots, u_m) := \{(v_1, \dots, v_m) \in \mathbf{R}^m \mid v_1 u_1 + \cdots + v_m u_m = 0\}.$$

Definition 145. Let $\mathbf{R}$ be a ring, $m \geq 2$, and $u = (u_1, \ldots, u_m) \in \mathbf{R}^m$. An $s \in \mathbf{R}^m$ is said to be an *obvious syzygy* of u if $s = u_j e_i - u_i e_j =: s_{i,j}$ with $1 \leq i < j \leq m$, where $(e_1, \ldots, e_m)$ stands for the canonical basis of $\mathbf{R}^m$. Note that we avoided the use of the terminology "trivial syzygy" as it is reserved for $(0, \ldots, 0)$. Of course there are $\binom{m}{2} = \frac{m(m-1)}{2}$ obvious syzygies for u, and we have:

$$\langle s_{i,j} \mid 1 \leq i < j \leq m \rangle \subseteq \mathrm{Syz}(u).$$

But, in general, there is no equality. For example, for $\alpha \notin \mathbf{R}^\times$ and $\alpha \neq 0$, $s = (-\alpha, 1) \in \mathrm{Syz}(\alpha, \alpha^2)$ but $s \notin \langle s_{1,2} \rangle = \mathbf{R}(\alpha^2, -\alpha)$. Note that $\alpha s \in \langle s_{1,2} \rangle$. This is a general fact that will be explained in Proposition 146 below. As another example, if $\mathbf{R}$ is a noncoherent domain, then $\mathrm{Syz}(u)$ is not finitely-generated for some $m \geq 2$ and $u \in \mathbf{R}^m$, and, thus, $\langle s_{i,j} \mid 1 \leq i < j \leq m \rangle \subsetneq \mathrm{Syz}(u)$.

Proposition 146. *Let $\mathbf{R}$ be a ring, $m \geq 2$, and $u = (u_1, \ldots, u_m) \in \mathbf{R}^m$. Then*

$$\langle u_1, \ldots, u_m \rangle \mathrm{Syz}(u) \subseteq \langle s_{i,j} \mid 1 \leq i < j \leq m \rangle \subseteq \mathrm{Syz}(u),$$

where the $s_{i,j}$'s are the obvious syzygies of u.

In particular, if u is unimodular, i.e., $1 \in \langle u_1, \ldots, u_m \rangle$, then

$$\mathrm{Syz}(u) = \langle s_{i,j} \mid 1 \leq i < j \leq m \rangle.$$

Proof. It suffices to prove that $u_m \mathrm{Syz}(u) \subseteq \langle s_{i,j} \mid 1 \leq i < j \leq m \rangle$. Letting $v = (v_1, \ldots, v_m) \in \mathrm{Syz}(u)$, as $v_m u_m = -(v_1 u_1 + \cdots + v_{m-1} u_{m-1})$, we have

$$u_m v = v_1 s_{1,m} + \cdots + v_{m-1} s_{m-1,m}.$$

$\square$

Corollary 147. *Let $\mathbf{R}$ be a Bezout domain, $m \geq 2$, and $u = (u_1, \ldots, u_m) \in \mathbf{R}^m \setminus \{0\}$. Then, denoting by $d = \gcd(u_1 \ldots, u_m)$, $\mathrm{Syz}(u)$ is generated as $\mathbf{R}$-module by the obvious syzygies of $\frac{1}{d}u$.*

Proof. This follows from Proposition 146 since $\mathrm{Syz}(u) = \mathrm{Syz}(\frac{1}{d}u)$ and $\frac{1}{d}u$ is unimodular.

$\square$

2.5.2 $E_2(\mathbf{R})$ as a Subgroup of $SL_2(\mathbf{R})$

This subsection is devoted to Park's algorithm [132] with which one can test whether a given matrix in $SL_2(\mathbf{R}[X_1, \ldots, X_k])$, where $\mathbf{R}$ is a Euclidean domain, allows a factorization into elementary matrices (i.e., $\in E_2(\mathbf{R}[X_1, \ldots, X_k])$), and if it does, gives its explicit factorization.

Definition 148. Let $\mathbf{R}$ be ring and $n \geq 2$. A square $n \times n$ matrix A with entries in $\mathbf{R}$ is said to be *realizable* if it can be written as a product of elementary matrices, i.e., if it is in $E_n(\mathbf{R})$.

Example 149.

(1) If $\mathbf{R}$ is a Euclidean domain then every matrix in $SL_n(\mathbf{R})$ $(n \geq 2)$ is realizable, i.e., $SL_n(\mathbf{R}) = E_n(\mathbf{R})$. In more details, considering $A \in SL_n(\mathbf{R})$, as its first column C_1 is unimodular, we can by successive applications of the Euclidean algorithm, transform C_1 using elementary operations to ${}^t(1, 0, \ldots, 0)$. Thus, there exists $E \in E_n(\mathbf{R})$ such that $EA = \begin{pmatrix} 1 & 0 \\ 0 & A' \end{pmatrix}$ with $A' \in SL_{n-1}(\mathbf{R})$. We finish by induction on n.

(2) *Suslin's stability theorem* (see Theorem 178): If $\mathbf{K}$ is a discrete field and $n \geq 3$, then every matrix in $SL_n(\mathbf{K}[X_1, \ldots, X_k])$ is realizable, i.e.,

$$SL_n(\mathbf{K}[X_1, \ldots, X_k]) = E_n(\mathbf{K}[X_1, \ldots, X_k]).$$

Definition and notation 150. Let $\mathbf{R}$ be a discrete ring and fix a monomial order on $\mathbf{R}[X_1, \ldots, X_k]$ (see Definition and Notation 199).

(1) As in Definition 214, for a nonzero polynomial f, $LT(f)$ denotes its leading term and $mdeg(f)$ its multidegree. We convene that $LT(0) = 0$ and $mdeg(0) = (0, \ldots, 0)$.

Note that if $\mathbf{R}$ is a domain, then $LT(fg) = LT(f)LT(g)$ for $f, g \in \mathbf{R}[X_1, \ldots, X_k]$. This is no longer true if $\mathbf{R}$ is not a domain. For example, taking $\mathbf{R} = \mathbb{Z}/4\mathbb{Z}$, $f = \bar{1} + \bar{2}X$, and $g = \bar{1} - \bar{2}X$, we have $LT(f) = \bar{2}X$, $LT(g) = -\bar{2}X$, while $LT(fg) = \bar{1}$ since $fg = \bar{1}$.

(2) For a matrix $A = (f_{i,j})$ with entries in $\mathbf{R}[X_1, \ldots, X_k]$, we define the matrix of its leading terms as

$$LT(A) := (LT(f_{i,j})).$$

We also define the multidegree of A to be

$$mdeg(A) := \max_{i,j}(mdeg(f_{i,j})).$$

The following result was given by Park [132] in the case where $\mathbf{R}$ is a Euclidean domain but, in fact, it holds for any domain $\mathbf{R}$. It gives a key property that satisfies every nonconstant realizable matrix in $SL_2(\mathbf{R}[X_1, \ldots, X_k])$.

Theorem 151. *Let $\mathbf{R}$ be a domain, fix a monomial order $>$ on $\mathbf{R}[X_1, \ldots, X_k]$, and consider a nonconstant realizable matrix $A \in SL_2(\mathbf{R}[X_1, \ldots, X_k])$. Then either A has a zero entry, or one of the rows of $LT(A)$ is a term multiple of the other row.*

Proof. Write $A = \begin{pmatrix} q_1 & q_2 \\ q_3 & q_4 \end{pmatrix} = E_1 \cdots E_\ell$ a realization of A, that is, the E_r's have the form $E_{i,j}(p) = I_2 + p e_{i,j}$, where $p \in \mathbf{R}[X_1, \ldots, X_k]$ and $e_{i,j}$ is the matrix in $M_2(\mathbf{R}[X_1, \ldots, X_k])$ with 1 on position (i, j) and 0s elsewhere $(i \neq j)$. We proceed by induction on ℓ.

If $\ell = 1$, then $A = E_{1,2}(p)$ or $E_{2,1}(p)$ for some $p \in \mathbf{R}[X_1,\ldots,X_k]$, and in both cases, it has a zero entry.

Now, suppose that $\ell \geq 2$ and denote by $B = E_1 \cdots E_{\ell-1} = \begin{pmatrix} p_1 & p_2 \\ p_3 & p_4 \end{pmatrix}$. Two cases may occur:

Case 1: B does not have a zero entry. In that case, by induction hypothesis, we may assume that

$$(LT(p_1), LT(p_2)) = cX^\alpha (LT(p_3), LT(p_4))$$

for some $c \in \mathbf{R}$ and some monomial $X^\alpha = X_1^{\alpha_1} \ldots X_k^{\alpha_k}$, $\alpha_i \in \mathbb{N}$.
We have $A = BE_\ell$ with $E_\ell = E_{1,2}(q)$ or $E_{2,1}(q)$ for some $q \in \mathbf{R}[X_1,\ldots,X_k]$. As the two cases are similar, we will treat the case $E_\ell = E_{1,2}(q)$. In that case, we have

$$A = \begin{pmatrix} q_1 & q_2 \\ q_3 & q_4 \end{pmatrix} = \begin{pmatrix} p_1 & p_2 \\ p_3 & p_4 \end{pmatrix} E_{1,2}(q) = \begin{pmatrix} p_1 & p_2 + qp_1 \\ p_3 & p_4 + qp_3 \end{pmatrix}.$$

As A is nonconstant, either $q_1 q_4$ or $q_3 q_2$ is nonconstant. Furthermore, since $\det A = q_1 q_4 - q_3 q_2 = 1$ is constant and $\mathbf{R}$ is a domain, we have

$$LT(q_1 q_4) = LT(q_1)LT(q_4) = LT(q_3 q_2) = LT(q_3)LT(q_2).$$

As $LT(q_1) = cX^\alpha LT(q_3)$, we infer that $LT(q_2) = cX^\alpha LT(q_4)$, and thus,

$$(LT(q_1), LT(q_2)) = cX^\alpha (LT(q_3), LT(q_4)),$$

as desired.

Case 2: B has a zero entry. We can, without loss of generality, suppose that $p_3 = 0$.
If $A = BE_{1,2}(q) = \begin{pmatrix} p_1 & p_2 + qp_1 \\ 0 & p_4 \end{pmatrix}$, then we are done as A has a zero entry.

Else, $A = BE_{2,1}(q) = \begin{pmatrix} p_1 + qp_2 & p_2 \\ qp_4 & p_4 \end{pmatrix}$ with $\det B = p_1 p_4 = 1$ (in particular, p_1 and p_4 are constant). Since A is nonconstant with constant determinant ($= 1$), necessarily qp_2 is nonconstant, and thus, $LT(p_1 + qp_2) = LT(qp_2) = LT(q)LT(p_2)$, and
$(LT(p_1 + qp_2), LT(p_2)) = LT(p_2)p_1 (LT(q)p_4, p_4) = LT(p_2 p_1)(LT(qp_4), LT(p_4))$,

as desired.

$\square$

Remark 152.

(i) Theorem 151 is not valid for constant realizable matrices. To see this, consider the matrix $\begin{pmatrix} 7 & 2 \\ 3 & 1 \end{pmatrix} \in SL_2(\mathbb{Z}) = E_2(\mathbb{Z})$.

(2) Theorem 151 is not valid if $\mathbf{R}$ is not a domain. To see this, consider the matrix $A = \begin{pmatrix} \bar{1} + \bar{2}X & \bar{2} \\ \bar{2} & \bar{1} - \bar{2}X \end{pmatrix} \in SL_2((\mathbb{Z}/4\mathbb{Z})[X])$. No entry of A is zero, and none of the rows $L_1 = \bar{2}(X, \bar{1})$ and $L_2 = \bar{2}(\bar{1}, -X)$ of $LT(A)$ is a term multiple of the other row.

(3) As Theorem 151 is true for any monomial order, and $\mathrm{LT}(A)$ usually changes when monomial order changes, one may get several relationships between different term vectors (i.e., vectors whose entries are terms).

Example 153. Let $\mathbf{K}$ be a field and consider the matrix

$$C := \begin{pmatrix} 1+xy & x^2 \\ -y^2 & 1-xy \end{pmatrix} \in \mathrm{SL}_2(\mathbf{K}[x,y]).$$

(1) Recall that the matrix C, called Cohn's matrix, appeared in [30] and was one of the earliest examples for $\mathrm{E}_2(\mathbf{R}) \subsetneq \mathrm{SL}_2(\mathbf{R})$. To see that C is not realizable, choosing any monomial order, we have

$$\mathrm{LT}(C) = \begin{pmatrix} xy & x^2 \\ -y^2 & -xy \end{pmatrix},$$

and we see that no entry of C is zero, and none of the rows $L_1 = x(y,x)$ and $L_2 = -y(y,x)$ of $\mathrm{LT}(C)$ is a term multiple of the other row. We conclude that $C \notin \mathrm{E}_2(\mathbf{K}[x,y])$, and thus, Suslin's stability theorem (see Theorem 178) does not hold for 2×2 matrices, i.e.,

$$\mathrm{E}_2(\mathbf{K}[x,y]) \subsetneq \mathrm{SL}_2(\mathbf{K}[x,y]).$$

(2) The unimodular row $(1+xy, x^2)$ is completable (as it is the first row of C) but is not elementarily completable, that is, cannot be brought to $(1,0)$ by elementary transformations. This is because this would imply that $C \in \mathrm{E}_2(\mathbf{K}[x,y])$.

Remark 154. (Lam [92], page 56) Let us consider again Cohn's matrix

$$C := \begin{pmatrix} 1+xy & x^2 \\ -y^2 & 1-xy \end{pmatrix} \in \mathrm{SL}_2(\mathbf{K}[x,y]) \setminus \mathrm{E}_2(\mathbf{K}[x,y]).$$

(1) As $C \notin \mathrm{E}_2(\mathbf{K}[x,y])$, so is the matrix

$$M := E_{1,2}(y) C E_{1,2}(x) = \begin{pmatrix} 1+xy-y^3 & x+y+x^2+x^2y-xy^2-xy^3 \\ -y^2 & 1-xy-xy^2 \end{pmatrix}.$$

The matrix M is not realizable. Choosing the lexicographic monomial order with $y > x$ (see Example 200), one has

$$\mathrm{LT}(M) = \begin{pmatrix} -y^3 & -xy^3 \\ -y^2 & -xy^2 \end{pmatrix}.$$

One can see that the first row of $\mathrm{LT}(M)$ is y times its second row. We conclude that the converse of Theorem 151 does not hold.

(2) Consider the matrix

$$N := C \begin{pmatrix} 0 & 1 \\ -1 & 0 \end{pmatrix} C^{-1} = \begin{pmatrix} -x^2 + x^3 y + y^2 + xy^3 & 1 + x^4 + 2xy + x^2 y^2 \\ -1 + 2xy - y^2 x^2 - y^4 & x^2 - x^3 y - y^2 - xy^3 \end{pmatrix}.$$

Choosing the lexicographic monomial order with $y > x$, one has

$$LT(N) = \begin{pmatrix} xy^3 & x^2 y^2 \\ -y^4 & -xy^3 \end{pmatrix}$$

(the leading terms matrix given by Lam in page 56 of [92] is not correct). One can see that no entry of N is zero, and none of the rows $L_1 = xy^2(y, x)$ and $L_2 = -y^3(y, x)$ of $LT(N)$ is a term multiple of the other row. So, $N \notin E_2(\mathbf{K}[x, y])$. But, by virtue of Proposition 164.(2), the matrix $\begin{pmatrix} 0 & 1 \\ -1 & 0 \end{pmatrix} \in E_2(\mathbf{K}[x, y])$. We infer that, $E_2(\mathbf{K}[x, y])$ is not a normal subgroup of $SL_2(\mathbf{K}[x, y])$, or in short,

$$E_2(\mathbf{K}[x, y]) \ntriangleleft SL_2(\mathbf{K}[x, y]).$$

By contrast, Suslin's Normality Theorem 163 says that for any ring $\mathbf{R}$ and $m \geq 3$, we have $E_m(\mathbf{R}) \triangleleft GL_m(\mathbf{R})$.

Before presenting Park's algorithm, it is worth recalling that for a ring $\mathbf{R}$, when we say that $SL_2(\mathbf{R}) = E_2(\mathbf{R})$ this means that we have an algorithm which expresses every matrix in $SL_2(\mathbf{R})$ as a product of elementary matrices. For example, any Euclidean domain R satisfies $SL_2(\mathbf{R}) = E_2(\mathbf{R})$ (see Example 149.(1)).

Algorithm 155. (Park's Realization Algorithm for $SL_2(\mathbf{R}[X_1, \ldots, X_k])$ [132])

Input: A matrix $A = \begin{pmatrix} q_1 & q_2 \\ q_3 & q_4 \end{pmatrix} \in SL_2(\mathbf{R}[X_1, \ldots, X_k])$ with $\mathbf{R}$ a domain equipped with an algorithm (call it, AlgorithmConstant) realizing $SL_2(\mathbf{R}) = E_2(\mathbf{R})$.

Output: An answer to the question "$A \in$? $E_2(\mathbf{R}[X_1, \ldots, X_k])$", and in case of positive answer, a factorization of A into elementary matrices.

Let us first fix a monomial order $>$ on $\mathbf{R}[X_1, \ldots, X_k]$.

Step 1: Check the following cases. If one of them occurs, then we are done. Else, go to **Step 2**.

Case 1: If A is constant then the answer is "yes", and use AlgorithmConstant to factorize A into elementary matrices.

Case 2: If A is nonconstant and has a zero entry then the answer is "yes". Transform A using elementary operations into a constant matrix A' and then use AlgorithmConstant to factorize A' into elementary matrices.

For example, if $q_3 = 0$, then $\det A = q_1 q_4 = 1$, and

$$E_{1,2}(-q_4^{-1} q_2) A = \begin{pmatrix} q_1 & 0 \\ 0 & q_4 \end{pmatrix} =: A'$$

is constant. The three other cases are similar.

Case 3: If A is nonconstant, has no zero entry, and none of the rows of $\mathrm{LT}(A)$ is a term multiple of the other row, then the answer is "no" by virtue of Theorem 151.

Step 2: If one of the rows of $\mathrm{LT}(A)$ is a term multiple of the other row.

Assume, without loss of generality, that $(\mathrm{LT}(q_1), \mathrm{LT}(q_2)) = cX^\alpha$ $(\mathrm{LT}(q_3), \mathrm{LT}(q_4))$ with $c \in \mathbf{R}$ and $\alpha \in \mathbb{N}^k$. Set $A := E_{1,2}(-cX^\alpha)A$ and go to **Step 1**.

Theorem 156. *Algorithm 155 terminates and is correct.*

Proof. Theorem 151 ensures the correctness of Algorithm 155. To prove the termination, first note that at each passage at **Step 2**, the multidegree of one of the two rows of the current matrix decreases. Thus, as the set of monomials is well-ordered (see Theorem 209), Algorithm 155 must terminate. $\qquad\square$

Example 157. Consider the matrix

$$A_1 = \begin{pmatrix} 1 - xy^2 & -x^2y^2 \\ 1 + xy + y^2 - xy^2 - x^2y^3 & 1 + xy^2 - x^2y^2 - x^3y^3 \end{pmatrix} \in \mathrm{SL}_2(\mathbb{Z}[x,y]).$$

Suppose that we want to answer the question $A_1 \in ? \, E_2(\mathbb{Z}[x,y])$. Fixing the lexicographic order with $y > x$ as monomial order, we have:

$$\mathrm{LT}(A_1) = \begin{pmatrix} -xy^2 & -x^2y^2 \\ -x^2y^3 & -x^3y^3 \end{pmatrix}.$$

We will list below the successive matrices computed with Algorithm 155 as well as their leading terms matrices:

$$A_2 := E_{2,1}(-xy)A_1 = \begin{pmatrix} 1 - xy^2 & -x^2y^2 \\ 1 + y^2 - xy^2 & 1 + xy^2 - x^2y^2 \end{pmatrix},$$

$$\mathrm{LT}(A_2) = \begin{pmatrix} -xy^2 & -x^2y^2 \\ -xy^2 & -x^2y^2 \end{pmatrix}$$

$$A_3 := E_{2,1}(-1)A_2 = \begin{pmatrix} 1 - xy^2 & -x^2y^2 \\ y^2 & 1 + xy^2 \end{pmatrix}, \quad \mathrm{LT}(A_3) = \begin{pmatrix} -xy^2 & -x^2y^2 \\ y^2 & xy^2 \end{pmatrix}$$

$$A_4 := E_{1,2}(x)A_3 = \begin{pmatrix} 1 & x \\ y^2 & 1 + xy^2 \end{pmatrix}, \quad \mathrm{LT}(A_4) = \begin{pmatrix} 1 & x \\ y^2 & xy^2 \end{pmatrix}$$

$$A_5 := E_{2,1}(-y^2)A_4 = \begin{pmatrix} 1 & x \\ 0 & 1 \end{pmatrix} = E_{1,2}(x).$$

Thus, $A_1 = E_{2,1}(xy) E_{2,1}(1) E_{1,2}(-x) E_{2,1}(y^2) E_{1,2}(x) \in E_2(\mathbb{Z}[x,y])$.

As pointed out by Park in [132], the problem of extending a realization algorithm to matrices of finite impulse response filters was raised in [172]. In mathematical terms, this is about realization of matrices over the multivariate Laurent polynomial ring $\mathbf{R}[x_1^\pm,\ldots,x_k^\pm]$. So, an interesting issue is to generalize Algorithm 155 to Laurent polynomials.

2.5.3 Suslin's Normality Theorem

This subsection is mainly extracted from [92]. The proofs given there are already constructive. These proofs are included here so that these notes become self-contained. The purpose is to prove the following wonderful general theorem due to Suslin [165]:

For any ring $\mathbf{R}$ and $n \geq 3$, the group $E_n(\mathbf{R})$ is normal in $GL_n(\mathbf{R})$, in short,

$$E_n(\mathbf{R}) \lhd GL_n(\mathbf{R}).$$

Notation 158. (i) Recall that for a ring $\mathbf{R}$ and n, $m \geq 1$, we denote by $M_{n,m}(\mathbf{R})$ the set of matrices of size $n \times m$ with entries in $\mathbf{R}$. The set $M_{n,n}(\mathbf{R})$ will simply be denoted by $M_n(\mathbf{R})$.

(ii) Recall that for any ring $\mathbf{R}$ and $n \geq 1$, an $n \times n$ elementary matrix $E_{i,j}(a)$ over $\mathbf{R}$, where $i \neq j$ and $a \in \mathbf{R}$, is the matrix in $M_n(\mathbf{R})$ with 1s on the diagonal, a on position (i,j) and 0s elsewhere. When multiplying on the left, that is, $M \to E_{i,j}(a)M$, for $M \in M_n(\mathbf{R})$, $E_{i,j}(a)$ corresponds to the elementary rows operation $L_i \to L_i + aL_j$. When multiplying on the right, that is, $M \to ME_{i,j}(a)$, for $M \in M_n(\mathbf{R})$, $E_{i,j}(a)$ corresponds to the elementary columns operation $C_j \to C_j + aC_i$. $E_n(\mathbf{R})$ will denote the subgroup of $SL_n(\mathbf{R})$ generated by elementary matrices.

The following is folklore. We will need it in Lemma 160.

Lemma 159. *Let $\mathbf{R}$ be a ring, n, $m \geq 1$, $M \in M_{n,m}(\mathbf{R})$, and $N \in M_{m,n}(\mathbf{R})$. Then*

$$I_n + MN \in GL_n(\mathbf{R}) \quad \Leftrightarrow \quad I_m + NM \in GL_m(\mathbf{R}).$$

In particular, if $u \in \mathbf{R}^{1 \times m}$ and $v \in \mathbf{R}^{m \times 1}$ are such that $uv = 0$ then

$$I_m + vu \in GL_m(\mathbf{R}) \quad \& \quad (I_m + vu)^{-1} = I_m - vu.$$

Proof. "$\Rightarrow$" Denoting by $A = (I_n + MN)^{-1}$, we have:

$$(I_m - NAM)(I_m + NM) = I_m + NM - NA(I_n + MN)M = I_m.$$

"$\Leftarrow$" This follows by symmetry. $\qquad\square$

Lemma 160. (Vaserstein [174]) Let $\mathbf{R}$ be a ring, n, $m \geq 1$, $M \in M_{n,m}(\mathbf{R})$, and $N \in M_{m,n}(\mathbf{R})$. If $I_n + MN \in GL_n(\mathbf{R})$ then

$$\begin{pmatrix} I_n + MN & 0 \\ 0 & (I_m + NM)^{-1} \end{pmatrix} \in E_{n+m}(\mathbf{R}).$$

In particular, if $u \in \mathbf{R}^{1 \times m}$ and $v \in \mathbf{R}^{m \times 1}$ are such that $uv = 0$ then

$$\begin{pmatrix} I_m + vu & 0 \\ 0 & 1 \end{pmatrix} \in E_{m+1}(\mathbf{R}).$$

Proof. Let us denote by $R = (I_m + NM)^{-1}$ and $S = (I_n + MN)^{-1}$. We will perform the following series of elementary block operations (multiplying on the right, i.e., operating on columns):

$$\begin{pmatrix} I_n + MN & 0 \\ 0 & R \end{pmatrix} \hookrightarrow \begin{pmatrix} I_n + MN & -M \\ 0 & R \end{pmatrix} \hookrightarrow \begin{pmatrix} I_n & -M \\ RN & R \end{pmatrix} \hookrightarrow$$

$$\begin{pmatrix} I_n & 0 \\ RN & R + RNM \end{pmatrix} = \begin{pmatrix} I_n & 0 \\ RN & I_m \end{pmatrix} \overset{\text{elementary operations on rows}}{\hookrightarrow} I_{n+m}.$$

$\square$

Example 161. Let $\mathbf{R} = \mathbf{A}[x,y]$, where $\mathbf{A}$ is a nontrivial ring. Take $n = 1$, $m = 2$, $u = (-y, -x)$, and $v = \begin{pmatrix} x \\ -y \end{pmatrix}$. We have $uv = 0$, $I_n + uv = 1$, and $C := (I_m + vu)^{-1} = I_m - vu = \begin{pmatrix} 1 + xy & x^2 \\ -y^2 & 1 - xy \end{pmatrix}$.

Recall that the matrix C, called Cohn's matrix, appeared in [30] and was one of the earliest examples for $E_2(\mathbf{R}) \subsetneq SL_2(\mathbf{R})$ (see Example 153 for a proof). Accordingly to the above proof, after the following series of elementary operations:

$$A := \begin{pmatrix} 1 & 0 & 0 \\ 0 & 1+xy & x^2 \\ 0 & -y^2 & 1-xy \end{pmatrix} \quad \begin{matrix} C_2 \to C_2 + yC_1 \\ C_3 \to C_3 + xC_1 \\ \hookrightarrow \end{matrix}$$

$$\begin{pmatrix} 1 & y & x \\ 0 & 1+xy & x^2 \\ 0 & -y^2 & 1-xy \end{pmatrix} \quad \begin{matrix} C_1 \to C_1 + xC_2 - yC_3 \\ \hookrightarrow \end{matrix}$$

$$\begin{pmatrix} 1 & y & -x \\ x & 1+xy & x^2 \\ -y & -y^2 & 1-xy \end{pmatrix} \quad \begin{matrix} C_2 \to C_2 - yC_1 \\ C_3 \to C_3 - xC_1 \\ \hookrightarrow \end{matrix}$$

$$\begin{pmatrix} 1 & 0 & 0 \\ x & 1 & 0 \\ -y & 0 & 1 \end{pmatrix} \quad \begin{matrix} L_2 \to L_2 - xL_1 \\ L_3 \to L_3 + yL_1 \\ \hookrightarrow \end{matrix} \quad I_3,$$

one finds the elementary factorization:
$$A = E_{2,1}(x)E_{3,1}(-y)E_{3,1}(x)E_{2,1}(y)E_{1,3}(y)E_{1,2}(-x)E_{3,1}(-x)E_{2,1}(-y).$$

Corollary 162. *Let* $\mathbf{R}$ *be a ring and* $m \geq 1$. *Consider* $u = (u_1, \ldots, u_m) \in \mathbf{R}^{1 \times m}$ *and* $v = {}^{t}(v_1, \ldots, v_m) \in \mathbf{R}^{m \times 1}$ *such that* $uv = 0$.

(1) *If* $v_{i_0} = 0$ *for some* $1 \leq i_0 \leq m$, *then* $\mathrm{I}_m + vu \in \mathrm{E}_m(\mathbf{R})$.

(2) *Let* $r \in \langle u_1, \ldots, u_m \rangle$ *and suppose that* $m \geq 3$. *Then* $\mathrm{I}_m + rvu \in \mathrm{E}_m(\mathbf{R})$.

In particular, if u *is unimodular, then* $\mathrm{I}_m + vu \in \mathrm{E}_m(\mathbf{R})$.

Proof. (1) We can suppose that $v_m = 0$. Write $v = {}^{t}(v', 0)$ and $u = (u', a)$ with $u' \in \mathbf{R}^{1 \times (m-1)}$, $v' \in \mathbf{R}^{(m-1) \times 1}$, and $a \in \mathbf{R}$. We have:

$$\mathrm{I}_m + vu = \begin{pmatrix} \mathrm{I}_{m-1} + v'u' & v'a \\ 0 & 1 \end{pmatrix} = \begin{pmatrix} \mathrm{I}_{m-1} & v'a \\ 0 & 1 \end{pmatrix} \begin{pmatrix} \mathrm{I}_{m-1} + v'u' & 0 \\ 0 & 1 \end{pmatrix},$$

where obviously $\begin{pmatrix} \mathrm{I}_{m-1} & v'a \\ 0 & 1 \end{pmatrix} \in \mathrm{E}_m(\mathbf{R})$ and also $\begin{pmatrix} \mathrm{I}_{m-1} + v'u' & 0 \\ 0 & 1 \end{pmatrix} \in \mathrm{E}_m(\mathbf{R})$
(this follows from Lemma 160 as $u'v' = 0$).

(2) By virtue of Proposition 146, $rv \in \langle {}^{t}s_{i,j} \mid 1 \leq i < j \leq m \rangle$, where the $s_{i,j}$'s are the obvious syzygies of u, that is, there exist $a_1, \ldots, a_k \in \mathbf{R}$ and $w_1, \ldots, w_k \in \{ {}^{t}s_{i,j} \mid 1 \leq i < j \leq m \}$ such that

$$rv = a_1 w_1 + \cdots + a_k w_k.$$

Since $uw_i = 0$, we have:

$$\mathrm{I}_m + rvu = (\mathrm{I}_m + a_1 w_1 u)(\mathrm{I}_m + a_2 w_2 u) \cdots (\mathrm{I}_m + a_k w_k u),$$

and, as $\mathrm{I}_m + a_i w_i u \in \mathrm{E}_m(\mathbf{R})$ by (1), we infer that $\mathrm{I}_m + rvu \in \mathrm{E}_m(\mathbf{R})$. $\qquad \square$

Now we are in position to prove Suslin's Normality Theorem. It is worth repeating that Suslin's Normality Theorem does not work in general for $m = 2$ (see Example 153).

Theorem 163. (Suslin's Normality Theorem [165]) *For any ring* $\mathbf{R}$ *and* $m \geq 3$, *the group* $\mathrm{E}_m(\mathbf{R})$ *is normal in* $\mathrm{GL}_m(\mathbf{R})$.

Proof. It suffices to prove that for $1 \leq i \neq j \leq m$, $a \in \mathbf{R}$, and $\gamma \in \mathrm{GL}_m(\mathbf{R})$, we have

$$\gamma E_{i,j}(a) \gamma^{-1} \in \mathrm{E}_m(\mathbf{R}).$$

Writing $E_{i,j}(a) = \mathrm{I}_m + a e_{i,j}$, where $e_{i,j}$ is the matrix in $\mathrm{M}_m(\mathbf{R})$ with 1 on position (i,j) and 0s elsewhere, we have

$$\gamma E_{i,j}(a) \gamma^{-1} = \mathrm{I}_m + a(\gamma e_{i,j} \gamma^{-1}).$$

Denoting by v the ith column of γ^{-1} and u the jth row of γ, as $\gamma \gamma^{-1} = \mathrm{I}_m$, we have $\gamma e_{i,j} \gamma^{-1} = vu$. Moreover, $uv = 0$ since $i \neq j$. Thus,

$$\gamma E_{i,j}(a) \gamma^{-1} = \mathrm{I}_m + avu = \mathrm{I}_m + (av)u.$$

The desired result follows from Corollary 162 (u being clearly unimodular). $\qquad \square$

2.5.4   Unimodular Rows and Elementary Operations

This subsection is mainly extracted from [92]. The proofs given there are already constructive. These proofs are included in here so that these notes become self-contained. The goal is to give some basic facts about elementary transformations and unimodular rows that we will need in Sects. 2.6.1 and 2.6.2.

Proposition 164. *Let* $\mathbf{R}$ *be a ring and* $n \geq 1$.

(1) *For any diagonal matrix* $D \in M_n(\mathbf{R})$ *and any* $E \in E_n(\mathbf{R})$, *we have*

$$DED^{-1} \in E_n(\mathbf{R}).$$

(2) $M_1 = \begin{pmatrix} 0 & -I_n \\ I_n & 0 \end{pmatrix} \in E_{2n}(\mathbf{R})$, $M_2 = \begin{pmatrix} 0 & I_n \\ -I_n & 0 \end{pmatrix} \in E_{2n}(\mathbf{R})$, *and for any* $a, b \in \mathbf{R}$,

$(\ldots, a, \ldots, b, \ldots)M_1 = (\ldots, b, \ldots, -a, \ldots)$ & $(\ldots, a, \ldots, b, \ldots)M_2 = (\ldots, -b, \ldots, a, \ldots)$.

(3) *(Whitehead's Lemma) For* $A, B \in GL_n(\mathbf{R})$, *we have*

$$\begin{pmatrix} AB & 0 \\ 0 & I_n \end{pmatrix} \in \begin{pmatrix} A & 0 \\ 0 & B \end{pmatrix} \cdot E_{2n}(\mathbf{R}).$$

(4) *For* $u_1, \ldots, u_n \in \mathbf{R}^\times$, *we have*

$$\mathrm{diag}(u_1, \ldots, u_n) \in \mathrm{diag}(\prod_{i=1}^{n} u_i, 1, \ldots, 1) \cdot E_n(\mathbf{R}).$$

(5) *Any diagonal matrix in* $SL_n(\mathbf{R})$ *belongs to* $E_n(\mathbf{R})$.

Proof. (1) We can suppose that $D = \mathrm{diag}(1, \ldots, 1, d, 1, \ldots, 1)$ where d is in kth position, and $E = I_m + a e_{i,j}$, where $e_{i,j}$ is the matrix in $M_m(\mathbf{R})$ with 1 on position (i, j) $(i \neq j)$ and 0s elsewhere. Then

$$DED^{-1} = \begin{cases} I_m + a e_{i,j} \text{ if } k \neq i \ \& \ k \neq j \\ I_m + d a e_{i,j} \text{ if } k = i \\ I_m + d^{-1} a e_{i,j} \text{ if } k = j. \end{cases}$$

(2) This follows from the following sequence of block elementary column transformations:

$$\begin{pmatrix} 0 & I_n \\ -I_n & 0 \end{pmatrix} \rightarrow \begin{pmatrix} I_n & I_n \\ -I_n & 0 \end{pmatrix} \rightarrow \begin{pmatrix} I_n & 0 \\ -I_n & I_n \end{pmatrix} \rightarrow \begin{pmatrix} I_n & 0 \\ 0 & I_n \end{pmatrix}.$$

(3) This follows from the following block column transformations:

$$\begin{pmatrix} A & 0 \\ 0 & B \end{pmatrix} \rightarrow \begin{pmatrix} A & 0 \\ I_n & B \end{pmatrix} \rightarrow \begin{pmatrix} A & -AB \\ I_n & 0 \end{pmatrix} \overset{\text{by (1)}}{\longrightarrow} \begin{pmatrix} AB & A \\ 0 & I_n \end{pmatrix} \rightarrow \begin{pmatrix} AB & 0 \\ 0 & I_n \end{pmatrix}.$$

(4) Use (3) and induct on n.

(5) This follows from (4).

$\square$

Notation 165. Let $\mathbf{R}$ be a ring, $n \geq 1$, G a subgroup of $GL_n(\mathbf{R})$, and consider $u, u' \in \mathbf{R}^{1 \times n}$. We write

$$u \sim_G u'$$

if there exists $A \in G$ such that $uA = u'$.

Proposition 166 (Roitman-Suslin-Vaserstein [146, 175]). *Let $\mathbf{R}$ be a ring, and $n \geq 3$.*

(1) *If $u, u' \in \mathbf{R}^{1 \times n}$ and $v \in \mathbf{R}^{n \times 1}$ are such that $uv = u'v = 1$, then $u \sim_{E_n(\mathbf{R})} u'$.*

(2) *If u and u' are two elements of a free basis of $\mathbf{R}^n$, then $u \sim_{E_n(\mathbf{R})} u'$.*

(3) *Let $u = (u_1, \ldots, u_n)$ a unimodular row in $\mathbf{R}^n$.*

    (i) *For $v_1, v_2 \in \mathbf{R}$, if $u_1v_1 + u_2v_2$ is invertible modulo $\langle u_3, \ldots, u_n \rangle$, then*

$$u \sim_{E_n(\mathbf{R})} (v_1, v_2, u_3, \ldots, u_n).$$

    (ii) *If $a \in \mathbf{R}$ is invertible modulo $\langle u_3, \ldots, u_n \rangle$, then*

$$u \sim_{E_n(\mathbf{R})} (a^2 u_1, u_2, \ldots, u_n).$$

Proof. (1) As $(u - u')v = 0$, then, by virtue of Corollary 162, $E := I_n + v(u' - u) \in E_n(\mathbf{R})$ (because ${}^t E \in E_n(\mathbf{R})$). We have $uE = u + uv(u - u') = u + u' - u = u'$.

(2) u and u' can be seen as the first two rows of a matrix $A \in GL_n(\mathbf{R})$. Denoting by c and c' the first two columns of A^{-1}, we have

$$\begin{cases} uc = 1 \\ uc' = 0 \\ u'c = 0 \\ u'c' = 1, \end{cases}$$

and, thus, $u(c + c') = u'(c + c')$. The desired result follows from (1).

(3) (i) Writing $b(u_1v_1 + u_2v_2) + b_3u_3 + \cdots + b_nu_n = 1$ with $b, b_3, \ldots, b_n \in \mathbf{R}$, we have:

$$b(v_1 + u_2)u_1 + b(v_2 - u_1)u_2 + b_3u_3 + \cdots + b_nu_n$$
$$= b(v_1 + u_2)v_2 + b(v_2 - u_1)(-v_1) + b_3u_3 + \cdots + b_nu_n = 1.$$

Thus,

$$(u_1,\ldots,u_n) \sim_{E_n(\mathbf{R})} (v_2,-v_1,u_3,\ldots,u_n) \text{ (by (1))}$$
$$\sim_{E_n(\mathbf{R})} (v_1,v_2,u_3,\ldots,u_n) \text{ (by Proposition 164.(2))}.$$

(ii) Write $u_1 w_1 + \cdots + u_n w_n = 1$ with $w_i \in \mathbf{R}$. By (i), we have

$$(u_1,\ldots,u_n) \sim_{E_n(\mathbf{R})} (w_1,w_2,u_3,\ldots,u_n).$$

As $a = (au_1)w_1 + (au_2)w_2 + (au_3)w_3 + \cdots + (au_n)w_n$ is invertible modulo $\langle u_3,\ldots, u_n\rangle$, we deduce that $(au_1)w_1 + (au_2)w_2$ is invertible modulo $\langle u_3,\ldots,u_n\rangle$, and hence, again by (i), we have

$$(w_1,w_2,u_3,\ldots,u_n) \sim_{E_n(\mathbf{R})} (au_1,au_2,u_3,\ldots,u_n).$$

Now, by Proposition 164.(5), writing $ab \equiv 1 \bmod \langle u_3,\ldots,u_n\rangle$, that is, $\bar{a}\bar{b} = \bar{1}$ in the ring $\mathbf{A} := \mathbf{R}/\langle u_3,\ldots,u_n\rangle$, the diagonal matrix $\mathrm{diag}(\bar{a},\bar{b}) \in E_2(\mathbf{A})$. It follows that after finitely many elementary operations performed on the first two entries of the row $(au_1,au_2,u_3,\ldots,u_n)$ (that is, operations of the form $c_i \to c_i + \lambda c_j$, with $1 \le i \ne j \le 2$ and $\lambda \in \mathbf{R}$, where c_1 and c_2 are the first two entries of the considered row), we get a new row τ of the form

$$(a^2 u_1 + \lambda_3 u_3 + \cdots + \lambda_n u_n, u_2 + \mu_3 u_3 + \cdots + \mu_n u_n, u_3,\ldots,u_n),$$

with $\lambda_i, \mu_i \in \mathbf{R}$. The desired result follows since

$$\tau E_{3,1}(-\lambda_3)\cdots E_{n,1}(-\lambda_n)E_{3,2}(-\mu_3)\cdots E_{n,2}(-\mu_n) = (a^2 u_1, u_2,\ldots,u_n).$$

□

2.5.5 Local-Global Principle for Elementary Polynomial Matrices

This subsection is reproduced from [106]. The goal is to obtain a constructive version of the following Quillen Induction result:

Let $\mathbf{R}$ be a ring, $n \ge 3$, and $M \in SL_n(\mathbf{R}[X])$. If $M_{\mathscr{M}} \in E_n(\mathbf{R}_{\mathscr{M}}[X])$ for every maximal ideal $\mathscr{M}$ of $\mathbf{R}$, then $M \in E_n(\mathbf{R}[X])$.

Notation 167.   (i) Recall that for a ring $\mathbf{A}$ and n, $m \ge 1$, we denote by $M_{n,m}(\mathbf{A})$ the set of matrices of size $n \times m$ with entries in $\mathbf{A}$. The set $M_{n,n}(\mathbf{A})$ will simply be denoted by $M_n(\mathbf{A})$.

(ii) If J is an ideal of a ring $\mathbf{A}$ and $n \ge 1$, we define

$$GL_n(\mathbf{A}, J) := \mathrm{Ker}\left(GL_n(\mathbf{A}) \to GL_n(\mathbf{A}/J)\right).$$

Thus, $\mathrm{GL}_n(\mathbf{A}, J)$ is the normal subgroup of $\mathrm{GL}_n(\mathbf{A})$ consisting of matrices M which are $\equiv \mathrm{I}_n \bmod \mathrm{M}_n(J)$. $\mathrm{GL}_n(\mathbf{A}, \mathbf{A})$ is just $\mathrm{GL}_n(\mathbf{A})$. Also, we define

$$\mathrm{E}_n(\mathbf{A}, J) := \{\gamma^{-1} E_{i,j}(a)\, \gamma;\ \gamma \in \mathrm{E}_n(\mathbf{A}),\ a \in J,\ 1 \leq i \neq j \leq n\}$$

the normal subgroup of $\mathrm{E}_n(\mathbf{A})$ generated by the elementary matrices $\{E_{i,j}(a)$; $a \in J,\ 1 \leq i \neq j \leq n\}$. $\mathrm{E}_n(\mathbf{A}, \mathbf{A})$ is just $\mathrm{E}_n(\mathbf{A})$. We have

$$\mathrm{E}_n(\mathbf{A}, J) \subseteq \mathrm{E}_n(\mathbf{A}) \cap \mathrm{GL}_n(\mathbf{A}, J) \subseteq \mathrm{SL}_n(\mathbf{A}, J) := \mathrm{SL}_n(\mathbf{A}) \cap \mathrm{GL}_n(\mathbf{A}, J)$$

but, in general, $\mathrm{E}_n(\mathbf{A}, J) \neq \mathrm{E}_n(\mathbf{A}) \cap \mathrm{GL}_n(\mathbf{A}, J)$. However, in the particular case where $\mathbf{A} = \mathbf{R}[X]$ and $J = \langle X \rangle$, the two groups coincide as affirms the following lemma.

We need the following series of lemmas.

Lemma 168. *For any ring $\mathbf{R}$ and $n \geq 1$, we have*

$$\mathrm{E}_n(\mathbf{R}[X], \langle X \rangle) = \mathrm{Ker}\left(\mathrm{E}_n(\mathbf{R}[X]) \to \mathrm{E}_n(\mathbf{R}[X]/\langle X \rangle) = \mathrm{E}_n(\mathbf{R}) \right).$$

It is generated by matrices of type $\gamma E_{i,j}(Xg)\gamma^{-1}$ with $g \in \mathbf{R}[X]$ and $\gamma \in \mathrm{E}_n(\mathbf{R})$.

Proof. Let K be the above-mentioned kernel, take $M \in K$, and write

$$M = \prod_{k=1}^{m} E_{i_k, j_k}(a_k + Xg_k) = \prod_{k=1}^{m} E_{i_k, j_k}(a_k)\, E_{i_k, j_k}(Xg_k)$$

with $a_k \in \mathbf{R}$ and $g_k \in \mathbf{R}[X]$. Denoting by $M_k := \prod_{\ell=1}^{k} E_{i_\ell, j_\ell}(a_\ell) \in \mathrm{E}_n(\mathbf{R})$, we have

$$M = \prod_{k=1}^{m} \left(M_k\, E_{i_k, j_k}(Xg_k)\, M_k^{-1} \right) \cdot \prod_{k=1}^{m} E_{i_k, j_k}(a_k) = \prod_{k=1}^{m} M_k\, E_{i_k, j_k}(Xg_k)\, M_k^{-1}$$

as $\prod_{k=1}^{m} E_{i_k, j_k}(a_k) = M(0) = \mathrm{I}_n$. $\qquad\square$

Lemma 169. *Let $\mathbf{R}$ be a ring, $s \in \mathbf{R}$, $n \geq 3$, and $M = M(X) \in \mathrm{E}_n(\mathbf{R}_s[X], \langle X \rangle)$. Then, there exists $k \in \mathbb{N}$ such that $M(s^k X) \in \mathrm{E}_n(\mathbf{R}[X], \langle X \rangle)$.*

Proof. We can, by virtue of Lemma 168, suppose that $M = \gamma E_{i,j}(Xg)\gamma^{-1}$ for some $g \in \mathbf{R}_s[X]$, $\gamma \in \mathrm{E}_n(\mathbf{R}_s)$, and $1 \leq i \neq j \leq n$. Writing $E_{i,j}(Xg) = \mathrm{I}_m + Xg\, e_{i,j}$, where $e_{i,j}$ is the matrix in $\mathrm{M}_n(\mathbf{R})$ with 1 on position (i, j) and 0s elsewhere, we have

$$\gamma E_{i,j}(Xg)\, \gamma^{-1} = \mathrm{I}_n + Xg(\gamma\, e_{i,j}\, \gamma^{-1}).$$

Denoting by $v \in \mathbf{R}_s^{n \times 1}$ the ith column of γ and $u \in \mathbf{R}_s^{1 \times n}$ the jth row of γ^{-1}, as $\gamma \gamma^{-1} = I_n$, we have $\gamma \, e_{i,j} \, \gamma^{-1} = vu$. Moreover, $uv = 0$ since $i \neq j$, u is clearly unimodular, and

$$\gamma E_{i,j}(Xg) \, \gamma^{-1} = I_n + (Xg)vu.$$

By virtue of Proposition 146, $v \in \langle {}^t s_{i,j} \mid 1 \leq i < j \leq n \rangle$, where the $s_{i,j}$'s are the obvious syzygies of u, that is, there exist $a_1, \ldots, a_k \in \mathbf{R}_s$ and $w_1, \ldots, w_k \in \{ {}^t s_{i,j} \mid 1 \leq i < j \leq n \}$ such that

$$v = a_1 w_1 + \cdots + a_k w_k.$$

Since $uw_i = 0$, we have:

$$I_n + (Xg)vu = (I_n + (Xg)a_1 w_1 u)(I_n + (Xg)a_2 w_2 u) \cdots (I_n + (Xg)a_k w_k u),$$

with $I_n + (Xg)a_i w_i u \in E_n(\mathbf{R}_s[X])$. Writing $g = \frac{\tilde{g}}{s^k}$, $a_i w_i = \frac{\tau_i}{s^k}$, and $u = \frac{\tilde{u}}{s^k}$, with $\tilde{g} \in \mathbf{R}[X]$, $\tau_i \in \mathbf{R}^{n \times 1}$, $\tilde{u} \in \mathbf{R}^{1 \times n}$, and $k \in \mathbb{N}$, we have

$$M(s^{3k}X) = I_n + (s^{3k}Xg)vu = (I_n + (X\tilde{g})\tau_1 \tilde{u}) \cdots (I_n + (X\tilde{g})\tau_k \tilde{u}) \in E_n(\mathbf{R}[X], \langle X \rangle).$$

$\square$

Lemma 170. *Let $\mathbf{R}$ be a ring, $n \geq 3$, $s \in \mathbf{R}$, and $M = M(X) \in GL_n(\mathbf{R}[X])$ such that $M \in E_n(\mathbf{R}_s[X])$. Then, there exists $k \in \mathbb{N}$ such that for all $a, b \in \mathbf{R}$ which are congruent modulo s^k, the matrix $M^{-1}(aX)M(bX) \in E_n(\mathbf{R}[X], \langle X \rangle)$.*

Proof. Consider two new variables T, U and set

$$M'(X, T, U) := M^{-1}\left((T+U)X\right)M(TX).$$

Applying Lemma 169 with $\mathbf{R}[X, T]$ instead of $\mathbf{R}$ and U instead of X, there exists $k \in \mathbb{N}$ such that

$$M'(X, T, s^k U) =: N(X, T, U) \in E_n(\mathbf{R}[X, T, U], \langle U \rangle).$$

As $N(X, T, U) = M^{-1}\left((T + s^k U)X\right)M(TX)$, for $b = a + s^k c$, we have

$$M^{-1}(aX) \, M(bX) = N(X, a, c) \text{ over } \mathbf{R}_s.$$

We have $N(0, T, U) = I_n$ over $\mathbf{R}_s$ but not necessarily over $\mathbf{R}$. The matrix

$$A(X, T, U) := N^{-1}(0, T, U) N(X, T, U)$$

is equal to $N(X, T, U)$ over $\mathbf{R}_s$, with $N(0, T, U) = I_n$ over $\mathbf{R}$, and

$$M^{-1}(aX) M(bX) = A(X, a, c)$$

over $\mathbf{R}_s$, and $A(X, a, c) \in E_n(\mathbf{R}[X], \langle X \rangle)$.

$\square$

Lemma 171. *Let $\mathbf{R}$ be a ring, $n \geq 3$, $r, s \in \mathbf{R}$ with $\langle r, s \rangle = \mathbf{R}$, and $M \in GL_n(\mathbf{R}[X])$. If $M \in E_n(\mathbf{R}_r[X])$ and $M \in E_n(\mathbf{R}_s[X])$ then $M \in E_n(\mathbf{R}[X])$.*

Proof. By virtue of Lemma 170, there exists $k \in \mathbb{N}$ such that for all $a, b \in \mathbf{R}$ which are congruent modulo r^k or modulo s^k, the matrix $M^{-1}(aX)M(bX)$ is in $E_n(\mathbf{R}[X], \langle X \rangle)$. Writing $r^k a + s^k b = 1$ for some $a, b \in \mathbf{R}$, and considering $c = s^k b$, we have $c \equiv 0 \bmod s^k$ and $c \equiv 1 \bmod r^k$. The desired result follows from the equality

$$M = M(X) = M^{-1}(0.X) M(c.X) M^{-1}(c.X) M(1.X).$$

$\square$

As an immediate consequence, one obtains:

Theorem 172. (Propagation Lemma for Elementary Polynomial Matrices [76, 106])
Let $\mathbf{R}$ be a ring, $n \geq 3$, and $M = M(X) \in GL_n(\mathbf{R}[X])$.

(i) *If $M(0) = I_n$ then the set $\{s \in \mathbf{R} \mid M \in E_n(\mathbf{R}_s[X])\}$ is an ideal of $\mathbf{R}$.*

(ii) *The set $\{s \in \mathbf{R} \mid M(X) \sim_{E_n(\mathbf{R}_s[X])} M(0)\}$ is an ideal of $\mathbf{R}$.*

Theorem 172 can be rephrased to give the following concrete local-global principle:

Theorem 173. (Concrete Local-Global Principle for Elementary Polynomial Matrices [106])
Let $\mathbf{R}$ be a ring, $n \geq 3$, $S_1, \ldots, S_m$ comaximal monoids of $\mathbf{R}$, and $M = M(X) \in GL_n(\mathbf{R}[X])$ with $M(0) = I_n$. Then:

$$M \in E_n(\mathbf{R}[X]) \quad \Leftrightarrow \quad M \in E_n(\mathbf{R}_{S_i}[X]) \; \forall \, 1 \leq i \leq m.$$

2.5.6   A Realization Algorithm for $SL_3(\mathbf{R}[X])$

This subsection is reproduced from [76] and [135]. The proofs given there are already constructive.

Lemma 174. ([135]) *Let* $\mathbf{R}$ *be a ring, and* $a, a', b, c, d \in \mathbf{R}$ *such that* $aa'd - bc = 1$. *Then*

$$\begin{pmatrix} aa' & b & 0 \\ c & d & 0 \\ 0 & 0 & 1 \end{pmatrix} \equiv \begin{pmatrix} a & b & 0 \\ c & a'd & 0 \\ 0 & 0 & 1 \end{pmatrix} \cdot \begin{pmatrix} a' & b & 0 \\ c & ad & 0 \\ 0 & 0 & 1 \end{pmatrix} \bmod E_3(\mathbf{R})$$

in the sense that there exist $M, N, A \in E_3(\mathbf{R})$ *such that*

$$\begin{pmatrix} aa' & b & 0 \\ c & d & 0 \\ 0 & 0 & 1 \end{pmatrix} = M \begin{pmatrix} a & b & 0 \\ c & a'd & 0 \\ 0 & 0 & 1 \end{pmatrix} N \begin{pmatrix} a' & b & 0 \\ c & ad & 0 \\ 0 & 0 & 1 \end{pmatrix} A.$$

Proof. It suffices to take

$$M = E_{2,1}(-cd) E_{2,3}(ad - 1) E_{3,2}(1) E_{2,3}(-1),$$
$$N = E_{2,3}(1) E_{3,2}(-1) E_{2,3}(1),$$
$$\text{and } A = E_{2,3}(-1) E_{3,2}(1) E_{2,3}(a - 1) E_{3,1}(-a'c) E_{3,2}(-a'd).$$

$\square$

Theorem 175. ([76, 135]) *Let* $\mathbf{R}$ *be a residually discrete local ring, and*

$$M = \begin{pmatrix} p & q & 0 \\ r & s & 0 \\ 0 & 0 & 1 \end{pmatrix} \in SL_3(\mathbf{R}[X]),$$

where p *is monic. Then* M *is realizable, i.e.,* $M \in E_3(\mathbf{R}[X])$.

Proof. We proceed by induction on $\deg p$. The case $p = 1$ is immediate. Now, suppose that $\deg p = d \geq 1$. Using one elementary operation, we can replace q with its remainder on division by p, and, thus, we can suppose that $\deg q < d$. As $1 = ps - qr$ and $\mathbf{R}$ is local, then either $p(0) \in \mathbf{R}^\times$ or $q(0) \in \mathbf{R}^\times$.

Case 1: $q(0) \in \mathbf{R}^\times$. Again, using one elementary operation, we can replace p with $p - p(0)q(0)^{-1}q$, and, thus, we can suppose that $p(0) = 0$, that is, $p = Xp'$ for some monic polynomial $p' \in \mathbf{R}[X]$. By Lemma 174, we have

$$\begin{pmatrix} p & q & 0 \\ r & s & 0 \\ 0 & 0 & 1 \end{pmatrix} \equiv \begin{pmatrix} X & q & 0 \\ r & p's & 0 \\ 0 & 0 & 1 \end{pmatrix} \cdot \begin{pmatrix} p' & q & 0 \\ r & Xs & 0 \\ 0 & 0 & 1 \end{pmatrix} \bmod E_3(\mathbf{R}[X]).$$

The second matrix on the right hand side is realizable by induction hypothesis. On the other hand, using one elementary operation, we can replace q with $q(0) \in \mathbf{R}^\times$, and, thus, the first matrix on the right hand side is realizable.

Case 2: $q(0) \in \mathrm{Rad}(\mathbf{R})$. By virtue of Propositions 52 and 56, there exist $p', q' \in \mathbf{R}[X]$ with $\deg p' < \deg q < d$ and $\deg q' < \deg p = d$ such that $p'p - q'q = 1$. Also, note that, as $p'(0)p(0) - q'(0)q(0) = 1$ and $q(0) \in \mathrm{Rad}(\mathbf{R})$, we have $p'(0) \in \mathbf{R}^\times$ and, thus, $q(0) + p'(0) \in \mathbf{R}^\times$. Now, we have

$$
\begin{pmatrix} p & q & 0 \\ r & s & 0 \\ 0 & 0 & 1 \end{pmatrix} = E_{2,1}(rp' - sq') \begin{pmatrix} p & q & 0 \\ q' & p' & 0 \\ 0 & 0 & 1 \end{pmatrix}
$$

$$
= E_{2,1}(rp' - sq') E_{1,2}(-1) \begin{pmatrix} p + q' & q + p' & 0 \\ q' & p' & 0 \\ 0 & 0 & 1 \end{pmatrix}.
$$

The last matrix on the right hand side is realizable by Case 1 since $q(0) + p'(0) \in \mathbf{R}^\times$ and $\deg(p + q') = d$. $\qquad\square$

Corollary 176. *Let* $\mathbf{K}$ *be a discrete field, and*

$$
M = \begin{pmatrix} p & q & 0 \\ r & s & 0 \\ 0 & 0 & 1 \end{pmatrix} \in \mathrm{SL}_3(\mathbf{K}[X_1, \ldots, X_k]).
$$

Then M *is realizable, i.e.,* $M \in E_3(\mathbf{K}[X_1, \ldots, X_k])$.

Proof. Let $\mathbf{R} = \mathbf{K}[X_1, \ldots, X_{k-1}]$ and $X = X_k$. We may assume that p is monic at X by applying a change of variables (à la Nagata for example). Also, applying the General Local-Global Principle 35 and the Concrete local-global Principle for elementary polynomial matrices (Theorem 173), we can suppose that $\mathbf{R}$ is local. The desired result follows from Theorem 175. $\qquad\square$

2.5.7 Elementary Unimodular Completion

Recall that for any ring $\mathbf{B}$, when we say that a matrix $N \in M_n(\mathbf{B})$ ($n \geq 3$) is in $\mathrm{SL}_2(\mathbf{B})$ we mean that it is of the form

$$
\begin{pmatrix} N' & 0 & \cdots & 0 \\ 0 & 1 & & \\ \vdots & & \ddots & \\ 0 & & & 1 \end{pmatrix}
$$

with $N' \in \mathrm{SL}_2(\mathbf{B})$.

Lemma 177. (Elementary Unimodular Completion, General Case) *Let* $\mathbf{R}$ *be a ring,* $n \geq 3$, *and* $\mathscr{V} = \mathscr{V}(X) = {}^{\mathrm{t}}(v_1(X), \ldots, v_n(X)) \in \mathrm{Um}_n(\mathbf{R}[X])$ *such that* v_1 *is monic.* *Then there exist* $B_1 \in \mathrm{SL}_2(\mathbf{R}[X])$ *and* $B_2 \in E_n(\mathbf{R}[X])$ *such that* $B_1 B_2 \mathscr{V} = \mathscr{V}(0)$.

Proof. This follows from Algorithm 69 and Suslin's Normality Theorem 163.

$\square$

Theorem 178. (Elementary Unimodular Completion, Case $\mathbf{K}[X_1,\ldots,X_k]$; [76, 135])

Let $\mathbf{K}$ be a field, and $n \geq 3$. Then the group $E_n(\mathbf{K}[X_1,\ldots,X_k])$ acts transitively on the set $\mathrm{Um}_n(\mathbf{K}[X_1,\ldots,X_k])$.

Proof. The case $k = 1$ is clear via the Euclidean division algorithm. We assume that the results holds for $\mathbf{R} = \mathbf{K}[X_1,\ldots,X_{k-1}]$. Let $X = X_k$ and $\mathcal{V} = {}^t(v_1(X),\ldots,v_n(X)) \in \mathrm{Um}_n(\mathbf{R}[X])$. We may assume that v_1 is monic by applying a change of variables (à la Nagata for example). By virtue of Lemma 177, there exist $B_1 \in \mathrm{SL}_2(\mathbf{R}[X])$ and $B_2 \in E_n(\mathbf{R}[X])$ such that

$$B_1 B_2 \mathcal{V} = \mathcal{V}(0).$$

By the induction hypothesis, there exists $B \in E_n(\mathbf{R})$ such that

$$BB_1B_2\mathcal{V} = {}^t(0,\ldots,0,1) =: e_n, \text{ and thus}, \mathcal{V} = B_2^{-1}B_1^{-1}B^{-1}e_n.$$

Since $E_n(\mathbf{R}[X])$ is a normal subgroup of $\mathrm{SL}_n(\mathbf{R}[X])$ (see Suslin's Normality Theorem 163), one can find a matrix $B' \in E_n(\mathbf{R}[X])$ such that $B_1^{-1}B^{-1} = B'B_1^{-1}$. Note that B_1^{-1} has the form

$$\begin{pmatrix} p & q & 0 & \ldots & 0 \\ r & s & 0 & \ldots & 0 \\ 0 & 0 & 1 & & \\ \vdots & \vdots & & \ddots & \\ 0 & 0 & & & 1 \end{pmatrix},$$

and, thus, $B_1^{-1}e_n = e_n$ and $\mathcal{V} = B_2^{-1}B'e_n$, or also

$$B'^{-1}B_2\mathcal{V} = e_n.$$

$\square$

Now, we are in position to prove Suslin's stability theorem.

Theorem 179. (Suslin's Stability Theorem; [165])

Let $\mathbf{K}$ be a field, and $n \geq 3$. Then

$$\mathrm{SL}_n(\mathbf{K}[X_1,\ldots,X_k]) = E_n(\mathbf{K}[X_1,\ldots,X_k]).$$

Proof. Let $M \in \mathrm{SL}_n(\mathbf{K}[X_1,\ldots,X_k])$. As its last column c_n is unimodular, in virtue of Theorem 179, there exists $B \in E_n(\mathbf{K}[X_1,\ldots,X_k])$ such that $Bc_n = e_n$, and, thus,

$$BM = \begin{pmatrix} & & & 0 \\ & \tilde{M} & & \vdots \\ & & & 0 \\ m_1 & \cdots & m_{n-1} & 1 \end{pmatrix}, \text{ and,}$$

$$BME_{n,1}(-m_1)\cdots E_{n,n-1}(-m_{n-1}) = \begin{pmatrix} \tilde{M} & 0 \\ 0 & 1 \end{pmatrix}.$$

Repeating this process, by successive elementary operations, we get a matrix of the form

$$\begin{pmatrix} p & q & 0 \\ r & s & 0 \\ 0 & 0 & 1 \end{pmatrix} \in SL_3(\mathbf{K}[X_1,\dots,X_k]).$$

Such matrix is realizable by virtue of Corollary 176.

□

It is worth recalling that Suslin's stability theorem does not hold for $n = 2$ (see Cohn's Example 153).

2.6 The Hermite Ring Conjecture

2.6.1 The Hermite Ring Conjecture in Dimension One

This subsection is extracted from [182]. Recall that a ring $\mathbf{R}$ is said to be *Hermite* if any finitely-generated stably free $\mathbf{R}$-module is free (see Definition 21). Examples of Hermite rings are local rings (see Theorem 10), rings of Krull dimension ≤ 1 (see Corollary 93), univariate polynomial rings with coefficients in a ring of Krull dimension ≤ 1 (see Corollary 187; the goal of this subsection is to prove this result), polynomial rings over Bezout domains (see Corollary 143), and polynomial rings over zero-dimensional rings (see Theorem 50).

Recall also that for any ring $\mathbf{R}$, saying that any finitely-generated stably free $\mathbf{R}$-module is free amounts to saying that any unimodular row over $\mathbf{R}$ is completable (see Propositions 20 and 22), i.e., it can be written as the first row of an invertible matrix with entries in $\mathbf{R}$.

Quillen's and Suslin's proofs of Serre's problem on projective modules had a big effect on the subsequent development of the study of projective modules. Nevertheless, many old conjectures and open questions about projective modules over polynomial rings still wait for solutions. Our concern here is the following equivalent two conjectures (for a proof of the equivalence, see Propositions 20 and 22)).

Conjecture 180. (Hermite Ring Conjecture (1978) [91, 92]) *If $\mathbf{R}$ is an Hermite ring, then $\mathbf{R}[X]$ is also Hermite.*

Conjecture 181. *If $\mathbf{R}$ is a ring and $v = (v_0(X),\dots,v_n(X))$ is a unimodular row over $\mathbf{R}[X]$ such that $v(0) = (1,0,\dots,0)$, then v can be completed to a matrix in $GL_{n+1}(\mathbf{R}[X])$.*

In this subsection we will prove constructively that for any ring $\mathbf{R}$ of Krull dimension ≤ 1 and $n \geq 3$, the group $E_n(\mathbf{R}[X])$ acts transitively on $Um_n(\mathbf{R}[X])$ (Theorem 186). In particular, we obtain that for any ring $\mathbf{R}$ with Krull dimension ≤ 1, all finitely-generated stably free modules over $\mathbf{R}[X]$ are free (Corollary 187). This settles the long-standing Hermite ring conjecture for rings of Krull dimension ≤ 1.

Let us begin by giving a constructive and elementary proof of a lemma which was used by Roitman [152] in the proof of his Theorem 5. Roitman gave as reference [91] (Chapter III, Lemma 1.1) but the proof given by Lam in [91] is not constructive and relies on the "going-up" property of prime ideals in integral extensions.

Lemma 182. *Let $\mathbf{R}$ be a ring, and I an ideal in $\mathbf{R}[X]$ that contains a monic polynomial. Let J be an ideal in $\mathbf{R}$ such that $I + J[X] = \mathbf{R}[X]$. Then $(I \cap \mathbf{R}) + J = \mathbf{R}$.*

Proof. Let us denote by f a monic polynomial in I. Since $I + J[X] = \mathbf{R}[X]$, there exist $g \in I$ and $h \in J[X]$ such that $g + h = 1$. It follows that $\langle \bar{f}, \bar{g} \rangle = (\mathbf{R}/J)[X]$ where the classes are taken modulo $J[X]$. By virtue of Proposition 56, we obtain that $\mathrm{Res}(\bar{f}, \bar{g}) \in (\mathbf{R}/J)^{\times}$. As f is a monic polynomial, $\mathrm{Res}(\bar{f}, \bar{g}) = \overline{\mathrm{Res}(f,g)}$, and, thus, $\langle \mathrm{Res}(f,g) \rangle + J = \mathbf{R}$. The desired conclusion follows from the fact that $\mathrm{Res}(f,g) \in I \cap \mathbf{R}$. $\qquad\square$

The following three lemmas were already proved constructively by their authors.

Lemma 183. *(Roitman's Lemma [152]) Let $\mathbf{R}$ be a ring, and $f(X) \in \mathbf{R}[X]$ of degree $n > 0$, such that $f(0) \in \mathbf{R}^{\times}$. Then for any $g(X) \in \mathbf{R}[X]$ and $k \geq \deg g(X) - \deg f(X) + 1$ there exists $h_k(X) \in \mathbf{R}[X]$ of degree $< n$ such that $g(X) \equiv X^k h_k(X) \bmod \langle f(X) \rangle$.*

Proof. Let $f(X) = a_0 + \cdots + a_n X^n$, $g(X) = c_0 + \cdots + c_m X^m$. Let $g(X) - c_0 a_0^{-1} f(X) = X h_1(X)$. Then $g(X) \equiv X h_1(X) \bmod \langle f(X) \rangle$ and $\deg h_1(X) < \max(m,n)$. Similarly we obtain $h_2(X)$ such that $h_1(X) \equiv X h_2(X) \bmod \langle f(X) \rangle$, $g(X) \equiv X^2 h_2(X) \bmod \langle f(X) \rangle$, $\deg h_2(X) < \max(m-1,n)$, and so on. $\qquad\square$

Lemma 184. *(Bass' Lemma [13]) Let $k \in \mathbb{N}$, $\mathbf{R}$ a ring, $f_1, \ldots, f_r \in \mathbf{R}[X]$ with degrees $\leq k-1$, and $f_{r+1} \in \mathbf{R}[X]$ monic with degree k. If the coefficients of $f_1, \ldots, f_r$ generate the ideal $\mathbf{R}$ of $\mathbf{R}$, then $\langle f_1, \ldots, f_r, f_{r+1} \rangle$ contains a monic with degree $k-1$.*

Proof. Let us denote by $\mathfrak{a} = \langle f_1, \ldots, f_r, f_{r+1} \rangle$ and $\mathfrak{b}$ the ideal formed by the coefficients of X^{k-1} of the elements of $\mathfrak{a}$ having degree $\leq k-1$. It suffices to prove that $\mathfrak{b} = \mathbf{R}$. In fact we will prove that $\mathfrak{b}$ contains all the coefficients of $f_1, \ldots, f_r$. For $1 \leq i \leq r$, denoting by $f_i = b_0 + b_1 X + \cdots + b_{k-1} X^{k-1}$ and $f_{r+1} = a_0 + \cdots + a_{k-1} X^{k-1} + X^k$, we have $b_{k-1} \in \mathfrak{b}$ and $f_i' = X f_i - b_{k-1} f = b_0' + b_1' X + \cdots + b_{k-1}' X^{k-1} \in \mathfrak{a}$ with $b_j' \equiv b_{j-1} \bmod \langle b_{k-1} \rangle$. Thus, $b_{k-1}' = b_{k-2} - a_{k-1} b_{k-1} \in \mathfrak{b}$, $b_{k-2} \in \mathfrak{b}$, and so on until getting that all the b_i's are in $\mathfrak{b}$. $\qquad\square$

Now we're reaching a crucial stage in our objective to prove the Hermite ring conjecture for rings of Krull dimension ≤ 1.

Lemma 185. *Let $\mathbf{R}$ be a reduced local ring of dimension ≤ 1, $n \geq 2$, and let $v(X) = {}^{\mathrm{t}}(v_0(X), \ldots, v_n(X)) \in \mathrm{Um}_{n+1}(\mathbf{R}[X])$. Then there exists $E \in \mathbf{E}_{n+1}(\mathbf{R}[X])$ such that $E v(X) = {}^{\mathrm{t}}(v_0(X), v_1(X), \ldots, v_{n-1}(X), c_n)$, where $c_n \in \mathbf{R}$.*

Proof. By the local-global principle for elementary matrices (see Theorem 173), we can suppose that $\mathbf{R}$ is local.

We prove the claim by double induction on the number N of nonzero coefficients of $v_0(X),\ldots,v_n(X)$ and d, starting with $N = 1$ (in that case the result is immediate) and $d = 0$ (in that case the result is well-known).

Let $N > 1$ and $d > 0$. We may assume that $v_0(0) \in \mathbf{R}^\times$. Let us denote by a the leading coefficient of v_0 and $m_0 := \deg v_0$. If $a \in \mathbf{R}^\times$ then the result follows from Suslin's lemma (Theorem 57). So we may assume $a \in \mathrm{Rad}(\mathbf{R})$. By the induction hypothesis applied to the ring $\mathbf{R}/\langle a \rangle$, we can assume that $v(X) \equiv {}^t(1,0,\ldots,0) \bmod (a\mathbf{R}[X])^{n+1}$. By Lemma 183, we assume now $v_i = X^{2k}w_i$, where $\deg w_i < m_0$ for $1 \leq i \leq n$. By Proposition 166.(3).(ii), we assume $\deg v_i < m_0$.

If $m_0 \leq 1$, our first claim is established. Assume now that $m_0 \geq 2$. Let $(c_1,\ldots, c_{m_0(n-1)})$ be the coefficients of $1, X, \ldots, X^{m_0-1}$ in the polynomials $v_2(X),\ldots,v_n(X)$. By Lemma 182, the ideal generated in $\mathbf{R}_a$ by $\mathbf{R}_a \cap (v_0\mathbf{R}_a[X] + v_1\mathbf{R}_a[X])$ and the c_i's is $\mathbf{R}_a$. As $m_0(n - 1) \geq 2 > \dim \mathbf{R}_a$, by the Stable range Theorem 92 there exists

$$(c'_1,\ldots,c'_{m_0(n-1)}) \equiv (c_1,\ldots,c_{m_0(n-1)}) \bmod (v_0\mathbf{R}[X] + v_1\mathbf{R}[X]) \cap \mathbf{R}$$

such that $c'_1\mathbf{R}_a + \cdots + c'_{m_0(n-1)}\mathbf{R}_a = \mathbf{R}_a$. Assume that we have already $c_1\mathbf{R}_a + \cdots + c_{m_0(n-1)}\mathbf{R}_a = \mathbf{R}_a$. By Lemma 184, the ideal $\langle v_0, v_2,\ldots,v_n \rangle$ of $\mathbf{R}[X]$ contains a polynomial $w(X)$ of degree $m_0 - 1$ which is unitary in $\mathbf{R}_a$. Let us denote the leading coefficient of w by ua^k where $u \in \mathbf{R}^\times$ and that of v_1 by b. Using Proposition 166.(3).(ii), as a is invertible modulo $\langle v_0, v_1 \rangle$ (because $v(X) \equiv {}^t(1,0,\ldots,0) \bmod (a\mathbf{R}[X])^{n+1}$), we can by elementary operations make the following transformations

$${}^t(v_0,v_1,\ldots,v_n) \to {}^t(v_0,a^{2k}v_1,\ldots,v_n) \to {}^t(v_0,a^{2k}v_1 + (1-a^ku^{-1}b)w, v_2,\ldots,v_n).$$

Now, $a^{2k}v_1 + (1 - a^ku^{-1}b)w$ is unitary in $\mathbf{R}_a$, we can assume that v_1 unitary in $\mathbf{R}_a$, $\deg(v_1) := m_1 < m_0$. By Proposition 166.(3).(ii), as a is invertible modulo $\langle v_0 \rangle$ (since $v(X) \equiv {}^t(1,0,\ldots,0) \bmod (a\mathbf{R}[X])^{n+1}$), by elementary operations, ${}^t(v_0,v_1,v_2,\ldots,v_n)$ can be transformed into ${}^t(v_0,v_1,a^\ell v_2,\ldots,a^\ell v_n)$ for a suitable $\ell \in \mathbb{N}$ so that we can divide (like in Euclidean division) all $a^\ell v_2,\ldots,a^\ell v_n$ by v_1, and, thus, we can assume that $\deg v_i < m_1$ for $2 \leq i \leq n$.

Repeating the argument above we lower the degree of v_1 until reaching the desired result. $\qquad\square$

Theorem 186. *Let $\mathbf{R}$ be a ring of dimension ≤ 1, $n \geq 2$, and let $v(X) = {}^t(v_0(X),\ldots, v_n(X))$ in $\mathrm{Um}_{n+1}(\mathbf{R}[X])$. Then there exists $E \in E_{n+1}(R[X])$ such that $E\,v(X) = {}^t(1,0,\ldots,0)$.*

Proof. By virtue of the Stable range theorem (see Theorem 92), it suffices to prove that there exists $E \in E_{n+1}(R[X])$ such that $E\,v(X) = v(0)$. By the local-global principle for elementary matrices (see Theorem 173), we can suppose that $\mathbf{R}$ is local. Moreover, it is clear that we can suppose that $\mathbf{R}$ is reduced. By virtue of Lemma 185, there exists $E \in E_{n+1}(R[X])$ such that $Ev(X) = {}^t(v_0(X), v_1(X), c_2, \ldots, c_n)$, where $c_i \in \mathbf{R}$. So we can without loss of generality suppose that $v_0 = a$ is constant.

Now, let us consider the ring $\mathbf{T} := \mathbf{R}/\mathscr{I}(a)$. Since dim $\mathbf{T} \leq 0$ (see Theorem 80), we have that $\mathbf{T}\langle X \rangle = \mathbf{T}(X)$ (see Remark 138) and thus $\mathbf{T}\langle X \rangle$ is a local ring. It follows that one among $v_1, \ldots, v_n$, say v_1, divides a monic polynomial in $\mathbf{T}[X]$. This means that there exist a monic polynomial $u \in \mathbf{R}[X]$, $w, h_1, h_2 \in \mathbf{R}[X]$ with $ah_2 = 0$, such that

$$wv_1 = u + ah_1 + h_2.$$

This means that $1 \in \langle v_1, a, h_2 \rangle$ in the ring $\mathbf{R}\langle X \rangle$ and thus $1 \in \langle v_1, a + h_2 \rangle$ by Lemma 90. That is, $\exists w_1, w_2 \in \mathbf{R}[X] \mid v_1 w_1 + (a + h_2)w_2 =: \tilde{u}$ is a monic polynomial.

Let $d \in \mathbb{N}$ and denote by $u_0, \ldots, u_n$ polynomials in $\mathbf{R}[X]$ such that $u_0 v_0 + \cdots + u_n v_n = 1$. Denoting by

$$\gamma_1 := E_{1,2}(h_2 u_1) \cdots E_{1,n+1}(h_2 u_n),$$
$$\gamma_2 := E_{3,2}(X^d w_1) E_{3,1}(X^d w_2),$$
$$\gamma := \gamma_2 \gamma_1,$$

we have

$$\gamma_1 v = {}^t(a + h_2, v_1, \ldots, v_n),$$

and

$$\gamma v = {}^t(a + h_2, v_1, v_2 + X^d \tilde{u}, v_3, \ldots, v_n).$$

So, for sufficiently large d, the third entry of γv becomes a monic polynomial. Thus, as already seen in Theorem 61, we have an algorithm transforming γv into ${}^t(1, 0, \ldots, 0)$ using elementary operations. $\qquad \square$

Corollary 187. ([182]) *For any ring $\mathbf{R}$ of Krull dimension ≤ 1, all finitely-generated stably free modules over $\mathbf{R}[X]$ are free. In other words, if $\mathbf{R}$ is a ring of Krull dimension ≤ 1 then $\mathbf{R}[X]$ is Hermite.*

Proof. We know that if $\mathbf{R}$ has Krull dimension ≤ 1 then all stably free modules over $\mathbf{R}$ are free (see Corollary 93 and Theorem 92). So, we have only to prove that all finitely-generated stably free modules over $\mathbf{R}[X]$ are extended from $\mathbf{R}$. For this, let $v = {}^t(v_0(X), \ldots, v_n(X)) \in \mathbf{R}[X]^{n+1}$ $(n \geq 2)$ be a unimodular vector. Our task amounts to proving that there exists $\Gamma \in GL_{n+1}(\mathbf{R}[X])$ such that $\Gamma V = {}^t(1, 0, \ldots, 0)$. This follows from Theorem 186. $\qquad \square$

Corollary 188. *The Hermite ring conjecture is true for rings of Krull dimension ≤ 1.*

Corollary 187 encourages us to set Conjecture 189. It is worth pointing out, that one cannot use the Quillen Induction Theorem 48 nor the constructive version of the Lequain-Simis Induction Theorem (Theorem 140) in order to settle affirmatively this conjecture because the class of rings with Krull dimension ≤ 1 is not stable by passage to none of the formations $\mathbf{R}\langle X \rangle$ and $\mathbf{R}(X)$. As a matter of fact, we have $\dim \mathbf{R}\langle X \rangle = \dim \mathbf{R}(X) = \dim \mathbf{R}[X] - 1$ [27] (we don't have yet a constructive proof of this fact; see Exercise 373 for a partial constructive proof), and thus, to see this, it suffices to consider a ring $\mathbf{R}$ such $\dim \mathbf{R} = 1 < \dim_v \mathbf{R} = \dim \mathbf{R}\langle X \rangle = \dim \mathbf{R}(X) = 2$ (for example $\mathbf{R} = \mathbb{Q} + y\mathbb{Q}(x)[y] = \{f(y) \in \mathbb{Q}(x)[y] \mid f(0) \in \mathbb{Q}\}$ where x, y are two independent indeterminates over the field of rationals $\mathbb{Q}$ [26]). The following conjecture is a generalization of Serre's conjecture to rings of Krull dimension ≤ 1.

Conjecture 189. (The One-Dimension Conjecture) *For any ring $\mathbf{R}$ of Krull dimension ≤ 1, and $k \in \mathbb{N}$, all finitely-generated stably free modules over $\mathbf{R}[X_1,\ldots,X_k]$ are free. In other words, if $\mathbf{R}$ is a of Krull dimension ≤ 1, then for any $k \in \mathbb{N}$, $\mathbf{R}[X_1,\ldots,X_k]$ is Hermite.*

Also, Corollary 187 raises the $\mathbf{K}_1$-analogue question. We will state it as a conjecture.

Conjecture 190. *Let $\mathbf{R}$ be a ring of Krull dimension ≤ 1 and $n \geq 3$. Then every matrix $M \in \mathrm{SL}_n(\mathbf{R}[X])$ is congruent to $M(0)$ modulo $\mathrm{E}_n(\mathbf{R}[X])$.*

In fact, by virtue of Theorem 186 and the local-global principle for elementary matrices (see Theorem 173), Conjecture 190 is equivalent to the following conjecture.

Conjecture 191. *Suppose $\mathbf{R}$ is a residually discrete local ring of Krull dimension ≤ 1, and*

$$M = \begin{pmatrix} p & q & 0 \\ r & s & 0 \\ 0 & 0 & 1 \end{pmatrix} \in \mathrm{SL}_3(\mathbf{R}[X]).$$

Then $M \in \mathrm{E}_3(\mathbf{R}[X])$.

2.6.2 Stably Free Modules Over $\mathbf{R}[X]$ of Rank $> \dim \mathbf{R}$ are Free

This subsection is extracted from [184]. The purpose of this subsection is to extend the results obtained in the one-dimensional case to the general case and of course always without supposing that the base ring is Noetherian.

Theorem 192. *Let $\mathbf{R}$ be a ring of dimension $\leq d$, $n \geq d+1$, and let*

$$v(X) = {}^t(v_0(X),\ldots,v_n(X)) \in \mathrm{Um}_{n+1}(\mathbf{R}[X]).$$

Then there exists $E \in \mathrm{E}_{n+1}(\mathbf{R}[X])$ such that $E v(X) = {}^t(1,0,\ldots,0)$.

Proof. By the Stable range Theorem 92, for any $w \in \mathrm{Um}_{n+1}(\mathbf{R})$, there exists $M \in \mathrm{E}_{n+1}(\mathbf{R})$ such that $Mw = {}^t(1,0,\ldots,0)$. So, it suffices to prove that there exists $E \in \mathrm{E}_{n+1}(\mathbf{R}[X])$ such that $E v(X) = v(0)$. For this aim, by the local-global principle for elementary matrices (see Theorem 173), we can suppose that $\mathbf{R}$ is local. Moreover, it is clear that we can suppose that $\mathbf{R}$ is reduced.

We prove the claim by double induction on the number N of nonzero coefficients of $v_0(X), \ldots, v_n(X)$ and d, starting with $N = 1$ (in that case the result is immediate) and $d = 0$ (in that case the result is well-known).

We will prove a first claim: $v(X)$ can be transformed by elementary operations into a vector with one constant entry.

Let $N > 1$ and $d > 0$. We may assume that $v_0(0) \in \mathbf{R}^\times$. Let us denote by a the leading coefficient of v_0 and $m_0 := \deg v_0$. If $a \in \mathbf{R}^\times$ then the result follows from Suslin's lemma (Theorem 57). So we may assume $a \in \mathrm{Rad}(\mathbf{R})$. By the induction hypothesis applied to the ring $\mathbf{R}/\langle a \rangle$, we can assume that $v(X) \equiv {}^t(1, 0, \ldots, 0) \bmod (a\mathbf{R}[X])^{n+1}$. By Lemma 183, we assume now $v_i = X^{2k} w_i$, where $\deg w_i < m_0$ for $1 \le i \le n$. By Proposition 166.(3).(ii), we assume $\deg v_i < m_0$.

If $m_0 \le 1$, our first claim is established. Assume now that $m_0 \ge 2$. Let $(c_1, \ldots, c_{m_0(n-1)})$ be the coefficients of $1, X, \ldots, X^{m_0-1}$ in the polynomials $v_2(X), \ldots, v_n(X)$. By Lemma 182, the ideal generated in $\mathbf{R}_a$ by $\mathbf{R}_a \cap (v_0 \mathbf{R}_a[X] + v_1 \mathbf{R}_a[X])$ and the c_i's is $\mathbf{R}_a$. As $m_0(n-1) \ge 2d > \dim \mathbf{R}_a$, by the Stable range Theorem 92 there exists

$$(c'_1, \ldots, c'_{m_0(n-1)}) \equiv (c_1, \ldots, c_{m_0(n-1)}) \bmod (v_0 \mathbf{R}[X] + v_1 \mathbf{R}[X]) \cap \mathbf{R}$$

such that $c'_1 \mathbf{R}_a + \cdots + c'_{m_0(n-1)} \mathbf{R}_a = \mathbf{R}_a$. Assume that we have already $c_1 \mathbf{R}_a + \cdots + c_{m_0(n-1)} \mathbf{R}_a = \mathbf{R}_a$. By Lemma 184, the ideal $\langle v_0, v_2, \ldots, v_n \rangle$ of $\mathbf{R}[X]$ contains a polynomial $w(X)$ of degree $m_0 - 1$ which is unitary in $\mathbf{R}_a$. Let us denote the leading coefficient of w by ua^k where $u \in \mathbf{R}^\times$ and that of v_1 by b. Using Proposition 166, as a is invertible modulo $\langle v_0, v_1 \rangle$ (because $v(X) \equiv {}^t(1, 0, \ldots, 0) \bmod (a\mathbf{R}[X])^{n+1}$), we can by elementary operations make the following transformations

$$ {}^t(v_0, v_1, \ldots, v_n) \to {}^t(v_0, a^{2k} v_1, \ldots, v_n) \to {}^t(v_0, a^{2k} v_1 + (1 - a^k u^{-1} b) w, v_2, \ldots, v_n). $$

Now, $a^{2k} v_1 + (1 - a^k u^{-1} b) w$ is unitary in $\mathbf{R}_a$, we can assume that v_1 unitary in $\mathbf{R}_a$, $\deg(v_1) := m_1 < m_0$. By Proposition 166.(3).(ii), as a is invertible modulo $\langle v_0 \rangle$ (since $v(X) \equiv {}^t(1, 0, \ldots, 0) \bmod (a\mathbf{R}[X])^{n+1}$), by elementary operations, ${}^t(v_0, v_1, v_2, \ldots, v_n)$ can be transformed into ${}^t(v_0, v_1, a^\ell v_2, \ldots, a^\ell v_n)$ for a suitable $\ell \in \mathbb{N}$ so that we can divide (like in Euclidean division) all $a^\ell v_2, \ldots, a^\ell v_n$ by v_1, and, thus we can assume that $\deg v_i < m_1$ for $2 \le i \le n$.

Repeating the argument above we lower the degree of v_1 until reaching the desired form of our first claim.

Assume now that $v_0 = a \in \mathbf{R}$. Let us consider the ring $\mathbf{T} := \mathbf{R}/\mathscr{I}(a)$. Since $\dim \mathbf{T} \le d - 1$ (see Theorem 80) and $(\bar{v}_1, \ldots, \bar{v}_n) \in \mathrm{Um}_n(\mathbf{T}[X])$, there exists $E_1 \in \mathrm{E}_n(\mathbf{R}[X])$ such that

$$ E_1 {}^t(v_1, \ldots, v_n) = {}^t(1 + ah_1 + y_1 \tilde{h}_1, ah_2 + y_2 \tilde{h}_2, \ldots, ah_n + y_n \tilde{h}_n), $$

where $h_i, \tilde{h}_i \in \mathbf{R}[X]$, $y_i \in \mathbf{R}$ with $a y_i = 0$.

Denoting by $E_2 = \begin{pmatrix} 1 & 0 \\ 0 & E_1 \end{pmatrix} \in E_{n+1}(\mathbf{R}[X])$, we have

$$E_2 v = {}^t(a, 1 + ah_1 + y_1\tilde{h}_1, ah_2 + y_2\tilde{h}_2, \ldots, ah_n + y_n\tilde{h}_n).$$

Thus,

$$E_{1,2}(-a)E_{2,1}(-h_1)\cdots E_{n+1,1}(-h_n)E_2 v = {}^t(0, 1 + y_1\tilde{h}_1, y_2\tilde{h}_2, \ldots, y_n\tilde{h}_n) =: \tilde{v},$$

and we can easily find $E_3 \in E_{n+1}(\mathbf{R}[X])$ such that $E_3 \tilde{v} = {}^t(1, 0, \ldots, 0)$. $\qquad \square$

Corollary 193. ([184]) *For any ring* $\mathbf{R}$ *with Krull dimension* $\leq d$, *all finitely-generated stably free modules over* $\mathbf{R}[X]$ *of rank* $> d$ *are free.*

Proof. By the Stable range theorem (Corollary 93), all finitely-generated stably free modules over $\mathbf{R}$ of rank $> d$ are free. So, we have only to prove that all stably free modules over $\mathbf{R}[X]$ are extended from $\mathbf{R}$.

For this, let $v = {}^t(v_0(X), \ldots, v_n(X)) \in \mathbf{R}[X]^{n+1}$ $(n \geq d + 1)$ be a unimodular vector. Our task amounts to proving that there exists $\Gamma \in \mathrm{GL}_{n+1}(\mathbf{R}[X])$ such that $\Gamma V = {}^t(1, 0, \ldots, 0)$. This follows from Theorem 192. $\qquad \square$

Corollary 193 encourages us to set the following conjecture.

Conjecture 194. *For any ring* $\mathbf{R}$ *with Krull dimension* $\leq d$, *all finitely-generated stably free modules over* $\mathbf{R}[X_1, \ldots, X_k]$ *of rank* $> d$ *are free.*

2.6.3 Two New Conjectures

The field with two elements will be denoted by $\mathbb{F}_2$. The goal of this subsection is to present a "serious" counterexample candidate to the Hermite ring conjecture, i.e., a ring $\mathbf{R}$ and a vector $V(X)$ in $\mathrm{Um}_n(\mathbf{R}[X])$ which is not equivalent to $V(0)$ by the action of $\mathrm{GL}_n(\mathbf{R}[X])$. The obstacle is to prove the following new conjecture:

Conjecture 195. (One Square Conjecture) *The unimodular vector* ${}^t(x_1, x_2, x_3)$ *is not completable over the ring*

$$\mathbb{F}_2[X_1, X_2, X_3, Y_2, Y_3]/\langle X_1^2 + X_2 Y_2 + X_3 Y_3 - 1\rangle = \mathbb{F}_2[x_1, x_2, x_3, y_2, y_3].$$

Or more generally:

Conjecture 196. (One Square Conjecture, bis) *If* $\mathbf{K}$ *is a field then the unimodular vector* ${}^t(x_1, x_2, x_3)$ *is not completable over the ring*

$$\mathbf{K}[X_1, X_2, X_3, Y_2, Y_3]/\langle X_1^2 + X_2 Y_2 + X_3 Y_3 - 1\rangle = \mathbf{K}[x_1, x_2, x_3, y_2, y_3].$$

Conjecture 195 above is true over the reals as the unimodular vector ${}^t(x_1, x_2, x_3)$ is not completable over the ring $\mathbb{R}[X_1, X_2, X_3]/\langle X_1^2 + X_2^2 + X_3^2 - 1\rangle = \mathbb{R}[x_1, x_2, x_3]$ (the real sphere counterexample, see for example pages 33 and 34 of [92]).

2.6.3.1   First Possible Application in Case of a Positive Answer to the One Square Conjecture

A positive answer to the One Square Conjecture (Conjecture 195) will give a negative answer to *Murthy's (a, b, c)-Problem* [91, 92]. Consider a vector ${}^t(1+a, b, c)$ over a ring **R**, where a is nilpotent modulo the ideal $\langle b, c \rangle$. Clearly, ${}^t(1+a, b, c) \in \mathrm{Um}_3(\mathbf{R})$. If $2 \in \mathbf{R}^\times$ then it is classical (it is a consequence of Suslin's $n!$-theorem, see Exercise 378), that ${}^t(1+a, b, c)$ is completable. If $2 \notin \mathbf{R}^\times$, for example if $2 = 0$ in **R**, this is a long-standing open problem set by Murthy [92] (page 323). It can be formulated as follows:

Murthy's (a, b, c)-Problem [91, 92].
*If **R** is a ring of characteristic 2 and $a^n \in \langle b, c \rangle$ for some $n \geq 2$, is the unimodular vector ${}^t(1+a, b, c)$ completable?*

Quoting from [92] (page 323): "although this question looks simple, it remained unanswered for more than 25 years. Even the case $n = 2$ seems wide open!".
If the One Square conjecture (Conjecture 195) is confirmed, then the answer to Murthy's (a, b, c)-Problem is negative. As a matter of fact, with notation of Conjecture 195, we have

$$(1+x_1)^2 \in \langle x_2, x_3 \rangle,$$

while the vector ${}^t(x_1, x_2, x_3)$ is not completable.

2.6.3.2   Second Possible Application in Case of a Positive Answer to the One Square Conjecture

A positive answer to the One Square Conjecture (Conjecture 195) will also give a negative answer to the Hermite Ring Conjecture (Conjecture 180).

Let us consider five independent variables A, B, C, D, E over $\mathbb{F}_2$ and set

$$\mathbf{R} := \mathbb{F}_2[A, B, C, D, E]/\langle A^2 - DB - CE \rangle = \mathbb{F}_2[a, b, c, d, e].$$

A positive answer to the One Square Conjecture will imply that the polynomial vector $v(X) := {}^t(1+aX, b, c)$ is not completable over $\mathbf{R}[X]$ as $v(1)$ is not elementarily completable over **R** (by the negative answer to Murthy's (a, b, c)-Problem), while $v(0) := {}^t(1, b, c)$ is easily completable to a matrix in $\mathrm{GL}_3(\mathbf{R})$. Thus, in case of a positive answer to the One Square Conjecture, $M(X) := \mathrm{Syz}(1+aX, b, c)$ (the first syzygy module) will be an example of a rank 2 stably free modules over $\mathbf{R}[X]$ which is not extended from $M(0) \equiv \mathbf{R}^2$.

It is worth pointing out that Rao and Swan had given in [147] an example of a vector in $\mathrm{Um}_3(\mathbf{R}[X])$ that is completable to a matrix in $\mathrm{GL}_3(\mathbf{R}[X])$, but not completable to one in $\mathrm{E}_3(\mathbf{R}[X])$, using topological arguments.

Before stating our second conjecture, we give the following simple result.

Proposition 197. *Let* **K** *be a field,* $n \geq 2$, *and* $m \geq 0$ *with* $n + m \geq 3$. *If* -1 *is the sum of* $n - 1$ *squares in* **K** *then the unimodular vector* ${}^{t}(x_1, \ldots, x_n, y_1, \ldots, y_m)$ *is elementarily completable over the ring*

$$\mathbf{K}[X_1, \ldots, X_n, Y_1, \ldots, Y_m, Z_1, \ldots, Z_m] /$$
$$\langle X_1^2 + \cdots + X_n^2 + Y_1 Z_1 + \cdots + Y_m Z_m - 1 \rangle$$
$$= \mathbf{K}[x_1, \ldots, x_n, y_1, \ldots, y_m, z_1, \ldots, z_m].$$

For example, in any finite field, -1 is the sum of two squares. Our intuition is that the converse of Proposition 197 holds. We state it as a conjecture.

Conjecture 198. (The Sum of Squares Conjecture) *Let* **K** *be a field,* $n \geq 2$, *and* $m \geq 0$ *with* $n + m \geq 3$. *The unimodular vector* ${}^{t}(x_1, \ldots, x_n, y_1, \ldots, y_m)$ *is elementarily completable over the ring*

$$\mathbf{K}[X_1, \ldots, X_n, Y_1, \ldots, Y_m, Z_1, \ldots, Z_m] / \langle X_1^2 + \cdots + X_n^2 + Y_1 Z_1 + \cdots + Y_m Z_m - 1 \rangle$$
$$= \mathbf{K}[x_1, \ldots, x_n, y_1, \ldots, y_m, z_1, \ldots, z_m]$$

if and only if -1 *is the sum of* $n - 1$ *squares in* **K**.

A positive answer to Conjecture 198 will, amongst other things, give new examples of completable unimodular vectors which are not elementary completable. As a matter of fact, it is well-known [92] that the unimodular vector ${}^{t}(x_1, \ldots, x_n)$ is completable over the ring

$$\mathbb{R}[X_1, \ldots, X_n] / \langle X_1^2 + \cdots + X_n^2 - 1 \rangle = \mathbb{R}[x_1, \ldots, x_n]$$

if and only if $n = 1, 2, 4$ or 8.

Chapter 3

Dynamical Gröbner Bases

3.1 Dickson's Lemma and the Division Algorithm

The Euclidean division algorithm plays a key role when dealing with univariate polynomials with coefficients in a field $\mathbf{K}$. For example, if one wants to divide a $f(X) = a_n X^n + a_{n-1} X^{n-1} + \cdots$ ($a_i \in \mathbf{K}$, $a_n \neq 0$) by a polynomial $g(X) = b_{n-r} X^{n-r} + b_{n-r-1} X^{n-r-1} + \cdots$ ($b_i \in \mathbf{K}$, $r \geq 0$, $b_{n-r} \neq 0$), one should subtract $a_n b_{n-r}^{-1} X^r g$ from f to cancel the leading term $a_n X^n$ of f, and then to resume the same process with $f - a_n b_{n-r}^{-1} X^r g$ until one obtains a polynomial of degree less than $n - r$. Note that the monomials of the polynomials above are written in decreasing order by degree in X:

$$\cdots > X^{k+1} > X^k > \cdots > X^3 > X^2 > X > 1.$$

In order to generalize the division algorithm to the multivariate case, as a first step, one has to define a total order on monomials which is compatible with multiplication and which is a well-ordering, in the sense that any nonincreasing sequence of monomials pauses (Corollary 210).

Definition and notation 199. Let $\mathbf{R}$ be a ring, $n \geq 1$, and consider n independent variables $X_1, \ldots, X_n$ on $\mathbf{R}$.

(1) For $\alpha = (\alpha_1, \ldots, \alpha_n) \in \mathbb{N}^n$, we denote by $X^\alpha := X_1^{\alpha_1} \cdots X_n^{\alpha_n}$.

(2) We denote by

$$\mathbb{M}_n := \{ X^\alpha \mid \alpha \in \mathbb{N}^n \}$$

the set of monomials at $X_1, \ldots, X_n$. Of course, there is a one-to-one correspondence between $\mathbb{M}_n$ and $\mathbb{N}^n$ given by $X^\alpha \leftrightarrow \alpha$, with $1 \leftrightarrow (0, \ldots, 0)$.

© Springer International Publishing Switzerland 2015
I. Yengui, *Constructive Commutative Algebra*, Lecture Notes in Mathematics 2138,
DOI 10.1007/978-3-319-19494-3_3

(3) A monomial order on $\mathbf{R}[X_1,\ldots,X_n]$ is a relation $>$ on $\mathbb{M}_n$ satisfying:

    (i) $>$ is a total order on $\mathbb{M}_n$.

    (ii) For $\alpha, \beta, \gamma \in \mathbb{N}^n$, if $X^\alpha > X^\beta$ then $X^\alpha X^\gamma > X^\beta X^\gamma$.

    (iii) $X^\alpha \geq 1$ for all $\alpha \in \mathbb{N}^n$.

Example 200. Let $\alpha, \beta \in \mathbb{N}^n$.

(1) **Lexicographic order** with $X_1 > X_2 > \cdots > X_n$: $X^\alpha >_{\text{lex}} X^\beta$ if the left-most nonzero entry of $\alpha - \beta$ is positive. For example, $X_1^2 X_2 X_3^2 >_{\text{lex}} X_1^2 >_{\text{lex}} X_1 X_2^2 X_3^2 >_{\text{lex}} X_1 X_2 X_3^3 >_{\text{lex}} X_3^7$.

(2) **Graded lexicographic order** with $X_1 > X_2 > \cdots > X_n$: $X^\alpha >_{\text{grlex}} X^\beta$ if $\sum_{i=1}^n \alpha_i > \sum_{i=1}^n \beta_i$ or ($\sum_{i=1}^n \alpha_i = \sum_{i=1}^n \beta_i$ and $X^\alpha >_{\text{lex}} X^\beta$). For example, $X_3^7 >_{\text{grlex}} X_1^2 X_2 X_3^2 >_{\text{grlex}} X_1 X_2^2 X_3^2 >_{\text{grlex}} X_1 X_2 X_3^3 >_{\text{grlex}} X_1^2$.

(3) **Graded reverse lexicographic order** with $X_1 > X_2 > \cdots > X_n$: $X^\alpha >_{\text{grevlex}} X^\beta$ if $\sum_{i=1}^n \alpha_i > \sum_{i=1}^n \beta_i$ or ($\sum_{i=1}^n \alpha_i = \sum_{i=1}^n \beta_i$ and the right-most nonzero entry of $\alpha - \beta$ is negative). For example, $X_3^7 >_{\text{grevlex}} X_1^2 X_2 X_3^2 >_{\text{grevlex}} X_1 X_2^2 X_3^2 >_{\text{grevlex}} X_1 X_2 X_3^3 >_{\text{grevlex}} X_1^2$.

Of course, for $n = 1$, all monomial orders coincide.

Remark 201. *(Monomial Orders on $\mathbf{R}[X_1,\ldots,X_n]^m$, [3])*

Let $\mathbf{R}$ be a ring, n, $m \geq 1$, consider n independent variables $X_1,\ldots,X_n$ on $\mathbf{R}$, and denote by

$$(e_1 = (1,0,\ldots,0),\ e_2 = (0,1,0,\ldots,0),\ldots,\ e_m = (0,\ldots,0,1))$$

the standard basis of $\mathbf{R}[X_1,\ldots,X_n]^m$.

(1) By a monomial in $\mathbf{R}[X_1,\ldots,X_n]^m$ we mean a vector of type $X^\alpha e_i$ ($1 \leq i \leq m$), where X^α is a monomial in $\mathbf{R}[X_1,\ldots,X_n]$. For example, $(0, X_1 X_2^3, 0)$ is a monomial in $\mathbf{R}[X_1, X_2]^3$, but $(0, X_1 + X_2^3, 0)$ and $(0, X_1, X_2^3)$ are not. If $M = X^\alpha e_i$ and $N = X^\beta e_j$, we say that M divides N if $i = j$ and X^α divides X^β. For example, $(X_1, 0, 0)$ divides $(X_1 X_2, 0, 0)$, but does not divide $(0, X_1 X_2, 0)$. Note that, is case M divides N, there exists a monomial X^γ in $\mathbf{R}[X_1,\ldots,X_n]$ such that $N = X^\gamma M$. In this case, we define

$$\frac{N}{M} := X^\gamma.$$

For example $\frac{(X_1 X_2, 0, 0)}{(X_1, 0, 0)} = X_2$. We denote by $\mathbb{M}_n^m$ the set of monomials in $\mathbf{R}[X_1,\ldots,X_n]^m$, with $\mathbb{M}_n^1 = \mathbb{M}_n$.

(2) A monomial order on $\mathbf{R}[X_1,\ldots,X_n]^m$ is a relation $>$ on $\mathbb{M}_n^m$ satisfying:

    (i) $>$ is a total order on $\mathbb{M}_n^m$.

    (ii) $X^\alpha M > M$ for all $M \in \mathbb{M}_n^m$ and $X^\alpha \in \mathbb{M}_n \setminus \{1\}$.

    (iii) $M > N \Rightarrow X^\alpha M > X^\alpha N$ for all M, $N \in \mathbb{M}_n^m$ and $X^\alpha \in \mathbb{M}_n$.

Note that, when specialized to the case $m = 1$, this definition coincides with the definition of a monomial order on $\mathbf{R}[X_1,\ldots,X_n]$ given in Definition and notation 199.

(3) Starting from a monomial order on $\mathbf{R}[X_1,\ldots,X_n]$, there are two natural ways of obtaining a monomial order on $\mathbf{R}[X_1,\ldots,X_n]^m$:

    (i) For monomials $M = X^\alpha e_i$, $N = X^\beta e_j \in \mathbb{M}_n^m$, we say that

$$M > N \text{ if } \begin{cases} X^\alpha > X^\beta \\ \text{or} \\ X^\alpha = X^\beta \text{ and } i > j. \end{cases}$$

This monomial order is called TOP for "term over position" as it gives more importance to the monomial order on $\mathbf{R}[X_1,\ldots,X_n]$ than to the position in the vector. For example, in case $X_1 > X_2$, we have

$$(0,X_1) > (X_1,0) > (0,X_2) > (X_2,0).$$

    (ii) For monomials $M = X^\alpha e_i$, $N = X^\beta e_j \in \mathbb{M}_n^m$, we say that

$$M > N \text{ if } \begin{cases} i > j \\ \text{or} \\ i = j \text{ and } X^\alpha > X^\beta. \end{cases}$$

This monomial order is called POT for "position over term" as it gives more importance to the position in the vector than to the monomial order on $\mathbf{R}[X_1,\ldots,X_n]$. For example, in case $X_1 > X_2$, we have

$$(0,X_1) > (0,X_2) > (X_1,0) > (X_2,0).$$

(4) Most of the results given in this book related to Gröbner bases for finitely-generated ideals in $\mathbf{R}[X_1,\ldots,X_n]$ can fairly be generalized to finitely-generated submodules of $\mathbf{R}[X_1,\ldots,X_n]^m$ using monomials order on $\mathbf{R}[X_1,\ldots,X_n]^m$, namely, Dickson' lemma (Theorem 209), the division algorithm, Buchberger's algorithm, the ideal membership test, computation of syzygies, and so on. We chose to limit ourselves to ideals simply to make the text easier to read.

Definition 202. Let $\mathbf{R}$ be a strongly discrete coherent ring, $f = \sum_\alpha a_\alpha X^\alpha$ a nonzero polynomial in $\mathbf{R}[X_1, \ldots, X_n]$, and $>$ a monomial order on $\mathbf{R}[X_1, \ldots, X_n]$.

(1) The X^α (resp. the $a_\alpha X^\alpha$) are called the monomials (resp. the terms) of f.

(2) The multidegree of f is $\mathrm{mdeg}(f) := \max\{\alpha \in \mathbb{N}^n : a_\alpha \neq 0\}$.

(3) The leading coefficient of f is $\mathrm{LC}(f) := a_{\mathrm{mdeg}(f)} \in \mathbf{R}$.

(4) The leading monomial of f is $\mathrm{LM}(f) := X^{\mathrm{mdeg}(f)}$.

(5) The leading term of f is $\mathrm{LT}(f) := \mathrm{LC}(f)\mathrm{LM}(f)$.

(6) For $g, h \in \mathbf{R}[X_1, \ldots, X_n] \setminus \{0\}$, we say that $\mathrm{LT}(g)$ divides $\mathrm{LT}(h)$ if $\mathrm{LM}(g)$ divides $\mathrm{LM}(h)$ and $\mathrm{LC}(g)$ divides $\mathrm{LC}(h)$.

In order to give a constructive proof of Dickson's lemma, following [105], we give the following definition.

Definition 203. A partially ordered set $(E, \leq)$ is said to satisfy the *descending chain condition* (in short, DCC) if for every nonincreasing sequence $(u_n)_{n \in \mathbb{N}}$ in E, there exists $n \in \mathbb{N}$ such that $u_n = u_{n+1}$. A partially ordered set $(E, \leq)$ is said to satisfy the *ascending chain condition* (in short, ACC) if for every nondecreasing sequence $(u_n)_{n \in \mathbb{N}}$ in E, there exists $n \in \mathbb{N}$ such that $u_n = u_{n+1}$.

Example 204. $\mathbb{N}$ with the usual order satisfies DCC.

Let $(E, \leq)$ be a partially ordered set. We will denote by $\leq_d$ the order on E^d defined by $(x_1, \ldots, x_d) \leq_d (y_1, \ldots, y_d)$ if and only if $x_i \leq y_i$ for all $1 \leq i \leq d$. We shall write $\leq$ instead of $\leq_d$ when there is no risk of confusion.

Lemma 205. ([105]) *If a partially ordered set $(E, \leq)$ satisfies DCC (resp., ACC), then so does $(E^d, \leq_d)$ (resp., ACC).*

Proof. It suffices to prove the result in the case $d = 2$. The same reasoning can be used to prove the general case by induction. Let $(u_n, v_n)_{n \in \mathbb{N}}$ be a nonincreasing sequence of elements of E^2. It is easy to see that, since the sequence $(u_n)_{n \in \mathbb{N}}$ is nonincreasing, one can find $n_1 < n_2 < \cdots$ such that $u_{n_i} = u_{n_i+1}$ for all $i \in \mathbb{N}$. The sequence $(v_{n_i})_{i \in \mathbb{N}}$ being nonincreasing, there exists $j \in \mathbb{N}$ such that $v_{n_j} = v_{n_j+1}$. But, as $v_{n_j} \geq v_{n_j+1} \geq v_{n_{j+1}}$, we have $v_{n_j} = v_{n_j+1}$, and, thus, $(u_{n_j}, v_{n_j}) = (u_{n_j+1}, v_{n_j+1})$. For the ACC case, consider the reverse order. $\qquad \square$

Definition and notation 206. Let $(E, \leq)$ be a partially ordered set.

(1) For $Y \in E$, we define
$$Y^{\uparrow} := \{Z \in E \mid Z \geq Y\},$$

and for $Y_1, \ldots, Y_m \in E$, we define

$$\mathcal{M}_E^+(Y_1, \ldots, Y_m) := \cup_{i=1}^m Y_i^{\uparrow} = \{Z \in E \mid Z \geq Y_1 \vee \cdots \vee Z \geq Y_m\}.$$

$\mathcal{M}_E^+(Y_1, \ldots, Y_m)$ is called a *final subset of finite type* of E (generated by $Y_1, \ldots, Y_m$). The set of final subsets of finite type of E, including the empty subset considered as generated by the empty family, will be denoted by $\mathcal{F}(E)$.

(2) In the particular case $E = \mathbb{N}^d$, for $Y = (y_1, \ldots, y_d) \in \mathbb{N}^d$, we have

$$Y^{\uparrow} := \{Z = (z_1, \ldots, z_d) \in \mathbb{N}^d \mid z_i \geq y_i \forall 1 \leq i \leq d\} = (y_1, \ldots, y_d) + \mathbb{N}^d.$$

The set $\mathcal{F}(\mathbb{N}^d) \setminus \{\emptyset\}$ will be denote by $\mathcal{M}_d$. So, $\mathcal{F}(\mathbb{N}^d)$ is isomorphic to $\mathcal{M}_d \cup \{-\infty\}$.

Proposition 207. ([105])

(1) *Every $A \in \mathcal{M}_d$ is generated by a unique minimal family (for $\subseteq$). This family can be obtained by taking the minimal elements (for $\leq_d$) of any family of generators of A.*

(2) *Given $A, B \in \mathcal{M}_d$, one can decide whether $A \subseteq B$ or not.*

(3) *The ordered set $(\mathcal{M}_d, \subseteq)$ satisfies ACC.*

Proof. (1) It is clear that for any $Y, Y_1, \ldots, Y_n \in \mathbb{N}^d$, we have

$$Y^{\uparrow} \subseteq Y_1^{\uparrow} \cup \cdots \cup Y_n^{\uparrow} \iff Y \in Y_1^{\uparrow} \cup \cdots \cup Y_n^{\uparrow} \iff Y_1 \leq_d Y \text{ or } \cdots \text{ or } Y_n \leq_d Y.$$

So, starting from a finite family of generators of A, to obtain a minimal family of generators of A, one has only to keep the minimal elements (for $\leq_d$) of the considered family. This proves the existence part of (1).

If $Y_1, \ldots, Y_n$ and $Z_1, \ldots, Z_m$ are minimal families of generators of A, then for each $1 \leq i \leq n$, there exists $1 \leq j \leq m$ such that $Z_j \leq_d Y_i$ and vice versa. By minimality of the two families, we deduce that $\{Y_1, \ldots, Y_n\} = \{Z_1, \ldots, Z_m\}$.

(2) This is straightforward.

(3) We will induct on d. The case $d = 1$ is clear. Suppose that $d \geq 2$ and consider
a nondecreasing sequence $(A_m)_{m \in \mathbb{N}}$ in $\mathcal{M}_d$. Let $a = (a_1, \ldots, a_d) \in A_0$. For all
$1 \leq i \leq d$ and $r \in \mathbb{N}$, let

$$H_{i,d}^r := \{(x_1, \ldots, x_d) \in \mathbb{N}^d \mid x_i = r\}.$$

There is an obvious order isomorphism between $(H_{i,d}^r, \leq_d)$ and $(\mathbb{N}^{d-1}, \leq_{d-1})$.
So $(\mathscr{F}(H_{i,d}^r), \subseteq)$ satisfies ACC by induction hypothesis (it is isomorphic to
$\mathcal{M}_{d-1} \cup \{-\infty\}$). The crucial point in the proof is the following observation:

$$\mathbb{N}^d \setminus a^{\uparrow} = \cup_{i=1}^d \cup_{r < a_i} H_{i,d}^r \text{ (a finite union).}$$

It follows that for all $m \in \mathbb{N}$, we have

$$A_m = a^{\uparrow} \bigcup \cup_{i=1}^d \cup_{r < a_i} (A_m \cap H_{i,d}^r).$$

The desired result follows since all the nondecreasing sequences $(A_m \cap H_{i,d}^r)_{m \in \mathbb{N}}$ pause by induction hypothesis. $\square$

Definition 208. Consider d independent variables $X_1, \ldots, X_d$ over a field K. As
usual, for $\alpha = (\alpha_1, \ldots, \alpha_d) \in \mathbb{N}^d$, X^{α} denotes the monomial $X_1^{\alpha_1} \cdots X_d^{\alpha_d}$.
A monomial ideal of $\mathbf{K}[X_1, \ldots, X_d]$ is an ideal generated by a family of monomials
at $X_1, \ldots, X_d$. Clearly, two monomial ideals are equal if and only if they contain the
same monomials, and the set of finitely-generated monomial ideals is in one-to-one
correspondence with $\mathcal{M}_d$.

The third assertion of Proposition 207 is equivalent to Dickson's lemma.

Theorem 209. (Dickson's Lemma, Constructive Version, Lombardi and Perdry
[105]) *The set of finitely-generated monomial ideals of* $\mathbf{K}[X_1, \ldots, X_d]$*, ordered with*
$\subseteq$*, satisfies ACC.*

Corollary 210. *Let* $\mathbf{R}$ *be a ring,* $>$ *be a monomial order on* $\mathbf{R}[X_1, \ldots, X_d]$*, and*
denote by $\mathbb{M}_d := \{X^{\alpha} = X_1^{\alpha_1} \cdots X_d^{\alpha_d} \mid \alpha \in \mathbb{N}^d\}$*. Then* $(\mathbb{M}_d, \leq)$ *satisfies DCC. In*
other words, any monomial order is a well-ordering.

Proof. There is a one-to-one correspondence between $\mathbb{M}_d$ and $\mathbb{N}^d$ given by $X^{\alpha} \leftrightarrow$
α. A nonincreasing sequence $(u_n)_{n \in \mathbb{N}}$ in $(\mathbb{N}^d, >)$ pauses at step n if and only if
the nondecreasing sequence $(\cup_{i=0}^n u_i^{\uparrow})_{n \in \mathbb{N}}$ in $(\mathcal{M}_d, \subseteq)$ pauses at step n. The desired
result follows from Proposition 207. $\square$

Now we have all the necessary tools to give a generalization of the Euclidean
division algorithm to multivariate polynomials.

Algorithm 211. (Division Algorithm in $\mathbf{K}[X_1,\ldots,X_d]$)

Input: $f_1,\ldots,f_s,f \in \mathbf{K}[X_1,\ldots,X_d]$, and $>$ a monomial order on $\mathbf{K}[X_1,\ldots,X_n]$, where $\mathbf{K}$ is a discrete field.
Output: $q_1,\ldots,q_s,r \in \mathbf{K}[X_1,\ldots,X_d]$ such that

$$f = q_1 f_1 + \cdots + q_s f_s + r,$$

$\mathrm{mdeg}(q_i f_i) \leq \mathrm{mdeg}(f)$, and either $r = 0$ or r is a linear combination, with coefficients in $\mathbf{K}$, of monomials, none of which is divisible by any of $\mathrm{LT}(f_1),\ldots,\mathrm{LT}(f_s)$. The polynomial r is called *a remainder* of f on division by $F := [f_1,\ldots,f_s]$, and is denoted by $\bar{f}^F$.

Initialization: $q_1 := 0;\cdots; q_s := 0; r := 0; p := f$
WHILE $p \neq 0$ DO
  $i := 1$;
  div := false
  WHILE $i \leq s$ AND div = false DO
    IF $\mathrm{LT}(f_i)$ divides $\mathrm{LT}(p)$ THEN
      $q_i := q_i + \frac{\mathrm{LT}(p)}{\mathrm{LT}(f_i)}$
      $p := p - \frac{\mathrm{LT}(p)}{\mathrm{LT}(f_i)} f_i$
      div := true
    ELSE
      $i := i+1$
  IF div = false THEN
    $r := r + \mathrm{LT}(p)$
    $p := p - \mathrm{LT}(p)$

Proposition 212. *Algorithm 211 terminates and is correct.*

Proof. The fact that Algorithm 211 terminates is constructively proven by Theorem 209 since $\mathrm{LM}(p)$ decreases until reaching $p = 0$. The correctness of Algorithm 211 is obvious. $\square$

Example 213. Let $f = X^2Y^2$, $f_1 = 2 + 12XY$, $f_2 = 8Y^2 \in \mathbb{Q}[X,Y]$, and fix any monomial order $>$ on $\mathbb{Q}[X,Y]$. Then

$$\bar{f}^{[f_1,f_2]} = \frac{1}{36} \text{ while } \bar{f}^{[f_2,f_1]} = 0.$$

We conclude that, contrary to the univariate case, the remainder is not unique. Moreover, from the second division, we infer that $f \in \langle f_1, f_2 \rangle$ despite that the remainder of the first division is not null (is a unit). It can be seen that the division algorithm (Algorithm 211) is not a satisfactory generalization of its univariate counterpart. The remedy for that is the "notion of Gröbner basis" (see Proposition 228).

Definition 214. Let $\mathbf{R}$ be a strongly discrete coherent ring, and $>$ a monomial order on $\mathbf{R}[X_1,\ldots,X_n]$.

(1) For a nonzero ideal I of $\mathbf{R}[X_1,\ldots,X_n]$, $\mathrm{LT}(I) := \langle \mathrm{LT}(g), g \in I \setminus \{0\} \rangle$. It is an ideal of $\mathbf{R}[X_1,\ldots,X_n]$.

(2) Let I be a nonzero ideal of $\mathbf{R}[X_1,\ldots,X_n]$, and $f_1,\ldots,f_s \in I$. We say that $G = \{f_1,\ldots,f_s\}$ is a *Gröbner basis* for I if

$$\mathrm{LT}(I) = \langle \mathrm{LT}(f_1),\ldots,\mathrm{LT}(f_s) \rangle.$$

We convene that $\emptyset$ is a Gröbner basis for $\{0\}$.

It was Buchberger [22] that first constructed Gröbner bases over fields (i.e., in $\mathbf{K}[X_1,\ldots,X_n]$, where $\mathbf{K}$ is a field). Rings over which one can construct Gröbner bases will be called Gröbner rings and are the subject of the following section.

We will explain how to construct Gröbner bases over discrete fields in the more general setting of coherent Notherian valuation rings (see Sect. 3.3.1) or, more generally, coherent archimedean valuation rings (they are valuation domains of Krull dimension ≤ 1 or coherent valuation rings with zero-divisors of Krull dimension ≤ 0, see Proposition 265 and Theorem 272).

3.2 Gröbner Rings

We first give the following series of definitions with the aim of extending Gröbner bases to multivariate polynomial rings with coefficients in a ring which is not a field.

Definition 215. Let $\mathbf{R}$ be a strongly discrete coherent ring.

(1) We say that $\mathbf{R}$ is an *n-Gröbner ring* if for every finitely-generated ideal I of $\mathbf{R}[X_1,\ldots,X_n]$, fixing the lexicographic order as monomial order on $\mathbf{R}[X_1,\ldots,X_n]$, $\mathrm{LT}(I)$ is also finitely generated; or equivalently, such that every finitely-generated ideal of $\mathbf{R}[X_1,\ldots,X_n]$ has a Gröbner basis with respect to the lexicographic monomial order.

(2) We say that $\mathbf{R}$ is a *Gröbner ring* if it is n-Gröbner for all positive integer n.

It is worth noting that, in the definition above, one can demand that Gröbner bases can be constructed for any monomial order but the fact that this can be done for at least the lexicographic order will suffice for dealing with the main issues we are interested in, namely the ideal membership question, computing elimination ideals, computing intersection of finitely-generated ideals, computing syzygies, etc. The second reason behind this restriction is that when $\mathbf{R}$ is a valuation domain of Krull dimension ≤ 1, we only know that Gröbner bases can be constructed in $\mathbf{R}[X_1,\ldots,X_n]$ with respect to the lexicographic monomial order (see Theorem 256). We will conjecture (see Conjecture 225) that over any strongly discrete coherent ring $\mathbf{R}$, if Gröbner bases can be constructed with respect to the lexicographic order, then they can be constructed with respect to any other monomial order.

The " n-Gröbner" property is inherited by localization since for any finitely-generated ideal I of $\mathbf{R}[X_1,\ldots,X_n]$ and any monoid S of $\mathbf{R}$, we have $\mathrm{LT}(S^{-1}I) =$

$S^{-1}\mathrm{LT}(I)$. In opposition to this, if $\mathbf{R}$ is n-Gröbner, and $\mathfrak{a}$ is an ideal of $\mathbf{R}$, then $\mathbf{R}/\mathfrak{a}$ need not be n-Gröbner (see Example 262). We don't know whether the implication "1-Gröbner $\Rightarrow$ Gröbner" is true but the obtained results about valuation domains (see for example Theorem 256) lead us to believe in this implication (see Conjecture 225).

Example 216. • A strongly discrete coherent Noetherian ring is Gröbner [3, 120, 159].

- If $\mathbf{K}$ is a discrete field then $\mathbf{K}[X_1, X_2, \ldots]$ is a non-Noetherian Gröbner ring (see Corollary 221).

- A valuation domain is 1-Gröbner if and only if its Krull dimension is ≤ 1, or equivalently, if its valuation group is archimedean (see Theorem 255).

- A Prüfer domain is 1-Gröbner if and only if its Krull dimension is ≤ 1 (this is a consequence of the previous bullet using the notion of dynamical Gröbner bases).

- More generally, a Prüfer domain is Gröbner if and only if its Krull dimension is ≤ 1 (consequence of Theorem 256).

- A valuation ring containing a zero-divisor is Gröbner if and only if it is zero-dimensional and coherent (see Proposition 272).

- An example of a one-dimensional Gröbner domain which is neither Noetherian nor Prüfer will be given in Example 222 (but this example is proven only within classical mathematics).

The following lemma is immediate.

Lemma 217. *Let $\mathbf{A}$ be a strongly discrete ring. A term aX^k (where $a \in \mathbf{A}$ and $k \in \mathbb{N}$) belongs to an ideal of $\mathbf{A}[X]$ of the form $\langle b_\lambda X^{k_\lambda}; \lambda \in \Lambda \rangle$, where $b_\lambda \in \mathbf{A}$ and $k_\lambda \in \mathbb{N}$, if and only if $a \in \langle b_\lambda; k_\lambda \leq k \rangle$.*

Definition 218. (i) Let $\mathbf{R}$ be a strongly discrete ring. For $n \in \mathbb{N}$ and I an ideal of $\mathbf{R}[X]$, we denote by $\mathrm{LC}_n(I)$ the ideal of $\mathbf{R}$ generated by the leading coefficients of the elements of I of degree n. In particular, $\mathrm{LC}_0(I) = I \cap \mathbf{R}$. The sequence $(\mathrm{LC}_n(I))_{n \in \mathbb{N}}$ is obviously nondecreasing and so $\mathrm{LC}_\infty(I) := \cup_{n \in \mathbb{N}} \mathrm{LC}_n(I)$ is an ideal of $\mathbf{R}$.

(ii) Following [79], we say that a ring $\mathbf{R}$ satisfies the *Kaplansky property*, if for any finitely-generated ideal I of $\mathbf{R}[X_1]$ and $n \in \mathbb{N} \cup \{\infty\}$, the ideal $\mathrm{LC}_n(I)$ is finitely-generated.

By the following proposition one can see that the 1-Gröbner property implies the Kaplansky property.

Proposition 219. *Let $\mathbf{R}$ be a strongly discrete ring and I a finitely-generated ideal of $\mathbf{R}[X_1]$. If $\mathrm{LT}(I)$ is finitely-generated then, $\forall n \in \mathbb{N} \cup \{\infty\}$, $\mathrm{LC}_n(I)$ is finitely-generated.*

Proof. By virtue of Lemma 217, denoting by $\mathrm{LT}(I) = \langle b_i X_1^{k_i} : 1 \leq i \leq s \rangle$, where $b_i \in \mathbf{R}$ and $k_i \in \mathbb{N}$, we have

$$\mathrm{LC}_n(I) = \langle b_i : k_i \leq n \rangle.$$

□

Note that the converse of Proposition 219 does not hold. As a matter of fact, we will give in Example 253, an example of a valuation domain $\mathbf{V}$ and a finitely-generated ideal I of $\mathbf{V}[X_1]$ such that $\mathrm{LT}(I)$ is not finitely-generated while $\mathrm{LC}_n(I)$ is principal for all $n \in \mathbb{N} \cup \{\infty\}$.

In order to construct an example of a one-dimensional Gröbner domain which is neither Noetherian nor Prüfer (as announced in Example 216), we first need the following technical result.

Proposition 220. *Let* $(\mathbf{R}_n)_{n \in \mathbb{N}}$ *be an increasing sequence of m-Gröbner (resp. Gröbner) rings* $(m \geq 1)$ *such that* $\mathbf{R}_{n+1}$ *is a free* $\mathbf{R}_n$*-module and* R_0 *is not trivial. Then* $\mathbf{R} = \bigcup_{n \in \mathbb{N}} \mathbf{R}_n$ *is a non-Noetherian m-Gröbner (resp. Gröbner) ring.*

Proof. The ring $\mathbf{R}[X_1, \ldots, X_m]$ will be denoted by $\mathbf{R}[\overline{X}]$. For all $n \geq 0$ considering $f_n \in \mathbf{R}_{n+1} \setminus \mathbf{R}_n$, the non-decreasing sequence $(\langle f_0, \ldots, f_n \rangle)_{n \in \mathbb{N}}$ of ideals of $\mathbf{R}$ does not pause and, thus, $\mathbf{R}$ is not Noetherian. Moreover, it is clear that for $i \geq j$, $\mathbf{R}_i$ is a free $\mathbf{R}_j$-module.

Letting $I = g_1 \mathbf{R}[\overline{X}] + \ldots + g_s \mathbf{R}[\overline{X}]$ be a finitely-generated ideal of $\mathbf{R}[\overline{X}]$, there exists $r \in \mathbb{N}$ such that $g_1, \ldots, g_s \in \mathbf{R}_r[\overline{X}]$. For $q \geq r$, we will denote by I_q the ideal of $\mathbf{R}_q[\overline{X}]$ generated by the g_i's.

Now, letting $f \in I$, there is $k \geq r$ such that $f \in I_k$. Denoting by $\{u_\lambda, \ \lambda \in \Lambda\}$ a basis of the free $\mathbf{R}_r$-module $\mathbf{R}_k$, f can be rewritten as

$$f = f_1 u_{\lambda_1} + \cdots + f_d u_{\lambda_d},$$

where $\lambda_i \in \Lambda$, all the f_i's are elements of I_r, and such that

$$\mathrm{mdeg}(f) = \max\{\mathrm{mdeg}(f_i u_{\lambda_i}), 1 \leq i \leq d\}.$$

Thus, $\mathrm{LT}(f) \in \mathrm{LT}(I_r)\mathbf{R}[\overline{X}]$. As $\mathrm{LT}(I_r)\mathbf{R}[\overline{X}] \subseteq \mathrm{LT}(I)$, we deduce that $\mathrm{LT}(I) = \mathrm{LT}(I_r)\mathbf{R}[\overline{X}]$ and the result clearly follows.

□

Corollary 221. *If* $\mathbf{K}$ *is a discrete field then* $\mathbf{K}[X_1, X_2, \ldots] = \bigcup_{n \geq 1} \uparrow \mathbf{K}[X_1, \ldots, X_n]$ *is a non-Noetherian Gröbner ring.*

Refining Hochster's example [70, page 225], one obtains an example of a one-dimensional Gröbner domain which is neither Noetherian nor Prüfer. It is worth pointing out that Example 222 is written within classical mathematics since for a discrete field $\mathbf{K}$, the formal series ring $\mathbf{K}[[X]]$ is not discrete. The "purely ideal" result proposed in Example 222 does not correspond to any known algorithm for computing Gröbner bases.

Example 222. Let $\mathbf{K}$ be a field. Setting $\mathbf{R}_n = \mathbf{K}[[X^{\frac{2}{3^n}}, X^3]]$ and $\mathbf{R} = \bigcup_{n \in \mathbb{N}} \uparrow \mathbf{R}_n$, we will prove that the ring $\mathbf{R}$ is a Gröbner local domain of Krull dimension 1 which is neither Noetherian nor Prüfer.

First, as $\mathbf{R}_n$ is isomorphic to $\mathbf{K}[[T,Z]]/\mathrm{Ker}\phi$ where $\phi : \mathbf{K}[[T,Z]] \longrightarrow \mathbf{R}_n$ is such that $\phi(f(T,Z)) = f(X^{\frac{2}{3^n}}, X^3)$, one can see easily that $\mathbf{R}_n$ is a one-dimensional Noetherian (and, thus, Gröbner) domain (recall that $\dim \mathbf{K}[[T,Z]] = 2$).

Now, let us prove that $\dim \mathbf{R} \leq 1$. Considering $a, b \in \mathbf{R}$, there exists $m \in \mathbb{N}$ such that $a, b \in \mathbf{R}_m$. As $\mathbf{R}_m$ has Krull dimension 1, there is a collapse of the form $a^k(b^k(1 + xb) + ya) = 0$ for some $x, y \in \mathbf{R}_m \subseteq \mathbf{R}$ and $k \in \mathbb{N}$. We infer that $\dim \mathbf{R} \leq 1$. As $\mathbf{R}$ is not a field, its Krull dimension must be equal to 1.

The following argument is extracted from [70]. Setting $j = X^{\frac{2}{3^{n+1}}}$, we have for all $k \in \mathbb{N}$, $j^{3k} = (X^{\frac{2}{3^n}})^k \in \mathbf{R}_n$, $j^{3k+1} = (X^{\frac{2}{3^n}})^k j \in j\mathbf{R}_n$, and $j^{3k+2} = (X^{\frac{2}{3^n}})^k j^2 \in j^2 \mathbf{R}_n$. Hence $\mathbf{R}_{n+1} = 1.\mathbf{R}_n + j\mathbf{R}_n + j^2 \mathbf{R}_n$. Let $f, g, h \in \mathbf{R}_n$ such that $f + jg + j^2 h = 0$. The monomials of $\mathbf{R}_n$ are of the form $X^{\frac{2l}{3^n}} X^{3l'}$, those of $j\mathbf{R}_n$ are of the form $X^{\frac{2}{3^{n+1}}} X^{\frac{2m}{3^n}} X^{3m'}$, and those of $j^2 \mathbf{R}_n$ are of the form $X^{\frac{4}{3^{n+1}}} X^{\frac{2t}{3^n}} X^{3t'}$. Hence, if we set $f + jg + j^2 h = 0$ with $(f, g, h) \neq (0, 0, 0)$, one of the following equalities will hold:

$$X^{\frac{2l}{3^n}} X^{3l'} = X^{\frac{2}{3^{n+1}}} X^{\frac{2m}{3^n}} X^{3m'},$$

$$X^{\frac{2l}{3^n}} X^{3l'} = X^{\frac{4}{3^{n+1}}} X^{\frac{2t}{3^n}} X^{3t'},$$

$$X^{\frac{2}{3^{n+1}}} X^{\frac{2m}{3^n}} X^{3m'} = X^{\frac{4}{3^{n+1}}} X^{\frac{2t}{3^n}} X^{3t'}.$$

This means that one of the following equalities will hold:

$$2l + 3^{n+1} l' = 2m + 3^{n+1} m' + \frac{2}{3},$$

$$2l + 3^{n+1} l' = 2t + 3^{n+1} t' + \frac{4}{3},$$

$$2m + 3^{n+1} m' + \frac{2}{3} = 2t + 3^{n+1} t' + \frac{4}{3}.$$

This is impossible because each of these equalities would imply that $\frac{2}{3}$ is an integer. We conclude that $\mathbf{R}_{n+1}$ is a free $\mathbf{R}_n$-module having $\{1, j, j^2\}$ as a basis.

The ring $\mathbf{R}$ is local because so is $\mathbf{K}[[T,Z]]$. As $X^{\frac{2}{3}}$ and X^3 are not comparable under division, $\mathbf{R}$ is not a valuation domain. The fact that $\mathbf{R}$ is Gröbner follows from Proposition 220.

Lemma 223. *Let $\mathbf{R}$ be a strongly discrete coherent ring and consider a monomial order $>$ on $\mathbf{R}[X,Y]$ such that $X > Y$. For any $a \in \mathbf{R}$, supposing that $\langle Y + aX \rangle$ has a Gröbner basis $G = \{g_1, \ldots, g_s\}$, then the annihilator of a is generated by the set $\{b \in \mathbf{R} \mid bY \in G\}$.*

Proof. Let $c \in \mathrm{Ann}(a)$. We have $cY = c(Y + aX) \in \langle Y + aX \rangle$. Thus, $cY \in \langle \mathrm{LT}(g_1), \ldots, \mathrm{LT}(g_s) \rangle$ and, obviously, it belongs to the ideal generated by $\{b \in \mathbf{R} \mid bY \in G\}$.

Conversely, let $b \in \mathbf{R}$ such that $bY \in G$. There exists $h \in \mathbf{R}[X,Y]$ such that $bY = (Y + aX)h$. Necessarily, h has the form $h = b + Yu(Y) + Xv(X) + XYw(X,Y)$. It follows that $abX + aX^2 v(X) \equiv 0 \bmod Y$, and hence $ab = 0$. $\qquad\square$

Proposition 224. ([143]) *A Gröbner ring is stably coherent.*

Proof. It suffices to prove that $\mathbf{R}[X]$ is coherent when $\mathbf{R}$ is a Gröbner ring. By virtue of Lemma 223 and Definition and Proposition 7, it remains to prove that the intersection of two finitely-generated ideals of $\mathbf{R}[X]$ is finitely-generated. For this, let $I = \langle f_1, \ldots, f_s \rangle$ and $J = \langle g_1, \ldots, g_r \rangle$ two finitely-generated ideals of $\mathbf{R}[X]$. As in the case where the base ring is a field,

$$I \cap J = \langle t f_1, \ldots, t f_s, (1-t)g_1, \ldots, (1-t)g_r \rangle \cap \mathbf{R}[X],$$

where t is an indeterminate over $\mathbf{R}[X]$. If G is a Gröbner basis for

$$\langle t f_1, \ldots, t f_s, (1-t)g_1, \ldots, (1-t)g_r \rangle$$

in $\mathbf{R}[X,t]$ accordingly to the lexicographic monomial order with $t > X$, then $G \cap \mathbf{R}[X]$ if a finite generating set for $I \cap J$. $\qquad\square$

We will see in Theorems 255 and 256 that, for a valuation domain $\mathbf{V}$ with explicit divisibility, the following assertions are equivalent:

(1) For any finitely-generated ideal I of $\mathbf{V}[X_1, \ldots, X_n]$, fixing the lexicographic order as monomial order on $\mathbf{R}[X_1, \ldots, X_n]$, the leading terms ideal $\mathrm{LT}(I)$ is also finitely-generated.

(2) For any finitely-generated ideal I of $\mathbf{V}[X]$, the leading terms ideal $\mathrm{LT}(I)$ is also finitely-generated.

(3) $\dim \mathbf{V} \leq 1$.

This encourages us to set the following conjecture.

Conjecture 225. (The Leading Terms Ideals Conjecture) *If $\mathbf{R}$ is a strongly discrete coherent ring then the following assertions are equivalent:*

(1) *For any finitely-generated ideal I of $\mathbf{R}[X_1, \ldots, X_n]$, and any monomial order on $\mathbf{R}[X_1, \ldots, X_n]$, the leading terms ideal $\mathrm{LT}(I)$ is also finitely-generated.*

(2) *For any finitely-generated ideal I of $\mathbf{R}[X_1, \ldots, X_n]$, fixing the lexicographic order as monomial order on $\mathbf{R}[X_1, \ldots, X_n]$, the leading terms ideal $\mathrm{LT}(I)$ is also finitely-generated.*

(3) *For any finitely-generated ideal I of $\mathbf{R}[X]$, the leading terms ideal $\mathrm{LT}(I)$ is also finitely-generated.*

3.3 Gröbner Bases Over Strongly Discrete Coherent Arithmetical Rings

3.3.1 Gröbner Bases Over a Coherent Valuation Ring

Recall that a coherent valuation ring $\mathbf{R}$ is nothing but a valuation ring such that given $a \in \mathbf{R}$, one can compute $b \in \mathbf{R}$ generating the annihilator of a, i.e., such that

$$\text{Ann}(a) = \{x \in \mathbf{R} \mid xa = 0\} = b\mathbf{R}.$$

Recall also (see Definition 8) that a coherent valuation ring $\mathbf{R}$ is integral if and only if it is without zero-divisors, and a reduced valuation ring is coherent if and only if it is integral.

Definition 226. Let $\mathbf{R}$ be a coherent valuation ring ring, $I = \langle f_1, \ldots, f_s \rangle$ a nonzero finitely-generated ideal of $\mathbf{R}[X_1, \ldots, X_n]$, and $>$ a monomial order on $\mathbf{R}[X_1, \ldots, X_n]$.

(1) As in the classical division algorithm in $\mathbf{F}[X_1, \ldots, X_n]$ ($\mathbf{F}$ a discrete field) (see Algorithm 211), for each polynomials $h, h_1, \ldots, h_m \in \mathbf{R}[X_1, \ldots, X_n]$, there exist $q_1, \ldots, q_m, r \in \mathbf{R}[X_1, \ldots, X_n]$ such that

$$h = q_1 h_1 + \cdots + q_m h_m + r,$$

with $\text{mdeg}(h) \geq \text{mdeg}(q_i h_i)$, and either $r = 0$ or r is a sum of terms none of which is divisible by any of $\text{LT}(h_1), \ldots, \text{LT}(h_m)$. The polynomial r is called a remainder of h on division by $H = \{h_1, \ldots, h_m\}$ and denoted $r = \overline{h}^H$.

(2) Let $g_1, \ldots, g_m \subset I$. Recall that $G = \{g_1, \ldots, g_m\}$ is said to be a *Gröbner basis* for I if $\text{LT}(I) = \langle \text{LT}(G) \rangle := \langle \text{LT}(g_1), \ldots, \text{LT}(g_m) \rangle$.

The following lemma gives a sufficient and necessary condition for a term to belong to an ideal generated by terms over a valuation ring.

Lemma 227. *Let $\mathbf{R}$ be a strongly discrete valuation ring and $I = \langle a_\alpha X^\alpha, \alpha \in A \rangle$ an ideal of $\mathbf{R}[X_1, \ldots, X_n]$ generated by a collection of terms. Then a term bX^β lies in I if and only if X^β is divisible by X^α and b is divisible by a_α for some $\alpha \in A$.*

Proof. It is obvious that the condition is sufficient. For proving the necessity, write $bX^\beta = \sum_{i=1}^s c_i a_{\alpha_i} X^{\gamma_i} X^{\alpha_i}$ for some $\alpha_1, \ldots, \alpha_s \in A$, $c_i, a_{\alpha_i} \in \mathbf{R} \setminus \{0\}$, and $\gamma_i \in \mathbb{N}^n$. Ignoring the superfluous terms, for each $1 \leq i \leq s$, $\gamma_i + \alpha_i = \beta$, and $b = \sum_{i=1}^s c_i a_{\alpha_i}$. It is clear that for each $1 \leq i \leq s$, X^β is divisible by X^{α_i}. Since all the coefficients are comparable under division, we can suppose that a_{α_1} divides all the a_{α_i} and thus divides b. $\square$

The following proposition shows that the rather "disappointing" behavior of the division algorithm (Algorithm 211) detected in Example 213 (nonuniqueness of the remainder) does not occur when one divides by a Gröbner basis.

Proposition 228. *Let* **R** *be a coherent valuation ring,* $>$ *a monomial order on* $\mathbf{R}[X_1,\ldots,X_n]$, *I a nonzero ideal of* $\mathbf{R}[X_1,\ldots,X_n]$, $f_1,\ldots,f_s \in I$ *such that* $G = \{f_1,\ldots,f_s\}$ *is a Gröbner basis for I, and* $f \in \mathbf{R}[X_1,\ldots,X_n]$. *Then:*

(1) *There is a unique* $r \in \mathbf{R}[X_1,\ldots,X_n]$ *with the following two properties:*

   (i) *No term of r is divisible by any of the* $\mathrm{LT}(f_i)$*'s.*

   (ii) *There is* $g \in I$ *such that* $f = g + r$.

In particular, r is the remainder on division of f by G regardless how the elements of G are listed when using the division algorithm (Algorithm 211).

(2) $f \in I \;\Leftrightarrow\; \bar{f}^G = 0$ (the ideal membership test).

(3) $I = \langle f_1,\ldots,f_s \rangle$.

Proof. (1) The division algorithm (Algorithm 211) gives $r = \bar{f}^G \in \mathbf{R}[X_1,\ldots,X_n]$ satisfying the property (i) and $q_1,\ldots,q_s \in \mathbf{R}[X_1,\ldots,X_n]$ such that $f = q_1 f_1 + \cdots + q_s f_s + r$. Then take $g = q_1 f_1 + \cdots + q_s f_s \in I$.

For the uniqueness, let $f = g_1 + r_1 = g_2 + r_2$ satisfying (i) and (ii). As $r_1 - r_2 = g_2 - g_1 \in I$ then either it is null or $\mathrm{LT}(r_1 - r_2)$ is divisible by one of the $\mathrm{LT}(f_i)$'s (because G is a Gröbner basis for I and taking into account Lemma 227). The latter case is impossible since no term of r_1, r_2 is divisible any of the $\mathrm{LT}(f_i)$'s. We conclude that $r_1 = r_2$.

(2) If $\bar{f}^G = 0$ then clearly $f \in \langle f_1,\ldots,f_s \rangle \subseteq I$. Conversely, if $f \in I$, then $f = f + 0$ satisfies the two conditions of (1), and thus $\bar{f}^G = 0$.

(3) This is an immediate consequence of (2).

$\qquad\qquad\qquad\qquad\qquad\qquad\qquad\qquad\qquad\qquad\qquad\qquad\qquad\qquad\quad$ $\square$

We now consider the problem of the construction of a Gröbner basis. A key tool introduced by Buchberger [22] in the case where the base ring is a discrete field is the notion of S-polynomial of two polynomials.

Definition 229. Let **R** be a coherent valuation ring, $f \neq g \in \mathbf{R}[X_1,\ldots,X_n] \setminus \{0\}$, and $>$ a monomial order on $\mathbf{R}[X_1,\ldots,X_n]$.

(1) If $\mathrm{mdeg}(f) = \alpha$ and $\mathrm{mdeg}(g) = \beta$ then let $\gamma = (\gamma_1,\ldots,\gamma_n)$, where $\gamma_i = \max(\alpha_i, \beta_i)$ for each i. Perform the test $\mathrm{LC}(f) \mid \mathrm{LC}(g)$ or $\mathrm{LC}(g) \mid \mathrm{LC}(f)$.

$$S(f,g) = \frac{X^\gamma}{\mathrm{LM}(f)} f - \frac{\mathrm{LC}(f)}{\mathrm{LC}(g)} \frac{X^\gamma}{\mathrm{LM}(g)} g \quad \text{if } \mathrm{LC}(g) \text{ divides } \mathrm{LC}(f).$$

$$S(f,g) = \frac{\mathrm{LC}(g)}{\mathrm{LC}(f)} \frac{X^\gamma}{\mathrm{LM}(f)} f - \frac{X^\gamma}{\mathrm{LM}(g)} g \quad \text{if } \mathrm{LC}(f) \text{ divides } \mathrm{LC}(g) \text{ and } \mathrm{LC}(g)$$
does not divide $\mathrm{LC}(f)$.[1]

$S(f,g)$ is called the *S-polynomial* of f and g. It is "designed" to produce cancellation of leading terms. Here, it is worth pointing out that $S(f,g)$ is not uniquely determined when **R** has zero-divisors. This minor technical issue will be repaired through the consideration in (2) of $S(f,f)$ and $S(g,g)$.

---

[1] That assumes that the ring **V** is residually discrete but this is not necessary for the algorithm.

(2) Let d be a generator of the annihilator of $\mathrm{LC}(f)$ (note that this annihilator is principal because **R** is a coherent valuation ring). We set

$$S(f,f) := df$$

(it is defined up to a unit). Note that $S(f,f)$ behaves exactly like usual S-polynomials in the sense that $\mathrm{mdeg}(S(f,f)) < \mathrm{mdeg}(f)$ and $S(X^\delta f, X^\delta f) = X^\delta S(f,f) \; \forall \delta \in \mathbb{N}^n$. In addition, if the leading coefficient of f is not a zero-divisor then automatically $S(f,f) = 0$ (as in the case where **R** is a field).

$S(f,f)$ is called the *auto-S-polynomial* of f. It is "designed" to cover the cancellation of the leading term of f produced by a multiplication of f by an element of the annihilator of $\mathrm{LC}(f)$.

Example 230. (Example 213 Continued) Let $f_1 = 2 + 12XY$, $f_2 = 8Y^2 \in \mathbb{Q}[X,Y]$, and fix any monomial order $>$ on $\mathbb{Q}[X,Y]$. Then

$$S(f_1, f_2) = Yf_1 - \frac{3}{2}Xf_2 = 2Y.$$

Of course, $S(f_1, f_1) = S(f_2, f_2) = 0$ as we are on an integral ground.

Example 231. (*S-Polynomials over* $\mathbb{F}_2[Y]/\langle Y^2 \rangle$, a Useful Ring in Coding Theory)
The ring $\mathbf{V} := \mathbb{F}_2[Y]/\langle Y^2 \rangle = \mathbb{F}_2[y]$ (where $y = \bar{Y}$) is a zero-dimensional coherent valuation ring with zero-divisors (as $y^2 = 0$).
Let $f \neq g \in \mathbf{V}[X_1, \ldots, X_n] \setminus \{0\}$, and $>$ a monomial order. Denoting by $\mathrm{mdeg}(f) = \alpha = (\alpha_1, \ldots, \alpha_n)$, $\mathrm{mdeg}(g) = \beta = (\beta_1, \ldots, \beta_n)$, $\gamma = (\gamma_1, \ldots, \gamma_n)$, where $\gamma_k = \max(\alpha_k, \beta_k)$ for each k, the only case where $S(f,g)$ is not equal to $\mathrm{LC}(g)\frac{X^\gamma}{X^\alpha}f - \mathrm{LC}(f)\frac{X^\gamma}{X^\beta}g$ (up to a unit) is when $\mathrm{LC}(f) = \mathrm{LC}(g) = y$. In that case $S(f,g)$ is simply equal to $\frac{X^\gamma}{X^\alpha}f - \frac{X^\gamma}{X^\beta}g$. On the other hand, for the computation of $S(f,f)$, two cases may arise:
If $\mathrm{LC}(f) = 1$ or $1 + y$ then $S(f,f) = 0$.
If $\mathrm{LC}(f) = y$ then $S(f,f) = yf$.
For example, fixing the lexicographic order with $X_1 > X_2$ as monomial order, and considering the polynomials $f_1 := yX_1 + X_2$ and $f_2 = y + yX_2$, we have:

$$S(f_1, f_2) = X_2 f_1 - X_1 f_2 = X_2^2 + yX_1,$$
$$S(f_1, f_1) = yf_1 = yX_2, \; S(f_2, f_2) = yf_2 = 0.$$

Example 232. (The Ring $\mathbb{Z}_{p\mathbb{Z}}$, Where p Is a Prime Number)

(1) Recall that the ring $\mathbb{Z}_{p\mathbb{Z}}$ is the following localization of $\mathbb{Z}$:

$$\mathbb{Z}_{p\mathbb{Z}} := \{\frac{a}{b} \in \mathbb{Q} \mid a \in \mathbb{Z} \text{ and } b \in \mathbb{Z} \setminus p\mathbb{Z}\}.$$

For $a \in \mathbb{Z} \setminus \{0\}$, we denote by $v_p(a) := \max\{k \in \mathbb{N} \mid p^k$ divides $a\}$ (the valuation of a at p), so that,

$$a = p^{v_p(a)} a' \text{ with } a' \wedge p = 1, \text{ that is, } a' \in \mathbb{Z} \setminus p\mathbb{Z}.$$

So, any element $x \in \mathbb{Z}_{p\mathbb{Z}}$ can be written in the form

$$x = p^k \frac{a'}{b'} \text{ with } k \in \mathbb{N} \text{ and } a', b' \in \mathbb{Z} \setminus p\mathbb{Z}.$$

The nonnegative integer k is called the *valuation of x at p* and is denoted by $v_p(x)$. It follows that for $x, y \in \mathbb{Z}_{p\mathbb{Z}} \setminus \{0\}$,

$$x \mid y \iff v_p(x) \leq v_p(y),$$

and, thus, $\mathbb{Z}_{p\mathbb{Z}}$ is a valuation domain. It follows also, that any nondecreasing sequence $(\langle x_n \rangle)_{n \in \mathbb{N}}$ of principal ideals of $\mathbb{Z}_{p\mathbb{Z}}$ (finitely-generated ideals of $\mathbb{Z}_{p\mathbb{Z}}$ are principal as $\mathbb{Z}_{p\mathbb{Z}}$ is valuation domain) pauses after at most $(v_p(x_0) + 1)$ iterations, and, thus, $\mathbb{Z}_{p\mathbb{Z}}$ is Noetherian. In classical literature, $\mathbb{Z}_{p\mathbb{Z}}$ is called a "discrete valuation domain" (discrete because its valuation group is $\mathbb{Z}$) but we prefer to call it a Noetherian valuation domain as, in this text following Richman, the terminology "discrete" is reserved for rings equipped with an equality test.

(2) In $\mathbb{Z}_{p\mathbb{Z}}[X_1, \ldots, X_n]$, for $x, y \in \mathbb{Z}_{p\mathbb{Z}} \setminus \{0\}$, a term xX^α divides a term yX^β if and only if $v_p(x) \leq v_p(y)$ and $X^\alpha \mid X^\beta$.

(3) We will specify the definition of S-polynomials given in Definition 226 to the case where the base ring is $\mathbb{Z}_{p\mathbb{Z}}$.

Let $f_i, f_j \in \mathbb{Z}_{p\mathbb{Z}}[X_1, \ldots, X_n] \setminus \{0\}$ $(i \neq j)$, and fix a monomial order $>$ on $\mathbb{Z}_{p\mathbb{Z}}[X_1, \ldots, X_n]$. Denote by $\text{mdeg}(f_i) = \beta = (\beta_1, \ldots, \beta_n)$, $\text{mdeg}(f_j) = \beta' = (\beta_1', \ldots, \beta_n')$, $\gamma = (\gamma_1, \ldots, \gamma_n)$, where $\gamma_k = \max(\beta_k, \beta_k')$ for each k.

Moreover, denoting by $\text{LC}(f_i) = c_i = p^{v_p(c_i)} \frac{a_i}{b_i}$, $\text{LC}(f_j) = c_j = p^{v_p(c_j)} \frac{a_j}{b_j}$, with $a_i, b_i, a_j, b_j \in \mathbb{Z} \setminus p\mathbb{Z}$, we have:

(i) $S(f_i, f_j) := \frac{X^\gamma}{\text{LM}(f_i)} f_i - \frac{a_i b_j}{b_i a_j} p^{v_p(c_i) - v_p(c_j)} \frac{X^\gamma}{\text{LM}(f_j)} f_j$ if $v_p(c_j) \leq v_p(c_i)$.

$S(f_i, f_j) := \frac{a_j b_i}{b_j a_i} p^{v_p(c_j) - v_p(c_i)} \frac{X^\gamma}{\text{LM}(f_i)} f_i - \frac{X^\gamma}{\text{LM}(f_j)} f_j$ if $v_p(c_j) > v_p(c_i)$.

(ii) $S(f_i, f_i) = S(f_j, f_j) = 0$ (as we are on an integral ground).

(4) Let $f_1 = 2 + 12XY$, $f_2 = 8Y^2 \in \mathbb{Z}_{2\mathbb{Z}}[X, Y]$. Fixing any monomial order $>$ on $\mathbb{Z}_{2\mathbb{Z}}[X, Y]$, we have:

$$S(f_1, f_2) = \frac{2}{3}Yf_1 - Xf_2 = \frac{4}{3}Y.$$

The following lemma will be of great use since it is a key result for the characterization of Gröbner bases by means of S-polynomials.

Lemma 233. *Let* $\mathbf{R}$ *be a valuation ring,* $>$ *a monomial order, and* $f_1,\ldots,f_s \in \mathbf{R}[X_1,\ldots,X_n]$ *such that* $\mathrm{mdeg}(f_i) = \gamma$ *for each* $1 \leq i \leq s$. *If* $\mathrm{mdeg}(\sum_{i=1}^{s} a_i f_i) < \gamma$ *for some* $a_1,\ldots,a_s \in \mathbf{R}$, *then* $\sum_{i=1}^{s} a_i f_i$ *is a linear combination with coefficients in* $\mathbf{R}$ *of the* S-*polynomials* $S(f_i,f_j)$ *for* $1 \leq i,j \leq s$. *Furthermore, each* $S(f_i,f_j)$ *has multidegree* $< \gamma$

Proof. Since $\mathbf{R}$ is a valuation ring, we can suppose that $\mathrm{LC}(f_s)/\mathrm{LC}(f_{s-1})/\cdots/\mathrm{LC}(f_1)$. Thus, for $i < j$, $S(f_i,f_j) = f_i - \frac{\mathrm{LC}(f_i)}{\mathrm{LC}(f_j)} f_j$.

$$
\begin{aligned}
\sum_{i=1}^{s} a_i f_i &= a_1 \left(f_1 - \frac{\mathrm{LC}(f_1)}{\mathrm{LC}(f_2)} f_2\right) + \left(a_2 + \frac{\mathrm{LC}(f_1)}{\mathrm{LC}(f_2)} a_1\right)\left(f_2 - \frac{\mathrm{LC}(f_2)}{\mathrm{LC}(f_3)} f_3\right) \\
&\quad + \cdots + \left(a_{s-1} + \frac{\mathrm{LC}(f_{s-2})}{\mathrm{LC}(f_{s-1})} a_{s-2}\right. \\
&\quad + \cdots + \left.\frac{\mathrm{LC}(f_1)}{\mathrm{LC}(f_{s-1})} a_1\right)\left(f_{s-1} - \frac{\mathrm{LC}(f_{s-1})}{\mathrm{LC}(f_s)} f_s\right) \\
&\quad + \left(a_s + \frac{\mathrm{LC}(f_{s-1})}{\mathrm{LC}(f_s)} a_{s-1} + \cdots + \frac{\mathrm{LC}(f_1)}{\mathrm{LC}(f_s)} a_1\right) f_s.
\end{aligned}
$$

But $\left(a_s + \frac{\mathrm{LC}(f_{s-1})}{\mathrm{LC}(f_s)} a_{s-1} + \cdots + \frac{\mathrm{LC}(f_1)}{\mathrm{LC}(f_s)} a_1\right) \mathrm{LC}(f_s) = 0$ since $\mathrm{mdeg}(\sum_{i=1}^{s} a_i f_i) < \gamma$, and, thus,

$$
\left(a_s + \frac{\mathrm{LC}(f_{s-1})}{\mathrm{LC}(f_s)} a_{s-1} + \cdots + \frac{\mathrm{LC}(f_1)}{\mathrm{LC}(f_s)} a_1\right) f_s \in \mathbf{R} S(f_s,f_s).
$$

$\square$

Using Lemma 227 and Lemma 233, we generalize some classical results about the existence and characterization of Gröbner bases for ideals in polynomial rings over coherent valuation rings.

Theorem 234. *Let* $\mathbf{R}$ *be a coherent valuation ring,* $I = \langle g_1,\ldots,g_s \rangle$ *an ideal of* $\mathbf{R}[X_1,\ldots,X_n]$, *and fix a monomial order* $>$ *on* $\mathbf{R}[X_1,\ldots,X_n]$. *Then,* $G = \{g_1,\ldots,g_s\}$ *is a Gröbner basis for* I *if and only if for all pairs* $i \leq j$, *the remainder on division of* $S(g_i,g_j)$ *by* G *is zero.*

Proof. "$\Rightarrow$" As $S(g_i,g_j) \in \langle g_i,g_j \rangle \subseteq I$, then, by virtue of Proposition 228.(2), $\overline{S(g_i,g_j)}^G = 0$.
"$\Leftarrow$" Instead of going through the details of the proof, we prefer to give the idea behind it. This is nicely explained in [43, page 83] in case the base ring is a discrete field. The same proof holds in our situation as we have all the necessary ingredients.

Letting $f \in I = \langle g_1, \ldots, g_s \rangle$, there are polynomials $h_i \in \mathbf{R}[X_1, \ldots, X_n]$ such that

$$f = \sum_{i=1}^{s} h_i g_i, \tag{3.1}$$

with $\mathrm{mdeg}(f) \leq \max_{1 \leq i \leq s}(\mathrm{mdeg}(h_i g_i))$.

Case 1: $\mathrm{mdeg}(f) = \max_{1 \leq i \leq s}(\mathrm{mdeg}(h_i g_i))$, say $\mathrm{mdeg}(f) = \mathrm{mdeg}(h_{i_0} g_{i_0})$ for some $i_0 \in \{1, \ldots, s\}$. As the leading coefficients of the $h_i g_i$'s such that $\mathrm{mdeg}(f) = \mathrm{mdeg}(h_i g_i)$ are comparable under division, we can suppose that all of them are divisible by the leading coefficient of $h_{i_0} g_{i_0}$. It follows that $\mathrm{LT}(f) \in \langle \mathrm{LT}(g_{i_0}) \rangle \subseteq \langle \mathrm{LT}(g_1), \ldots, \mathrm{LT}(g_s) \rangle$.

Case 2: $\mathrm{mdeg}(f) < \max_{1 \leq i \leq s}(\mathrm{mdeg}(h_i g_i))$. Then, roughly speaking, some cancellation must occur among the leading terms of (3.1). Using Lemma 233, we can rewrite this in terms of S-polynomials. Then, the assumption that S-polynomials have zero remainders modulo G allows to replace the S-polynomials by expressions involving less cancellation. Thus, we obtain an expression for f that has less cancellation of leading terms. An so on, as the set of monomials is well-ordered (by virtue of Corollary 210), we end up with a situation like that of Case 1.

$\square$

From Theorem 234 ensues the following algorithm for constructing Gröbner bases over coherent valuation rings.

Algorithm 235. (Buchberger's Algorithm for Coherent Valuation Rings)

Input: $g_1, \ldots, g_s \in \mathbf{R}[X_1, \ldots, X_n]$ where $\mathbf{R}$ is a coherent valuation ring, and a monomial order $>$ on $\mathbf{R}[X_1, \ldots, X_n]$
Output: a Gröbner basis G for $\langle g_1, \ldots, g_s \rangle$ with $\{g_1, \ldots, g_s\} \subseteq G$

> $G := \{g_1, \ldots, g_s\}$
> REPEAT
> $G' := G$
> For each pair f, g in G' DO
> $\quad S := \overline{S(f,g)}^{G'}$
> $\quad$ If $S \neq 0$ THEN $G := G' \cup \{S\}$
> UNTIL $G = G'$

The algorithm above is exactly the same algorithm as in the case where the base ring is a discrete field (Algorithm 211). The only modifications are in the definition of S-polynomials, in the consideration of the auto-S-polynomials, and in the divisions of terms. Note that we will see later in this chapter (Theorem 272) that the precise reason why Buchberger's Algorithm 235 terminates is that the valuation ring is archimedean, or equivalently, either it is a valuation domain of Krull dimension ≤ 1 or a zero-dimensional valuation ring containing zero-divisors. But for now, we will content ourselves with the following result.

Theorem 236. *If* **R** *is a coherent Noetherian valuation ring then Algorithm 235 terminates and is correct.*

Proof. This algorithm computes a nondecreasing sequence $G_1 \subseteq G_2 \subseteq \cdots$. First $G_0 = \{g_1, \ldots, g_s\}$ is a family of generators of $I = \langle g_1, \ldots, g_s \rangle$. If $G_i \subseteq I$, then for $f, g \in G_i$, we have $S(f, g) \in I$, and hence $\overline{S(f,g)}^{G_i} \in I$, and $G_{i+1} \subseteq I$. By induction, $G_m \subseteq I$ for all m.

If the algorithm ends, Theorem 234 guarantees that the computed family G is a Gröbner basis for I.

Hence we just need to prove that the algorithm ends. For each i, we denote by $\langle LT(G_i) \rangle$ the ideal of $\mathbf{R}[X_1, \ldots, X_n]$ generated by the leading terms of the elements of G_i, by $\langle LM(G_i) \rangle$ the ideal of $\mathbf{R}[X_1, \ldots, X_n]$ generated by the leading monomials of the elements of G_i, and by $\langle LC(G_i) \rangle$ the ideal of $\mathbf{R}$ generated by the leading coefficients of the elements of G_i. Since $G_i \subseteq G_{i+1}$, we have $\langle LT(G_i) \rangle \subseteq \langle LT(G_{i+1}) \rangle$. But if $G_i \subset G_{i+1}$, then there exists $f, g \in G_i$ such that $\overline{S(f,g)}^{G_i} \neq 0$, and hence $LT(\overline{S(f,g)}^{G_i}) \in \langle LT(G_{i+1}) \rangle \setminus \langle LT(G_i) \rangle$, and $\langle LT(G_i) \rangle \subset \langle LT(G_{i+1}) \rangle$. Since the sequence $(\langle LM(G_i) \rangle)_{i \in \mathbb{N}}$ is nondecreasing, one can find $n_1 < n_2 < \cdots$ such that $\langle LM(G_{n_i}) \rangle = \langle LM(G_{n_i+1}) \rangle$ for all $i \in \mathbb{N}$. The sequence $\langle LC(G_{n_i}) \rangle_{i \in \mathbb{N}}$ being nondecreasing, using the fact that $\mathbf{R}$ is Noetherian, there exists $j \in \mathbb{N}$ such that $\langle LC(G_{n_j}) \rangle = \langle LC(G_{n_j+1}) \rangle$. But, as $\langle LC(G_{n_j}) \rangle \subseteq \langle LC(G_{n_j+1}) \rangle \subseteq \langle LC(G_{n_j+1}) \rangle$, we have $\langle LC(G_{n_j}) \rangle = \langle LC(G_{n_j+1}) \rangle$, and, thus, $\langle LT(G_{n_j}) \rangle = \langle LT(G_{n_j+1}) \rangle$. Hence $G_i = G_{i+1}$, completing the proof. $\qquad\square$

Note that in classical mathematics, a Noetherian valuation domain which is not a field is called a DVR. From a constructive point of view, we don't know a priori the generator of the maximal ideal.

Example 237. Examples of coherent Noetherian valuation rings are:

- Discrete fields ($\mathbb{Q}$ for example).

- $\mathbb{Z}_{p\mathbb{Z}} = \{\frac{a}{b} \mid a \in \mathbb{Z} \,\&\, b \in Z \setminus p\mathbb{Z}\}$ (p a prime number). It is a domain.

- $\mathbf{D}/\langle a^k \rangle$ with $\mathbf{D}$ a PID (such as $\mathbb{Z}$), and a an irreducible element. When $k \geq 2$, it has zero-divisors ($\mathbb{F}_2[Y]/\langle Y^2 \rangle$ for example).

- Galois rings $GR(p^k, n) = (\mathbb{Z}/p^k\mathbb{Z})[t]/\langle f \rangle$, where f is a monic irreducible polynomial in $(\mathbb{Z}/p^k\mathbb{Z})[t]$ (p a prime number) of degree n whose image modulo p is irreducible. When $k \geq 2$, they have zero-divisors.

Keeping the above notation, it is obvious that if G is a Gröbner basis for I, then for any $p \in G$ such that $LT(p) \in \langle LT(G \setminus \{p\}) \rangle$, $G \setminus \{p\}$ is also a Gröbner basis for I. So, using Algorithm 235 and removing any p with $LT(p) \in \langle LT(G \setminus \{p\}) \rangle$, we can construct for I a Gröbner basis G such that

$$\forall p \in G, \text{LT}(p) \notin \langle \text{LT}(G \setminus \{p\}) \rangle.$$

Such a Gröbner basis will be called a *pseudo-minimal* Gröbner basis.
But even more, one can ask that the Gröbner basis G satisfies the following property:

$$\forall p \in G, \text{ no term of } p \text{ lies in } \langle \text{LT}(G \setminus \{p\}) \rangle. \tag{3.2}$$

Such a Gröbner basis will be called a *pseudo-reduced* Gröbner basis, and can be computed by the following algorithm.

Algorithm 238. Let $\mathbf{R}$ be a coherent Noetherian valuation ring, $G = \{g_1, \ldots, g_s\}$ a pseudo-minimal Gröbner basis for $I = \langle g_1, \ldots, g_s \rangle$, with $g_i \in \mathbf{R}[X_1, \ldots, X_n]$, accordingly to a monomial order $>$ on $\mathbf{R}[X_1, \ldots, X_n]$. Then, a pseudo-reduced Gröbner basis for I can be computed in a finite number of steps by the following algorithm:

Input: a pseudo-minimal Gröbner basis $G = \{g_1, \ldots, g_s\}$ for $\langle g_1, \ldots, g_s \rangle$
Output: a pseudo-reduced Gröbner basis $\tilde{G} = \{\tilde{g}_1, \ldots, \tilde{g}_s\}$ for $\langle g_1, \ldots, g_s \rangle$

  IF $s = 1$ THEN $\tilde{g}_1 := g_1$
  ELSE
    IF $s = 2$ THEN $\tilde{g}_1 := \overline{g_1}^{\{g_2\}}; \tilde{g}_2 := \overline{g_2}^{\{\tilde{g}_1\}}$
    ELSE
      $\tilde{g}_1 := \overline{g_1}^{\{g_2,\ldots,g_s\}}; \tilde{g}_2 := \overline{g_2}^{\{\tilde{g}_1,g_3,\ldots,g_s\}}$
      $i := 3$
      WHILE $i \leq s$ DO $\tilde{g}_i := \overline{g_i}^{\{\tilde{g}_1,\ldots,\tilde{g}_{i-1},g_{i+1},\ldots,g_s\}}$
      $i := i + 1$

Remark 239. In search of uniqueness of a pseudo-reduced Gröbner basis, we need a "normalization" of the elements of a pseudo-reduced Gröbner basis. In case the base ring is a discrete field, this is easily done by requiring the leading coefficients of the elements of a pseudo-reduced Gröbner basis to be 1. This gives birth to the notion of *reduced* Gröbner basis. For rings which are not fields, this has to be done on a case-by-case basis. Hereafter a few examples:

(1) Consider the case where the base ring is $\mathbb{Z}/p^\alpha \mathbb{Z}$, with p is a prime number and $\alpha \geq 2$. For $f \in (\mathbb{Z}/p^\alpha \mathbb{Z})[X_1, \ldots, X_n] \setminus \{0\}$, fixing a monomial order on $(\mathbb{Z}/p^\alpha \mathbb{Z})[X_1, \ldots, X_n]$, and denoting by $\text{LC}(f) = a$, where $a = p^m c$, $0 \leq m \leq \alpha - 1$ and $c \wedge p = 1$, the normalization of f, denoted by $\tilde{f}$, can be defined as

$$\tilde{f} := c^{-1} f, \text{ with } \text{LC}(\tilde{f}) = p^m, 0 \leq m \leq \alpha - 1.$$

(2) Consider the case where the base ring is $\mathbb{Z}_{p\mathbb{Z}} = \{\frac{a}{b} \mid a \in \mathbb{Z} \,\&\, b \in Z \setminus p\mathbb{Z}\}$ (p a prime number). Let $f \in \mathbb{Z}_{p\mathbb{Z}}[X_1, \ldots, X_n] \setminus \{0\}$, and fix a monomial order on $\mathbb{Z}_{p\mathbb{Z}}[X_1, \ldots, X_n]$. Denote by $\text{LC}(f) = \frac{a}{b}$, where $a = p^m c$, $b = p^r d$, $0 \leq r \leq m$, and $c, d \in \mathbb{Z}$ with $c \wedge p = d \wedge p = 1$. The normalization of f, denoted by $\tilde{f}$, can be defined as

$$\tilde{f} := \frac{d}{c} f, \text{ with } \text{LC}(\tilde{f}) \in p^{\mathbb{N}}.$$

(3) Consider the ring $\mathbf{V} := \mathbb{F}_2[Y]/\langle Y^2 \rangle = \mathbb{F}_2[y] = \{0, 1, 1 + y, y\}$, with $\mathbf{V}^{\times} = \{1, 1 + y\}$. For $f \in \mathbf{V}[X_1, \ldots, X_n] \setminus \{0\}$, fixing a monomial order, the normalization of $\tilde{f}$ of f can be defined as:

$$\tilde{f} = \begin{cases} f & \text{if } \mathrm{LC}(f) = y \text{ or } 1 \\ (1 + y)f & \text{if } \mathrm{LC}(f) = 1 + y, \end{cases}$$

with $\mathrm{LC}(\tilde{f}) = 1$ or y.

Example 240. (Example 230 Continued) Let $f_1 = 2 + 12XY$, $f_2 = 8Y^2 \in \mathbb{Q}[X, Y]$, and fix any monomial order $>$ on $\mathbb{Q}[X, Y]$. Then

$$S(f_1, f_2) = Yf_1 - \frac{3}{2}Xf_2 = 2Y \xrightarrow{\text{normalization}} Y =: f_3; \ f_1 \xrightarrow{f_3} 1.$$

Thus, $\{1\}$ is the reduced Gröbner basis of $\langle f_1, f_2 \rangle$ in $\mathbb{Q}[X, Y]$.

Example 241. (Example 232.(4) Continued) Let $f_1 = 2 + 12XY$, $f_2 = 8Y^2 \in \mathbb{Z}_{2\mathbb{Z}}[X, Y]$, and fix any monomial order $>$ on $\mathbb{Z}_{2\mathbb{Z}}[X, Y]$. Then

$$S(f_1, f_2) = \frac{2}{3}Yf_1 - Xf_2 = \frac{4}{3}Y \xrightarrow{\text{normalization}} 4Y =: f_3; \ f_1 \xrightarrow{f_3} 2, \ f_2 \xrightarrow{2} 0, f_3 \xrightarrow{2} 0.$$

Thus, $\{2\}$ is the reduced Gröbner basis of $\langle f_1, f_2 \rangle$ in $\mathbb{Z}_{2\mathbb{Z}}[X, Y]$.

Example 242. (Gröbner Bases Over $\mathbb{F}_2[Y]/\langle Y^2 \rangle$, Example 231 Continued)
Consider the ring $\mathbf{V} := \mathbb{F}_2[Y]/\langle Y^2 \rangle = \mathbb{F}_2[y]$, and the ideal $I = \langle f_1 = yX_1 + X_2, f_2 = y + yX_2 \rangle \subseteq \mathbf{V}[X_1, X_2]$. Let us compute a Gröbner basis for I accordingly to the lexicographic order with $X_1 > X_2$. We have:

$$S(f_1, f_1) = yX_2 \xrightarrow{f_2} y =: f_3; \ f_2 \xrightarrow{f_3} 0; \ f_1 \xrightarrow{f_3} X_2.$$

Thus, $\{y, X_2\}$ is the reduced Gröbner basis of I.

Gröbner bases are a powerful tool to eliminate variables. To see this, let us first define the elimination ideals of an ideal of $\mathbf{R}[X_1, \ldots, X_n]$.

Definition 243. Let $\mathbf{R}$ be a ring, $I = \langle f_1, \ldots, f_s \rangle$ an ideal of $\mathbf{R}[X_1, \ldots, X_n]$, and $1 \leq k \leq n$. The *kth elimination ideal* of I is

$$I_k := I \cap \mathbf{R}[X_{k+1}, \ldots, X_n].$$

It is an ideal of $\mathbf{R}[X_{k+1}, \ldots, X_n]$ consisting in all combinations of $f_1, \ldots, f_s$ eliminating the variables $X_1, \ldots, X_k$. Note that I_n is nothing but $I \cap \mathbf{R}$.

The following theorem shows that Gröbner bases with respect to the lexicographic monomial order allow the computation of the elimination ideals.

Theorem 244. *Let $\mathbf{R}$ be a coherent valuation ring, $G = \{g_1, \ldots, g_s\}$ a Gröbner basis for an ideal $I = \langle g_1, \ldots, g_s \rangle$ of $\mathbf{R}[X_1, \ldots, X_n]$ with respect to the lexicographic monomial order with $X_1 > X_2 > \cdots > X_n$. Then, for all $1 \leq k \leq n$,*

$$G_k := G \cap \mathbf{R}[X_{k+1}, \ldots, X_n]$$

is a Gröbner basis for the kth elimination ideal I_k of I.

Proof. As $G_k \subseteq I_k$, it suffices to show that $\mathrm{LT}(I_k) \subseteq \langle \mathrm{LT}(G_k) \rangle$. Let $f \in I_k \subseteq I$. Since G is a Gröbner basis for I, there exists $1 \leq i \leq s$ such that $\mathrm{LT}(g_i)$ divides $\mathrm{LT}(f)$. It follows that $\mathrm{LT}(g_i)$ involves only the variables $X_{k+1}, \ldots, X_n$. As any monomial involving one of the variables $X_1, \ldots, X_k$ is greater that $\mathrm{LT}(g_i)$, we infer that $g_i \in \mathbf{R}[X_{k+1}, \ldots, X_n]$, and thus, $g_i \in G_k$. $\square$

Example 245. (Example 241 Continued) Let $f_1 = 2 + 12XY$, $f_2 = 8Y^2 \in \mathbb{Z}_{2\mathbb{Z}}[X,Y]$, and fix any monomial order $>$ on $\mathbb{Z}_{2\mathbb{Z}}[X,Y]$. Then $\{2\}$ is the reduced Gröbner basis of $I = \langle f_1, f_2 \rangle$ in $\mathbb{Z}_{2\mathbb{Z}}[X,Y]$. It follows that:

$$I \cap \mathbb{Z}_{2\mathbb{Z}}[X] = \langle 2 \rangle = 2\mathbb{Z}_{2\mathbb{Z}}[X], \ \ I \cap \mathbb{Z}_{2\mathbb{Z}}[Y] = \langle 2 \rangle = 2\mathbb{Z}_{2\mathbb{Z}}[Y], \text{ and } I \cap \mathbb{Z}_{2\mathbb{Z}} = \langle 2 \rangle = 2\mathbb{Z}_{2\mathbb{Z}}.$$

Example 246. (Example 242 Continued)

Consider the ring $\mathbf{V} := \mathbb{F}_2[Y]/\langle Y^2 \rangle = \mathbb{F}_2[y]$, and the ideal $I = \langle f_1 = yX_1 + X_2, f_2 = y + yX_2 \rangle \subseteq \mathbf{V}[X_1, X_2]$. We know that $\{y, X_2\}$ is a Gröbner basis for I with respect to the lexicographic order with $X_1 > X_2$. Thus,

$$I \cap \mathbf{V}[X_2] = \langle y, X_2 \rangle \text{ and } I \cap \mathbf{V} = \langle y \rangle = \{0, y\}.$$

If one wants to compute $I \cap \mathbf{V}[X_1]$, then he has to consider the lexicographic order with $X_2 > X_1$. For this order, $\{y, X_2\}$ is again a Gröbner basis for I, and thus,

$$I \cap \mathbf{V}[X_1] = \langle y \rangle.$$

At the end of this subsection, let us point out that almost all the improvements that have been made in case where the base ring is a discrete field will prove to be easily adaptable to coherent valuation rings. Of course, in such case, one must take into account the considerable number of optimizations that have been made in recent years for the purpose of speeding up Buchberger's algorithm in case where the base ring is a discrete field (the faster version was given in [63]). The interested reader can refer to [73] for an introduction to this subject. Our goal is simply to introduce the main lines of the computation of Gröbner bases over coherent valuation rings.

3.3.2 Gröbner Bases Over $\mathbb{Z}/p^\alpha\mathbb{Z}$

Recently Gröbner bases techniques in polynomial rings over $\mathbb{Z}/m\mathbb{Z}$ and $(\mathbb{Z}/p^\alpha\mathbb{Z}) \times (\mathbb{Z}/p^\alpha\mathbb{Z})$ (in particular $\mathbb{Z}/2^\alpha\mathbb{Z}$ and $(\mathbb{Z}/2^\alpha\mathbb{Z}) \times (\mathbb{Z}/2^\alpha\mathbb{Z})$) have attracted some attention due to their potential applications in formal verification of data paths [21, 74, 162], and coding theory [25, 127, 128, 129, 130, 144] (see also the recent Ph.D thesis of Wienand [177]). Also, many authors [6, 54, 69, 136, 154, 169, 178] have been interested in computing Gröbner bases over $\mathbb{Z}/p^\alpha\mathbb{Z}$ (where p is a "lucky" prime number), because modular methods give a satisfactory way to avoid intermediate coefficients swell with Buchberger's algorithm for computing Gröbner bases over the rational numbers.

We will specify the definition of S-polynomials given in Definition 226 to the important case where the base ring is $\mathbb{Z}/p^\alpha\mathbb{Z}$, where p is a prime number and $\alpha \geq 2$. To lighten the notation, the class of $a \in \mathbb{Z}$ modulo $p^\alpha\mathbb{Z}$ will also be denoted by a.

Definition 247. Recall that the valuation of $a \in \mathbb{Z} \setminus \{0\}$ at p is $v_p(a) := \max\{k \in \mathbb{N} \mid p^k \text{ divides } a\}$, so that,

$$a = p^{v_p(a)}a' \text{ with } a' \wedge p = 1.$$

Note that, in $\mathbb{Z}/p^\alpha\mathbb{Z}$, a' is a unit and, writing $ua' + vp^\alpha = 1$ (Bezout identity) for some $u, v \in \mathbb{Z}$, we have $a'^{-1} = u$ in $\mathbb{Z}/p^\alpha\mathbb{Z}$.

Definition 248. (S-Polynomials Over $\mathbb{Z}/p^\alpha\mathbb{Z}$)

Let p be a prime number, $f_i, f_j \in (\mathbb{Z}/p^\alpha\mathbb{Z})[X_1,\ldots,X_n] \setminus \{0\}$ ($i \neq j$), and fix a monomial order $>$ on $(\mathbb{Z}/p^\alpha\mathbb{Z})[X_1,\ldots,X_n]$. Denote by $\mathrm{mdeg}(f_i) = \beta = (\beta_1,\ldots,\beta_n)$, $\mathrm{mdeg}(f_j) = \beta' = (\beta_1',\ldots,\beta_n')$, $\gamma = (\gamma_1,\ldots,\gamma_n)$, where $\gamma_k = \max(\beta_k, \beta_k')$ for each k.

Moreover, denote by $\mathrm{LC}(f_i) = a_i$, $a_i = p^{m_i}c_i$, $\mathrm{LC}(f_j) = a_j$, $a_j = p^{m_j}c_j$ with $0 \leq m_i, m_j \leq \alpha - 1$ and $c_i \wedge p = c_j \wedge p = 1$.

(i) $S(f_i,f_j) := \frac{X^\gamma}{\mathrm{LM}(f_i)}f_i - p^{m_i - m_j}c_i c_j^{-1}\frac{X^\gamma}{\mathrm{LM}(f_j)}f_j$ if $m_j \leq m_i$.

$S(f_i,f_j) := p^{m_j - m_i}c_j c_i^{-1}\frac{X^\gamma}{\mathrm{LM}(f_i)}f_i - \frac{X^\gamma}{\mathrm{LM}(f_j)}f_j$ if $m_j > m_i$.

(ii) $S(f_i,f_i) := p^{\alpha - m_i}f_i$.

Example 249. Let $\mathbf{V}[X] = (\mathbb{Z}/16\mathbb{Z})[X]$ and consider the ideal $I = \langle f_1 \rangle$, where $f_1 = 2 + 4X + 8X^2$.

$S(f_1,f_1) = 2f_1 = 4 + 8X =: f_2$,
$S(f_1,f_2) = 2 =: f_3$,
$S(f_2,f_2) = 2f_2 = 8 \xrightarrow{f_3} 0, \; S(f_3,f_3) = 0$,
$f_2 \xrightarrow{f_3} 0$.

Thus, $\mathcal{G} = \{2\}$ is the reduced Gröbner basis of I.

Example 250. ([25, Example 2.4.6]) Let $\mathbf{V}[X,Y] = (\mathbb{Z}/27\mathbb{Z})[X,Y]$ and consider $\mathcal{G} = \{g_i\}_{i=1}^4$, where $g_1 = 9, g_2 = X + 1, g_3 = 3Y^2, g_4 = Y^3 + 13Y^2 - 12$. Let us fix the lexicographic order as monomial order with $X > Y$.

$S(g_1,g_2) = Xg_1 - 9g_2 = -9 \xrightarrow{g_1} 0$,
$S(g_1,g_3) = Y^2 g_1 - 3g_3 = 0$,
$S(g_1,g_4) = -9Y^2 \xrightarrow{g_1} 0$,
$S(g_2,g_3) = 3Y^2 g_2 - Xg_3 = 3Y^2 \xrightarrow{g_3} 0$,
$S(g_2,g_4) = Y^3 g_2 - Xg_4 = -13XY^2 + 12X + Y^3 \xrightarrow{g_2} 12X + Y^3 + 13Y^2$
$\xrightarrow{g_2} Y^3 + 13Y^2 - 12 \xrightarrow{g_3} 0$,
$S(g_3,g_4) = Yg_3 - 3g_4 = -12Y^3 + 9 \xrightarrow{g_3} 9 \xrightarrow{g_1} 0$.

Thus, $\mathcal{G}$ is a Gröbner basis for $\langle g_1, g_2, g_3, g_4 \rangle$ in $\mathbf{V}[X,Y]$.

Example 251. Let $\mathbf{V}[X,Y] = (\mathbb{Z}/4\mathbb{Z})[X,Y]$ and consider the ideal $I = \langle f_1, f_2, f_3 \rangle$, where $f_1 = X^4 - X, f_2 = Y^3 - 1, f_3 = 2XY$. Let us fix the lexicographic order as monomial order with $X > Y$.

$$S(f_1, f_2) = Y^3 f_1 - X^4 f_2 = X^4 - XY^3 \xrightarrow{f_1} X - XY^3 \xrightarrow{f_2} 0,$$
$$S(f_1, f_3) = 2Y f_1 - X^3 f_3 = -2XY \xrightarrow{f_3} 0,$$
$$S(f_2, f_3) = 2X f_2 - Y^2 f_3 = -2X =: f_4,$$
$$S(f_2, f_4) = 2X f_2 + Y^3 f_4 = -2X \xrightarrow{f_4} 0,$$
$$S(f_1, f_4) = 2f_1 + X^3 f_4 = -2X \xrightarrow{f_4} 0,$$
$$f_3 \xrightarrow{f_4} 0.$$

Thus, $\mathscr{G} = \{f_1, f_2, f_4\}$ is a Gröbner basis for I in $\mathbf{V}[X,Y]$.

Example 252. Take $p = 2$ and $\alpha = 4$. Let $I = \langle f_1 \rangle \subseteq (\mathbb{Z}/16\mathbb{Z})[X,Y]$, where $f_1 = 8X + 2Y + 1$. Let us fix the lexicographic order as monomial order with $X > Y$.

$S(f_1, f_1) = 2f_1 = 4Y + 2 =: f_2$, $S(f_2, f_2) = 4f_2 = 8 =: f_3$, $S(f_2, f_3) = 4 =: f_4$, $S(f_2, f_4) = 2 =: f_5$, $S(f_2, f_5) = 1$. We conclude that $\{1\}$ is the reduced Gröbner of I.

Here, it is worth pointing out that, contrary to the integral case, in the presence of zero-divisors, $\{f\}$ need not be a Gröbner basis of $\langle f \rangle$.

3.3.3 When a Valuation Domain Is Gröbner?

A natural question arising is :

For a coherent valuation ring $\mathbf{R}$, is it always possible to compute a Gröbner basis for each finitely-generated nonzero ideal of $\mathbf{R}[X_1, \ldots, X_n]$ by Buchberger's Algorithm 235 in a finite number of steps ?

In fact, in the integral case, if the totally ordered group corresponding to the valuation is not archimedean, Buchberger's Algorithm 235 does not always work in a finite number of steps as can be seen by the following example.

Example 253. Let $\mathbf{V}$ be a valuation domain with a corresponding valuation v and group G. Suppose that G is not archimedean, that is there exist $a, b \in \mathbf{V}$ such that:

$$v(a) > 0, \text{ and } \forall\, n \in \mathbb{N}^*, \; v(b) > nv(a).$$

Denote by I the ideal of $\mathbf{V}[X]$ generated by $g_1 = aX + 1$ and $g_2 = b$.
Since $S(g_1, g_2) = (\frac{b}{a})g_1 - Xg_2 = \frac{b}{a}$ and $\frac{b}{a}$ is not divisible by b, then one must add $g_3 = \frac{b}{a}$ when executing Buchberger's Algorithm 235.
In the same way, $S(g_1, g_3) = (\frac{b}{a^2})g_1 - Xg_3 = \frac{b}{a^2}$ and $\frac{b}{a^2}$ is not divisible by b nor by $\frac{b}{a}$. Thus, one must add $g_4 = \frac{b}{a^2}$, and so on, we observe that Buchberger's Algorithm 235 does not terminate.

Let us take the particular case where $G = \mathbb{Z} \times \mathbb{Z}$ equipped with the lexicographic order, $a = (0,1)$, and $b = (1,0)$. We can prove $\langle \mathrm{LT}(I) \rangle$ is not finitely-generated despite that I is finitely-generated and that clearly $\langle \mathrm{LC}_n(I) \rangle = \langle a \rangle$ for all $n \in \mathbb{N} \cup \{\infty\}$ (there is no such example in the literature).

Proof. To check this, by way of contradiction, suppose that $\langle \mathrm{LT}(I) \rangle = \langle h_1, \ldots, h_s \rangle$, $h_i \in I \setminus \{0\}$, $s \in \mathbb{N}^*$. We can suppose that $h_1, \ldots, h_s$ are terms, that is $h_i = \mathrm{LT}(h_i)$ for each $1 \le i \le s$. From Lemma 227, it follows that for each $n \in \mathbb{N}$, there exists $i_n \in \{1, \ldots, s\}$ such that h_{i_n} divides $\frac{b}{a^n}$. We infer that there exists $1 \le i_0 \le s$ such that h_{i_0} is constant ($h_{i_0} \in \mathbf{V} \setminus \{0\}$) and such that

$$\forall n \in \mathbb{N}, \ h_{i_0} \text{ divides } \frac{b}{a^n}.$$

That is, $v(h_{i_0}) \le (1, -n) \ \forall n \in \mathbb{N}$. It follows that there exists $k \in \mathbb{N}$ such that $v(h_{i_0}) = (0, k)$ and hence there exists u invertible in $\mathbf{V}$ such that $h_{i_0} = ua^k$. Now,

$$\begin{cases} a^k \in I \\ aX + 1 \in I \end{cases} \Rightarrow \begin{cases} a^k \in I \\ a^{k-1}(aX + 1) \in I \end{cases} \Rightarrow a^{k-1} \in I \Rightarrow \cdots \Rightarrow a \in I \Rightarrow 1 \in I,$$

a contradiction.

$\square$

As a consequence of this example, keeping the notation above, we know that a necessary condition so that Buchberger's Algorithm 235 terminates in the integral case is that the group G is archimedean (this is in fact equivalent to $\dim \mathbf{V} \le 1$, see Exercise 372). Moreover, we already know that a sufficient condition is that $\mathbf{V}$ be Noetherian (see Theorem 236). This encouraged us to set the following conjecture.

Conjecture 254. (The Gröbner Ring Conjecture [77]) *For a valuation domain* $\mathbf{V}$, *the following assertions are equivalent:*

(i) *It is always possible to compute a Gröbner basis for each finitely-generated nonzero ideal of* $\mathbf{V}[X_1, \ldots, X_n]$ *by Buchberger's Algorithm 235, with respect to any monomial order on* $\mathbf{V}[X_1, \ldots, X_n]$, *in a finite number of steps.*

(ii) $\dim \mathbf{V} \le 1$.

The following result gives a solution to the Gröbner Ring Conjecture 254 in the univariate case.

Theorem 255. ([114]) *For a valuation domain* $\mathbf{V}$, *the following assertions are equivalent:*

(1) *For any finitely-generated ideal* I *of* $\mathbf{V}[X]$, *the leading terms ideal* $\mathrm{LT}(I)$ *is also finitely-generated.*

(2) *If* J *is a finitely-generated ideal of* $\mathbf{V}[X]$, *then* $J \cap \mathbf{V}$ *is a principal ideal of* $\mathbf{V}$.

(3) $\dim \mathbf{V} \le 1$.

Proof. "(1) $\Rightarrow$ (2)" Let J be a finitely-generated ideal of $\mathbf{V}[X_1,\ldots,X_m]$. As $\langle \mathrm{LT}(J)\rangle$ is finitely-generated, denote it by $\langle h_1,\ldots,h_s\rangle$ where $h_1,\ldots,h_s$ are terms. We can suppose that $h_1 \in \mathbf{V}$ and $h_2,\ldots,h_s \notin \mathbf{V}$. By virtue of Lemma 227, we infer that $J\cap \mathbf{V} = \langle h_1\rangle$.

" (2) $\Rightarrow$ (3)" Let us denote by v and G respectively the valuation and the valuation group associated with $\mathbf{V}$ and consider $a,b \in \mathrm{Rad}(\mathbf{V})$ (the radical of $\mathbf{V}$). Our goal is to find $n \in \mathbb{N}$ such that $v(b) \leq n v(a)$, or equivalently, such that b divides a^n.
Let us denote by I the ideal of $\mathbf{V}[X]$ generated by $g_1 = aX + 1$ and $g_2 = b$. Because I finitely-generated, $I\cap \mathbf{V}$ is principal, write $I\cap \mathbf{V} = \langle c\rangle$. As $c \in I$, it can be written in the form

$$c = U(X).(aX+1)+V(X).b,$$

with $U(X),V(X) \in \mathbf{V}[X]$. Supposing that $\deg V \leq k$ and evaluating X at $\frac{-1}{a}$, we obtain that $c = V(\frac{-1}{a})b$ and, thus, b divides $c a^k$. This means that $v(b) \leq v(c a^k)$, or equivalently, $v(c) \geq v(\frac{b}{a^k})$.

It is worth pointing out that for any $m \in \mathbb{N}$, if a^m divides b then $\frac{b}{a^m} \in I$ as $S(g_1,g_2) = (\frac{b}{a})g_1 - Xg_2 = \frac{b}{a} =: g_3 \in I,\ldots,g_{m+1} := \frac{b}{a^{m-1}} \in I, g_{m+2} := \frac{b}{a^m} = \frac{b}{a^m}(aX + 1) - Xg_{m+1} \in I$.

If a^k does not divide b, we are done by taking $n = k$; otherwise $v(c) = v(\frac{b}{a^k})$ because $c/\frac{b}{a^k}$ and necessarily $I\cap \mathbf{V} = \{x \in \mathbf{V} \mid v(x) \geq v(\frac{b}{a^k})\}$. Thus, $\frac{b}{a^{k+1}} \notin I$, b divides a^{k+1}, and we are done by taking $n = k+1$.

"(3) $\Rightarrow$ (1)" Let I be a finitely-generated nonzero ideal of $\mathbf{V}[X]$, say $I = \langle f_1,\ldots,f_s\rangle$. Denoting by $\mathbf{K}$ the quotient field of $\mathbf{V}$ and setting $\Delta := \gcd(f_1,\ldots,f_s)$ in $\mathbf{K}[X]$, we have $I = \langle f_1,\ldots,f_s\rangle = \langle \Delta h_1,\ldots,\Delta h_s\rangle$ for some coprime polynomials $h_1,\ldots,h_s \in \mathbf{K}[X]$. Replacing I by αI for an appropriate $\alpha \in \mathbf{V} \setminus \{0\}$, we may suppose that $\Delta, h_1,\ldots,h_s \in \mathbf{V}[X]$. As $\mathbf{V}$ is a valuation domain, there is one coefficient a of one of the h_i's which divides all the others. Thus, one can write $I = a\Delta \langle g_1,\ldots,g_s\rangle$ where $\Delta, g_1,\ldots,g_s \in \mathbf{V}[X]$, $\gcd(g_1,\ldots,g_s) = 1$ in $\mathbf{K}[X]$ and at least one of the g_i's is primitive. In particular, it follows that $\gcd(g_1,\ldots,g_s) = 1$ in $\mathbf{V}[X]$. As $\mathbf{V}\langle X\rangle$ (the localization of $\mathbf{V}[X]$ at monic polynomials) is a Bezout domain (see Theorem 132), the ideal $J = \langle g_1,\ldots,g_s\rangle$ contains a monic polynomial. Since proving that $\mathrm{LT}(I)$ is finitely-generated amounts to proving that $\mathrm{LT}(J)$ is finitely-generated, one may suppose that I contains a monic polynomial. The desired result follows from Theorem 75 (a valuation domain obviously being coherent). $\qquad\square$

Now we pass to the multivariate case. Recall that if a valuation domain $\mathbf{V}$ has Krull dimension ≤ 1 then $\mathbf{V}\langle X\rangle$ is a Bezout domain of Krull dimension ≤ 1 (see Theorem 132).

For any ring $\mathbf{R}$, one can define by induction the ring

$$\mathbf{R}\langle X_1,\ldots,X_n\rangle := (\mathbf{R}\langle X_1,\ldots,X_{n-1}\rangle)\langle X_n\rangle.$$

It is in fact the localization of the multivariate polynomial ring $\mathbf{R}[X_1,\ldots,X_n]$ at the monoid

$$S_n = \{p \in \mathbf{R}[X_1,\ldots,X_n] \mid \mathrm{LC}(p) = 1\},$$

where $\mathrm{LC}(p)$ denotes the leading coefficient of p with respect to the lexicographic monomial order on with $X_1 < X_2 < \cdots < X_n$.
As mentioned above, if $\mathbf{V}$ is a valuation domain with dimension ≤ 1, then $\mathbf{V}\langle X_1,\ldots,X_n\rangle$ is a Bezout domain with Krull dimension ≤ 1.

The following result gives a solution to the Gröbner Ring Conjecture 254 in the lexicographic monomial order case.

Theorem 256. ([185]) *For a valuation domain* $\mathbf{V}$, *fixing the lexicographic order as monomial order, the following assertions are equivalent:*

(1) *For any finitely-generated ideal* I *of* $\mathbf{V}[X_1,\ldots,X_n]$, *the leading terms ideal* $\mathrm{LT}(I)$ *is also finitely-generated.*

(2) *If* J *is a finitely-generated ideal of* $\mathbf{V}[X_1,\ldots,X_n]$, *then* $J \cap \mathbf{V}$ *is a principal ideal of* $\mathbf{V}$.

(3) $\dim \mathbf{V} \leq 1$.

Proof. For proving the implications "$(1) \Rightarrow (2) \Rightarrow (3)$" do as in the proof of Theorem 255.
"$(3) \Rightarrow (1)$" We suppose that $X_1 < X_2 < \cdots < X_n$. We proceed by induction on n. The result is obviously true for $n = 0$. Let I be a finitely-generated nonzero ideal of $\mathbf{V}[X_1,\ldots,X_n]$, say $I = \langle f_1,\ldots,f_s\rangle$. Denoting by $\mathbf{K}$ the quotient field of $\mathbf{V}$ and setting $\Delta := \gcd(f_1,\ldots,f_s)$ in $\mathbf{K}[X_1,\ldots,X_n]$, we have $I = \langle f_1,\ldots,f_s\rangle = \langle \Delta h_1,\ldots,\Delta h_s\rangle$ for some coprime polynomials $h_1,\ldots,h_s \in \mathbf{K}[X_1,\ldots,X_n]$. Replacing I by αI for an appropriate $\alpha \in \mathbf{V} \setminus \{0\}$, we may suppose that $\Delta, h_1,\ldots,h_s \in \mathbf{V}[X_1,\ldots,X_n]$. As $\mathbf{V}$ is a valuation domain, there is one coefficient a of one of the h_i's which divides all the others. Thus, one can write $I = a\Delta\langle g_1,\ldots,g_s\rangle$ where $\Delta, g_1,\ldots,g_s \in \mathbf{V}[X_1,\ldots,X_n]$, $\gcd(g_1,\ldots,g_s) = 1$ in $\mathbf{K}[X_1,\ldots,X_n]$ and at least one of the g_i's is primitive. In particular, it follows that $\gcd(g_1,\ldots,g_s) = 1$ in $\mathbf{V}[X_1,\ldots,X_n]$. As $\mathbf{V}\langle X_1,\ldots,X_n\rangle$ is a Bezout domain, denoting by $J = \langle g_1,\ldots,g_s\rangle$, we infer that

$$J \cap S_n \neq \emptyset.$$

Since proving that $\mathrm{LT}(I)$ is finitely-generated amounts to proving that $\mathrm{LT}(J)$ is finitely-generated, one may suppose that $I \cap S_n \neq \emptyset$. Moreover, by a change of variables "à la Nagata", we can suppose that I contains a monic polynomial at the variable X_n. Note that this change of variables does not "distort" our lexicographic monomial order. To see this, let us consider the case of two variables X and Y with $Y > X$ and denote the change of variables by

$$\varphi : Y \mapsto Y + X^r, \ X \mapsto X, \ \text{with } r \in \mathbb{N}^*.$$

We have for $n, n', m, m' \in \mathbb{N}$,

$$\mathrm{LM}(\varphi(X^n Y^m)) > \mathrm{LM}(\varphi(X^{n'} Y^{m'})) \Leftrightarrow \mathrm{LM}(X^n(Y + X^r)^m) > \mathrm{LM}(X^{n'}(Y + X^r)^{m'})$$

$$\Leftrightarrow X^n(\mathrm{LM}(Y + X^r))^m > X^{n'}(\mathrm{LM}(Y + X^r))^{m'} \Leftrightarrow X^n Y^m > X^{n'} Y^{m'}.$$

From now on, denoting by $\mathbf{A} = \mathbf{V}[X_1, \dots, X_{n-1}]$, the leading terms of polynomials in $\mathbf{V}[X_1, \dots, X_n]$ will be denoted using "LT" when considered as multivariate polynomials at the variables $X_1, \dots, X_n$ and using "L" when considered as univariate polynomials at the variable X_n (i.e., in $\mathbf{A}[X_n]$). By virtue of Theorem 75 and its proof, with $\mathbf{A}$ as above and $X = X_n$, ($\mathbf{A}$ being coherent, see Corollary 369), we have

$$\mathrm{L}(I) = \langle c_1(X_1, \dots, X_{n-1}) X_n^{\alpha_1}, \dots, c_\ell(X_1, \dots, X_{n-1}) X_n^{\alpha_\ell}, X_n^m \rangle,$$

for some $\alpha_1 \le \dots \le \alpha_\ell < m$ in $\mathbb{N}$ and $c_i \in \mathbf{A}$. One can rewrite $\{\alpha_1, \dots \alpha_\ell\} = \{\beta_1, \dots \beta_r\}$ with $\beta_1 < \dots < \beta_r$. For $1 \le j \le r$, we set

$$\mathfrak{I}_j := \mathrm{LT}(\langle c_i \mid \alpha_i \le \beta_j \rangle).$$

Now, for $f \in I$, let us denote by $\mathrm{LT}(f) = u X_1^{\gamma_1} \cdots X_{n-1}^{\gamma_{n-1}} X_n^{\gamma_n}$ and $\mathrm{L}(f) = (\dots + u X_1^{\gamma_1} \cdots X_{n-1}^{\gamma_{n-1}}) X_n^{\gamma_n}$ with $u \in \mathbf{V}$. If $\gamma_n \ge m$, then $\mathrm{LT}(f) \in \langle X_n^m \rangle$. Otherwise, as

$$\mathrm{LT}(f) = \mathrm{LT}(\mathrm{L}(f)),$$

then by writing $\mathrm{L}(f)$ as an element of $\langle c_1 X_n^{\alpha_1}, \dots, c_\ell X_n^{\alpha_\ell}, X_n^m \rangle$ and using Lemma 217, one easily obtains that

$$\mathrm{L}(f) \in \mathfrak{I}_1 \cdot \langle X_n^{\beta_1} \rangle \vee \dots \vee \mathfrak{I}_r \cdot \langle X_n^{\beta_r} \rangle.$$

Thus,

$$\mathrm{LT}(I) = \mathfrak{I}_1 \cdot \langle X_n^{\beta_1} \rangle + \dots + \mathfrak{I}_r \cdot \langle X_n^{\beta_r} \rangle + \langle X_n^m \rangle.$$

By the induction hypothesis, all the $\mathfrak{I}_j$'s are finitely-generated and, thus, so is $\mathrm{LT}(I)$. $\qquad \square$

Corollary 257. *A valuation domain is Gröbner if and only if its Krull dimension is* ≤ 1.

The following question was pointed to us by Henri Lombardi for further extension of Corollary 257 to domains whose divisors group have dimension ≤ 1.

Question 258. *Is a strongly discrete coherent unique factorization domain Gröbner?*

3.3.4 When a Coherent Valuation Ring with Zero-Divisors is Gröbner?

Notation 259. Let a be an element of a ring $\mathbf{R}$. Recall that the annihilator of a in $\mathbf{R}$ is the ideal

$$\text{Ann}(a) := \{x \in \mathbf{R} \mid xa = 0\}.$$

As the sequence $(\text{Ann}(a^n))_{n \in \mathbb{N}}$ is nondecreasing,

$$\text{Ann}(a^\infty) := \cup_{n \in \mathbb{N}} \text{Ann}(a^n)$$

is an ideal of $\mathbf{R}$.

For example, if a is regular then $\text{Ann}(a^\infty) = \{0\}$, and if it is nilpotent $\text{Ann}(a^\infty) = \mathbf{R}$.

Lemma 260. *Let $\mathbf{R}$ be a discrete ring. For any $a \in \mathbf{R}$, we have:*

$$\langle 1 + aX \rangle \cap \mathbf{R} = \text{Ann}(a^\infty) \quad \& \quad \text{LT}(\langle 1 + aX \rangle) = \text{Ann}(a^\infty)[X] + \langle aX \rangle.$$

In particular, $\text{LT}(\langle 1 + aX \rangle)$ is finitely-generated if and only if so is $\text{Ann}(a^\infty)$.

Proof. Letting $c \in \langle 1 + aX \rangle \cap \mathbf{R}$, there exists $g = \sum_{i=0}^{m} b_i X^i \in \mathbf{R}[X]$ such that

$$(1 + aX)g = c \in \mathbf{R}.$$

By identification, we have $ab_m = 0$, $b_m + ab_{m-1} = 0, \ldots, b_1 + ab_0 = 0$, $b_0 = c$, and thus $b_k = (-a)^k c \ \forall \ 0 \le k \le m$ and $a^{m+1}c = 0$.

Conversely, letting $b \in \text{Ann}(a^\infty)$, there exists $n \in \mathbb{N}$ such that $ba^n = 0$. It follows that

$$b(1 + aX)(1 - aX + \cdots + (-a)^{n-1}X^{n-1}) = b(1 - (-a)^n X^n) = b,$$

and, thus, $b \in \langle 1 + aX \rangle \cap \mathbf{R}$. We conclude that $\langle 1 + aX \rangle \cap \mathbf{R} = \text{Ann}(a^\infty)$ and necessarily $\text{Ann}(a^\infty)[X] + \langle aX \rangle \subseteq \text{LT}(\langle 1 + aX \rangle)$.

Letting $f = c_0 + c_1 X + \cdots + c_n X^n \in \langle 1 + aX \rangle$ (we suppose that $n \ge 1$), there exists $g = \sum_{i=0}^{m} b_i X^i \in \mathbf{R}[X]$ $(m + 1 \ge n)$ such that

$$(1 + aX)g = f.$$

By identification, we have

$$S : \begin{cases} ab_m = 0 \\ b_m + ab_{m-1} = 0 \\ \vdots \\ b_{n+1} + ab_n = 0 \\ b_n + ab_{n-1} = c_n \\ \vdots \\ b_1 + ab_0 = c_1 \\ b_0 = c_0, \end{cases}$$

and, thus, $b_n = c_n - ac_{n-1} + \cdots + (-a)^n c_0$ and $a^{m-n+1} b_n = 0$. It follows that $b_n \in$ $\text{Ann}(a^\infty)$ and $c_n \in \text{Ann}(a^\infty) + \langle a \rangle$, as desired.

The final particular affirmation easily follows by adapting the second members in the equalities of S.

$\square$

Proposition 261. *For any discrete ring* **R**, *we have* (i) $\Rightarrow$ (ii) $\Rightarrow$ (iii) *where:*

(i) **R** *is* 1-*Gröbner.*

(ii) *If J is a finitely-generated ideal of* **R**$[X]$, *then $J \cap$* **R** *is a finitely-generated ideal of* **R**.

(iii) *For any $a \in$* **R**, $\text{Ann}(a^\infty)$ *is a finitely-generated ideal of* **R**.

Proof. "(i) $\Rightarrow$ (ii)" Let J be a finitely-generated ideal of **R**$[X]$. As $\langle \text{LT}(J) \rangle$ is finitely-generated, denote it by $\langle h_1, \ldots, h_s \rangle$ where $h_1, \ldots, h_s$ are terms. We can suppose that $h_1, \ldots, h_r \in$ **R** and $h_{r+1}, \ldots, h_s \notin$ **R** where $1 \le r \le s$. It is clear that $J \cap$ **R** $= \langle h_1, \ldots, h_r \rangle$.

"(ii) $\Rightarrow$ (iii)" This follows immediately from Lemma 260.

$\square$

In the following, we give an example of a ring in which assertion (iii) of Proposition 261 fails. Moreover, this example shows also that if **R** is 1-Gröbner, and $\mathfrak{a}$ is an ideal of **R**, then **R**$/\mathfrak{a}$ need not be 1-Gröbner.

Example 262. Take $X_0, X_1, X_2, \ldots$ infinitely many independent indeterminates over a discrete field **K** and consider the ring **R** $:=$ **K**$[X_n : n \ge 0] / \langle X_0^k X_k : k \ge 1 \rangle$. Then, clearly

$$\text{Ann}(\bar{X}_0^\infty) = \langle \bar{X}_k : k \ge 1 \rangle,$$

which is not a finitely-generated ideal of **R**. It follows, by virtue of Proposition 261, that **R** is not 1-Gröbner, though **K**$[X_n : n \ge 0]$ is 1-Gröbner because a finitely-generated ideal of **K**$[X_n : n \ge 0][X]$ involves in its generators only a finite number of indeterminates among the X_i's.

Definition 263. (The Archimedean Property) We say that a valuation ring **V** is *archimedean* if

$$\forall a, b \in \text{Rad}(\mathbf{V}) \setminus \{0\} \ \exists n \in \mathbb{N} \mid a \text{ divides } b^n,$$

where $\text{Rad}(\mathbf{V})$ denotes the radical of **V**.

For valuation domains, the situation is clear (this is folklore):

Proposition 264. *For any valuation domain* **V**, *the following three assertions are equivalent:*

  (i) **V** *is archimedean.*

  (ii) *The valuation group of* **V** *is archimedean.*

  (iii) $\dim \mathbf{V} \leq 1$.

Proof. See Exercise 372.

                                                                          □

For a valuation ring with zero-divisors, the implication "(iii) $\Rightarrow$ (i)" in Proposition 264 is no longer true (see Example 268). The following proposition gives a characterization of archimedean valuation rings by means of Krull dimension.

Proposition 265. ([123]) *Let* **V** *be a valuation ring. Then,* **V** *is archimedean if and only if either* **V** *is a valuation domain of Krull dimension* ≤ 1, *or* **V** *contains a nonzero zero-divisor and is zero-dimensional.*

Proof. A *nonconstructive proof.* Denote by $\mathfrak{m}$ the maximal ideal of **V**. Assume that **V** is archimedean, let $\mathfrak{p}$ be any prime ideal of **V** and fix a nonzero element a of $\mathfrak{p}$. Since, for every $b \in \mathfrak{m}$, there exists n such that a divides b^n, $b \in \mathfrak{p}$, and hence, $\mathfrak{m} = \mathfrak{p}$. Conversely, if $\dim \mathbf{V} = 0$, every element of $\mathfrak{m}$ is nilpotent and **V** is archimedean.

A *constructive proof.* The proof above can be transformed into a constructive one as follows: assume that **V** is archimedean. If **V** is reduced then it is necessarily without zero-divisors (for $a, b, c \in \mathbf{V}$, $ab = 0$ & $b = ac \Rightarrow a^2 c = 0 \Rightarrow (ac)^2 = 0 \Rightarrow ac = 0 \Rightarrow b = 0$) and thus $\dim \mathbf{V} \leq 1$. Otherwise, there exists a nonzero nilpotent element a in **V** and hence, as above, any element in $\mathrm{Rad}(\mathbf{V})$ is nilpotent. Thus, $\dim \mathbf{V} = 0$.

                                                                          □

Remark 266. In order for the disjunction in Proposition 265 to become fully constructive, we have to suppose that we can test the existence of a nonzero nilpotent element in **V**.

Theorem 267. *For any valuation ring* **V**, *we have* (i) $\Rightarrow$ (ii) $\Rightarrow$ (iii) *where:*

  (i) **V** *is* 1-*Gröbner.*

  (ii) *If* J *is a finitely-generated ideal of* $\mathbf{V}[X]$, *then* $J \cap \mathbf{V}$ *is a principal ideal of* **V**.

  (iii) **V** *is archimedean (in particular,* $\dim \mathbf{V} \leq 1$).

Proof. By virtue of Proposition 261, we have only to prove that "(ii) $\Rightarrow$ (iii)". For this, by way of contradiction, suppose that **V** is not archimedean and take $a, b \in \mathrm{Rad}(\mathbf{V}) \setminus \{0\}$ such that b^n divides a for every $n \in \mathbb{N}$. Let us denote by J the ideal of

$\mathbf{V}[X]$ generated by $g_1 = bX + 1$ and $g_2 = a$. Because J is finitely-generated, $J \cap \mathbf{V}$ is principal, write $J \cap \mathbf{V} = \langle c \rangle$. As $c \in J$, it can be written in the form

$$c = U(X).(bX + 1) + V(X).a,$$

with $U(X), V(X) \in \mathbf{V}[X]$. Denoting by $U(X) = \sum_{i=0}^{k-1} u_i X^i$ and $V(X) = \sum_{j=0}^{k} v_j X^j$, with $u_i, v_j \in \mathbf{V}$ and $k \geq \max(1, \deg U + 1, \deg V)$, we have by identification:

$bu_{k-1} + av_k = 0 \Rightarrow bu_{k-1} = a(-v_k),$
$bu_{k-2} + u_{k-1} + v_{k-1}a = 0 \Rightarrow b^2 u_{k-2} = a(v_k - bv_{k-1}),$
$\vdots$
$b^k u_0 = a\gamma$, where $\gamma = \sum_{i=1}^{k} (-1)^{k-i} b^{k-i} v_i.$

Now, $c = u_0 + v_0 a \Rightarrow b^k c = b^k u_0 + b^k v_0 a = a(\gamma + b^k v_0) = ar$ where $r = \gamma + b^k v_0 \in \mathbf{V}$.

On the other hand, let $x \in \mathbf{V}$ be such that $a = xb^{k+1}$. For $1 \leq j \leq k+1$ let $x_j = xb^{k+1-j}$, so that $x_j b^j = a$. We have $x_1 g_1 - X g_2 = x_1 =: g_3 \in J, \dots, g_{k+2} := x_k \in J$, $g_{k+3} := x_{k+1} = x = x_{k+1} g_1 - X g_{k+2} \in J$. Thus, c divides x_{k+1}, and then $c b^{k+1}$ divides a. It follows that $a = c b^{k+1} s = arsb$ (for some $s \in \mathbf{V}$) and thus $(1 - rsb)a = 0$. As $1 - rsb \in \mathbf{V}^\times$, we infer that $a = 0$, a contradiction. $\qquad \square$

The example below, shows that, contrary to the case of valuation domains (Theorem 255), If $\mathbf{V}$ is a one-dimensional valuation ring with zero-divisors, there may exist a finitely-generated ideal J of $\mathbf{V}[X]$ whose leading terms ideal $\mathrm{LT}(J)$ is not finitely-generated.

Example 268. ([123]) Let $\mathbf{T}$ be a rank-two valuation domain explicitly given and take a nonzero element a in the height-one prime ideal of $\mathbf{T}$. For example, one can consider a valuation domain $\mathbf{T}$ whose valuation group is $\mathbb{Z} \times \mathbb{Z}$ equipped with the lexicographic order, and a an element in $\mathbf{T}$ whose valuation is $(0,1)$.
Then $\mathbf{V} := \mathbf{T}/\langle a \rangle$ is a one-dimensional valuation ring which is not archimedean (by virtue of Proposition 265), and hence, is not 1-Gröbner (by virtue of Theorem 267).

3.3.4.1 Buchberger's Algorithm for Strongly Discrete Coherent Archimedean Valuation Rings

Recall that a valuation ring $\mathbf{V}$ is coherent if for any $a \in \mathbf{V}$, $\mathrm{Ann}(a)$ is principal.

Algorithm 269. (Buchberger's Algorithm for Coherent Archimedean Valuation Rings)
Input: $g_1, \dots, g_s \in \mathbf{V}[X_1, \dots, X_n]$, where $\mathbf{V}$ is a coherent archimedean valuation ring
Output: a Gröbner basis G for $\langle g_1, \dots, g_s \rangle$ with respect to the lexicographic monomial order $>$, with $\{g_1, \dots, g_s\} \subseteq G$

$G := \{g_1, \dots, g_s\}$
REPEAT
$G' := G$

For each pair f_i, f_j in G' DO
$$S := \overline{S(f_i, f_j)}^{G'}$$
If $S \neq 0$ THEN $G := G' \cup \{S\}$
UNTIL $G = G'$

Theorem 270. *Algorithm 269 terminates and is correct.*

Proof. The termination proof of Algorithm 269 is included in the proof of Theorem 272. The correctness of Algorithm 269 ensues from Theorem 244. □

The following example shows that an archimedean valuation ring need not be coherent.

Example 271. Let $\mathbf{W}$ be a non-Noetherian valuation domain of Krull dimension 1, denote by $\mathfrak{m}$ its radical, and consider $\alpha \in \mathfrak{m} \setminus \{0\}$. The ring $\mathbf{V} := \mathbf{W}/\alpha\mathfrak{m}$ is a zero-dimensional (local with $\mathfrak{m}/\alpha\mathfrak{m}$ as radical) valuation ring, and hence, archimedean by virtue of Proposition 265. It is clear that in $\mathbf{V}$, $\mathrm{Ann}(\bar{\alpha}) = \mathfrak{m}/\alpha\mathfrak{m}$ which is not principal as $\mathfrak{m}$ is not principal as an ideal of $\mathbf{W}$.

Theorem 272. *Let $\mathbf{V}$ be a valuation ring. Then, $\mathbf{V}$ is Gröbner if and only if $\mathbf{V}$ is both coherent and archimedean, or also, if and only if either $\mathbf{V}$ is a valuation domain of Krull dimension ≤ 1, or $\mathbf{V}$ contains a nonzero zero-divisor, $\mathbf{V}$ is zero-dimensional and the annihilator of any element in $\mathbf{V}$ is finitely-generated.*

Proof. Let I be a finitely-generated ideal of $\mathbf{V}[X_1, \ldots, X_n]$ and fix the lexicographic order as monomial order. A finite basis for $\mathrm{LT}(I)$ can be obtained by executing Buchberger's Algorithm for coherent valuation rings (Algorithm 269). As a matter of fact, on the one hand, the hypothesis "$\mathbf{V}$ is a valuation ring" is needed for the computation of the S-polynomials of the form $S(f_i, f_j)$ with $i \neq j$, while the coherence hypothesis is needed for the computation of the auto-S-polynomials of the form $S(f_i, f_i)$. Thus, the hypothesis "$\mathbf{V}$ is a coherent valuation ring" ensures the correctness of the algorithm. On the other hand, the hypothesis "$\mathbf{V}$ is archimedean" (not all the powers of an element in $\mathrm{Rad}(\mathbf{V}) \setminus \{0\}$ can divide another element in $\mathrm{Rad}(\mathbf{V}) \setminus \{0\}$) ensures its termination (as in the integral case, Theorem 256) because it is the same algorithm, only the computation of the $S(f_i, f_i)$ is added. This latter does not affect the termination of the algorithm as $\mathrm{mdeg}(S(f_i, f_i)) < \mathrm{mdeg}(f_i)$. □

Remark 273. In order for the disjunction in Theorem 272 to become fully constructive, we have to suppose that we can test the existence of a nonzero nilpotent element in $\mathbf{V}$.

3.3.5   Dynamical Gröbner Bases Over Gröbner Arithmetical Rings

The concept of Gröbner basis was originally introduced by Buchberger in his Ph.D. thesis (1965) in order to solve the ideal membership problem for polynomial rings over a field [22]. The ideal membership problem has received considerable attention from the constructive algebra community resulting in algorithms that generalize the work of Buchberger [3, 6, 7, 65, 86]. A dynamical approach to Gröbner bases over PID was first introduced in [181, 183]. Our goal in this subsection is to extend the notion of dynamical Gröbner basis to Gröbner arithmetical rings.

First note that for a Dedekind domain $\mathbf{R}$ with field of fractions $\mathbf{F}$, a necessary condition so that $f \in \langle f_1, \ldots, f_s \rangle$ in $\mathbf{R}[X_1, \ldots, X_n]$ is: $f \in \langle f_1, \ldots, f_s \rangle$ in $\mathbf{F}[X_1, \ldots, X_n]$.

Suppose that this condition is fulfilled, that is, there exists $d \in \mathbf{R} \setminus \{0\}$ such that

$$df \in \langle f_1, \ldots, f_s \rangle \text{ in } \mathbf{R}[X_1, \ldots, X_n] \quad (0).$$

If the base ring $\mathbf{R}$ is a Dedekind domain in which complete prime factorization is feasible, we can write

$$\langle d \rangle = \prod_{i=1}^{\ell} \mathfrak{p}_i^{n_i},$$

where the $\mathfrak{p}_i$'s are nonzero distinct prime ideals of $\mathbf{R}$, and $n_i \in \mathbb{N} \setminus \{0\}$.
Other necessary conditions so that $f \in \langle f_1, \ldots, f_s \rangle$ in $\mathbf{R}[X_1, \ldots, X_n]$ is: $f \in \langle f_1, \ldots, f_s \rangle$ in $\mathbf{R}_{\mathfrak{p}_i}[X_1, \ldots, X_n]$ for each $1 \leq i \leq \ell$. Here the polynomial ring is over the Noetherian valuation domain $\mathbf{R}_{\mathfrak{p}_i}$. Write:

$$d_i f \in \langle f_1, \ldots, f_s \rangle \text{ in } \mathbf{R}[X_1, \ldots, X_n] \text{ for some } d_i \in \mathbf{R} \setminus \mathfrak{p}_i \quad (i).$$

Since no prime of $\mathbf{R}$ contains the ideal $\langle d, d_1, \ldots, d_\ell \rangle$, we obtain that $1 \in \langle d, d_1, \ldots, d_\ell \rangle$, that is, we can find an equality $\alpha d + \alpha_1 d_1 + \cdots + \alpha_\ell d_\ell = 1$, $\alpha, \alpha_i \in \mathbf{R}$. Using this Bezout identity, we can find an equality asserting that $f \in \langle f_1, \ldots, f_s \rangle$ in $\mathbf{R}[X_1, \ldots, X_n]$. Thus, the necessary conditions are sufficient and it suffices to treat the problem in case the base ring is a Noetherian valuation domain.

This method raises the following question:

How to avoid the obstacle of complete prime factorization when it is expensive or infeasible in the considered Dedekind ring?

The fact that the method explained above is based on gluing "local realizability" appeals to the use of dynamical methods, namely, the use of the notion of "dynamical Gröbner basis". Our goal is to mimic dynamically as much as we can the method explained above using a constructive theory of Dedekind rings, or more generally, Gröbner arithmetical rings.

3.3.5.1 How to Construct a Dynamical Gröbner Basis Over a Gröbner Arithmetical Ring?

Definition 274. A dynamical Gröbner basis of an ideal $I = \langle f_1, \ldots, f_s \rangle$ of $\mathbf{R}[X_1, \ldots, X_n]$ is a set $G = \{(G_1, S_1), \ldots, (G_\ell, S_\ell)\}$, where $S_1, \ldots, S_\ell$ are comaximal multiplicative subsets of $\mathbf{R}$, and, for each $1 \le i \le \ell$, G_i is a Gröbner basis for $S_i^{-1}I$ in the ring $(S_i^{-1}\mathbf{R})[X_1, \ldots, X_n]$.

Example 275. (Examples 241 and 252 Continued) Let $\mathbf{V}_1 = \mathbb{Z}_{2\mathbb{Z}} = \{\frac{a}{b} \in \mathbb{Q} \mid (a,b) \in \mathbb{Z} \times \mathbb{Z} \text{ and } b \text{ is odd}\}$, $\mathbf{V}_2 = \mathbb{Z}/16\mathbb{Z} = \mathbb{Z}/2^4\mathbb{Z}$, and $\mathbf{T} = \mathbf{V}_1 \times \mathbf{V}_2$. Note that $\mathbf{T}$ is a Gröbner arithmetical ring by Proposition 283.

Consider the ideal $I = \langle f_1 = (2,1) + (0,2)Y + (0,8)X + (12,0)XY, f_2 = (8,0)Y^2 \rangle \subseteq \mathbf{T}[X,Y]$, and fix the lexicographic order as monomial order with $X > Y$.

Let us denote by $\mathbf{e}_1 = (1,0)$, $\mathbf{e}_2 = (0,1)$, and $\mathbf{1} = (1,1) = \mathbf{e}_1 + \mathbf{e}_2$ the unit of $\mathbf{R}$. We know in advance that the monoids $S_1 = \mathbf{e}_1{}^{\mathbb{N}}$ and $S_2 = \mathbf{e}_2{}^{\mathbb{N}}$ are comaximal with

$$S_1^{-1}\mathbf{T} = \mathbf{T}[\frac{1}{\mathbf{e}_1}] \cong \mathbf{V}_1 = \mathbb{Z}_{2\mathbb{Z}} \text{ and } S_2^{-1}\mathbf{T} = \mathbf{T}[\frac{1}{\mathbf{e}_2}] \cong \mathbf{V}_2 = \mathbb{Z}/16\mathbb{Z}.$$

This can be represented as follows:

$$
\mathbf{T}
$$
$$
\swarrow \qquad \searrow
$$
$$
\mathbf{T}[\tfrac{1}{\mathbf{e}_1}] \cong \mathbf{V}_1 \qquad \mathbf{T}[\tfrac{1}{\mathbf{e}_2}] \cong \mathbf{V}_2
$$

Denoting by π_1 and π_2, the first and second projections, respectively, we have $\pi_1(\mathbf{T}) = \mathbf{V}_1$, $\pi_2(\mathbf{T}) = \mathbf{V}_2$,

$$\pi_1(I) = \langle 2 + 12XY, 8Y^2 \rangle, \text{ and } \pi_2(I) = \langle 1 + 2Y + 8X \rangle.$$

By Examples 241 and 252, we know that $G_1 = \{2\}$ is a Gröbner basis for $\pi_1(I)$ and $G_2 = \{1\}$ is a Gröbner basis for $\pi_2(I)$. So, denoting by $\mathscr{G}_i = \mathbf{e}_iG_i = \{\mathbf{e}_ig : g \in G_i\}$, $G = \{(\mathscr{G}_1, \mathbf{e}_1{}^{\mathbb{N}}), (\mathscr{G}_2, \mathbf{e}_2{}^{\mathbb{N}})\} = \{(\{(2,0)\}, \mathbf{e}_1{}^{\mathbb{N}}), (\{(0,1)\}, \mathbf{e}_2{}^{\mathbb{N}})\}$ is a dynamical Gröbner basis for I.

Let $\mathbf{R}$ be a Gröbner arithmetical ring (for example, a Dedekind ring), $I = \langle f_1, \ldots, f_s \rangle$ a nonzero finitely-generated ideal of $\mathbf{R}[X_1, \ldots, X_n]$, and fix a monomial order $>$. The purpose is to construct a dynamical Gröbner basis G for I.

3.3.5.2 Dynamical Version of Buchberger's Algorithm for Coherent Arithmetical Rings

This algorithm works like Buchberger's Algorithm 269 for coherent valuation rings. The only difference is when it has to handle two incomparable (under division) elements a, b in $\mathbf{R}$. In this situation, one should first compute $u, v, w \in \mathbf{R}$ such that

$$\begin{cases} ub = va \\ wb = (1-u)a. \end{cases}$$

Now, one opens two branches: the computations are pursued in $\mathbf{R}_u$ and $\mathbf{R}_{1-u}$.

- First possibility: the two incomparable elements a and b are encountered when performing the division algorithm (analogous to the division algorithm in the case of a valuation ring). Suppose that one has to divide a term $aX^\alpha = \mathrm{LT}(f)$ by another term $bX^\beta = \mathrm{LT}(g)$ with X^β divides X^α.

  In the ring $\mathbf{R}_{1-u}$: $f = \frac{w}{1-u}\frac{X^\alpha}{X^\beta}g + r$ ($\mathrm{mdeg}(r) < \mathrm{mdeg}(f)$) and the division is pursued with f replaced by r.

  In the ring $\mathbf{R}_u$: $\mathrm{LT}(f)$ is not divisible by $\mathrm{LT}(g)$ and thus $f = \overline{f}^{\{g\}}$.

- Second possibility: the two incomparable elements a and b are encountered when computing $S(f,g)$ with $\mathrm{LT}(f) = aX^\alpha$ and $\mathrm{LT}(g) = bX^\beta$. Denote $\gamma = (\gamma_1,\dots,\gamma_n)$, with $\gamma_i = \max(\alpha_i,\beta_i)$ for each i.

  In the ring $\mathbf{R}_{1-u}$: $S(f,g) = \frac{X^\gamma}{X^\alpha}f - \frac{w}{1-u}\frac{X^\gamma}{X^\beta}g$.

  In the ring $\mathbf{R}_u$: $S(f,g) = \frac{v}{u}\frac{X^\gamma}{X^\alpha}f - \frac{X^\gamma}{X^\beta}g$.

At each new branch, if $S = \overline{S(f,g)}^{G'} \neq 0$ where G' is the current Gröbner basis, then S must be added to G'.

3.3.5.3 The Ideal Membership Problem Over Gröbner Arithmetical Rings

Theorem 276. (Dynamical Gluing)
Let $\mathbf{R}$ be a Gröbner arithmetical ring, $I = \langle f_1,\dots,f_s \rangle$ a nonzero finitely-generated ideal of $\mathbf{R}[X_1,\dots,X_n]$, $f \in \mathbf{R}[X_1,\dots,X_n]$, and fix a monomial order. Suppose that $G = \{(G_1,S_1),\dots,(G_k,S_k)\}$ is a dynamical Gröbner basis for I in $\mathbf{R}[X_1,\dots,X_n]$. Then, $f \in I$ if and only if $\overline{f}^{G_i} = 0$ in $(S_i^{-1}\mathbf{R})[X_1,\dots,X_n]$ for each $1 \leq i \leq k$.

Proof. " $\Rightarrow$ " This follows from the fact that $f \in \langle f_1,\dots,f_s \rangle$ in the ring $(S_i^{-1}\mathbf{R})$ $[X_1,\dots,X_n]$ and from the valuation case.
" $\Leftarrow$ " Since $\overline{f}^{G_i} = 0$, then $f \in \langle f_1,\dots,f_s \rangle$ in $(S_i^{-1}\mathbf{R})[X_1,\dots,X_n]$, for each $1 \leq i \leq k$. This means that for each $1 \leq i \leq k$, there exist $s_i \in S_i$ and $h_{i,1},\dots,h_{i,s} \in \mathbf{R}[X_1,\dots,X_n]$ such that

$$s_if = h_{i,1}f_1 + \dots + h_{i,s}f_s.$$

Using the fact that $S_1,\dots,S_k$ are comaximal, there exist $a_1,\dots,a_k \in \mathbf{R}$ such that $\sum_{i=1}^{k} a_is_i = 1$. It follows that

$$f = (\sum_{i=1}^{k} a_ih_{i,1})f_1 + \dots + (\sum_{i=1}^{k} a_ih_{i,s})f_s \in I.$$

□

3.3.5.4 Locally Gröbner Rings

Definition 277. Let $\mathbf{R}$ ba a strongly discrete coherent ring. We say that $\mathbf{R}$ is *locally Gröbner* if, for every n, fixing the lexicographic order as monomial order on $\mathbf{R}[X_1, \ldots, X_n]$, every finitely-generated ideal of $\mathbf{R}[X_1, \ldots, X_n]$ has a dynamical Gröbner basis.

Within classical mathematics, this amount to saying that $\mathbf{R}_\mathfrak{m}$ is Gröbner for every maximal ideal $\mathfrak{m}$ of $\mathbf{R}$.

Lemma 278. *If $\mathbf{R}$ is a locally Gröbner arithmetical ring the so is $\mathbf{R}[\frac{1}{a}]$ for any $a \in \mathbf{R}$.*

Proof. We have only to prove that for any $b_1, \ldots, b_n \in \mathbf{R}$, the ideal $\langle b_1, \ldots, b_n \rangle : a^\infty$ is finitely generated. This follows from Exercise 387. □

The following result is not fully constructive. To be so, when using the "dynamical evaluation" one has to be able to ensure that one of the two situations mentioned by Corollary 279 applies.

Proposition 279. *Let $\mathbf{R}$ be an arithmetical ring. Then, $\mathbf{R}$ is locally Gröbner if and only if, locally, it is a valuation domain of Krull dimension ≤ 1 or a zero-dimensional coherent valuation ring. In particular, a Gröbner arithmetical ring has Krull dimension ≤ 1.*

Proof. This can be seen via Dynamical Gröbner bases. □

Examples of locally Gröbner arithmetical rings are valuation domains of Krull dimension ≤ 1, zero-dimensional coherent valuation rings, and Dedekind rings (i.e., arithmetical coherent Noetherian rings). Also, a finite product of Gröbner valuation rings is a Gröbner arithmetical ring (see Example 275 and Proposition 283).

Proposition 280. *A Dedekind ring is a locally Gröbner arithmetical ring. It is, locally, a valuation domain of Krull dimension ≤ 1 or a coherent zero-dimensional valuation ring. In particular, it has Krull dimension ≤ 1.*

Proof. Let $\mathbf{R}$ be a Dedekind ring. It is, by definition, an arithmetical coherent Noetherian ring. By virtue of Theorem 236 and using Dynamical Gröbner bases, we infer that $\mathbf{R}$ is locally Gröbner. The remaining desired results follow from Proposition 279. □

It is worth pointing out that Lombardi and Quitté, have proved constructively in [108, Theorem 7.8, page 752], but adopting another approach, that if an arithmetical pp-ring is Noetherian then its Krull dimension is ≤ 1.

Our goal now is to give a series of examples of dynamical Gröbner bases to illustrate our dynamical approach.

3.3.6 A Parallelisable Algorithm for Computing Dynamical Gröbner Bases Over $\mathbb{Z}/m\mathbb{Z}$ via the Chinese Remainder Theorem

The idea of using the Chinese remainder theorem for computing Gröbner bases over $\mathbb{Z}/m\mathbb{Z}$ via the Chinese remainder theorem is already introduced in [129]. Our objective here is to explain how this method can be seen as a particular case of computing dynamical Gröbner bases and to bring some simplifications.

Let $m \in \mathbb{N} \setminus \{0, 1, 2\}$ and suppose that we know the prime factorization $m = p_1^{\alpha_1} \cdots p_\ell^{\alpha_\ell}$ of m, where $\ell, \alpha_i \in \mathbb{N}^*$ and the p_i's are pairwise different prime numbers. The goal of this section is to present a simple way for constructing a dynamical Gröbner basis over the Dedekind ring $\mathbf{R} := \mathbb{Z}/m\mathbb{Z}$ whose leaves (i.e., comaximal localizations) are known in advance by the Chinese remainder theorem.

It is worth pointing out that if the prime factorization of m is not possible then one has to follow the general theory of dynamical Gröbner bases.

By the Chinese remainder theorem, we have the ring isomorphism

$$\mathbb{Z}/m\mathbb{Z} \cong (\mathbb{Z}/p_1^{\alpha_1}\mathbb{Z}) \times (\mathbb{Z}/p_2^{\alpha_2}\mathbb{Z}) \times \cdots \times (\mathbb{Z}/p_\ell^{\alpha_\ell}\mathbb{Z}).$$

So, we can assume that $\mathbf{R} = \prod_{i=1}^{\ell}(\mathbb{Z}/p_i^{\alpha_i}\mathbb{Z})$.

Our objective now is to explain how to construct a dynamical Gröbner basis over $\mathbf{R}$. The advantage of working with the ring $\mathbf{R} = \prod_{i=1}^{\ell}(\mathbb{Z}/p_i^{\alpha_i}\mathbb{Z})$ is that we know in advance that the binary tree we will construct when computing dynamically a Gröbner basis of an ideal of $\mathbf{R}[X_1, \ldots, X_n]$ is formed by only ℓ leaves as follows (we denote by $\mathbf{e_i} = (0, \ldots, 0, 1, 0, \ldots, 0)$ where 1 is on the ith position, and $\mathbf{1} = (1, \ldots, 1) = \mathbf{e_1} + \cdots + \mathbf{e_\ell}$):

$$\mathbf{R}$$
$$\swarrow \quad \cdots \quad \searrow$$
$$\mathbf{R}[\tfrac{1}{\mathbf{e_1}}] \quad \cdots \quad \mathbf{R}[\tfrac{1}{\mathbf{e_\ell}}]$$

Note that, as $\mathbf{R}[\tfrac{1}{\mathbf{e_i}}] \cong \mathbb{Z}/p_i^{\alpha_i}\mathbb{Z}$, in order to compute a dynamical Gröbner basis of an ideal I of $\mathbf{R}[X_1, \ldots, X_n]$, one only has to execute ℓ times (possibly in a parallel way) Buchberger's Algorithm over $\mathbb{Z}/p^\alpha\mathbb{Z}$ (Algorithm 269). Denoting by G_i ($1 \leq i \leq \ell$) the computed Gröbner basis for $\pi_i(I)$, where π_i is the ith canonical projection, and setting $\mathbf{e_i}G_i = \{\mathbf{e_i}g : g \in G_i\}$, $G = \{(\mathbf{e_1}G_1, \mathbf{e_1}^{\mathbb{N}}) \ldots, (\mathbf{e_\ell}G_\ell, \mathbf{e_\ell}^{\mathbb{N}})\}$ is a dynamical Gröbner basis for I.

Example 281. Take $\mathbf{A} = \mathbb{Z}/216\mathbb{Z}$, fix the lexicographic order as monomial order with $X > Y$, and suppose that we want to compute a Gröbner basis for the following ideal of $\mathbf{A}[X,Y]$:

$$\begin{aligned} J &= \langle u_1 = 144, u_2 = X + 162Y - 80, u_3 = 162X^2 + 81, u_4 = -24Y^2, \\ u_5 &= -80Y^3 + 40Y^2 - 120 \rangle. \end{aligned}$$

As $216 = 2^3 \times 3^3$, by the Chinese remainder theorem, we have $\mathbb{Z}/216\mathbb{Z} \overset{\varphi}{\cong} (\mathbb{Z}/8\mathbb{Z}) \times (\mathbb{Z}/27\mathbb{Z})$, where

$$\begin{aligned} \varphi : \mathbb{Z}/216\mathbb{Z} &\to (\mathbb{Z}/8\mathbb{Z}) \times (\mathbb{Z}/27\mathbb{Z}) \\ \bar{x} &\mapsto (\dot{x}, \tilde{x}), \end{aligned}$$

and $\bar{x}, \dot{x}, \tilde{x}$ denote the classes of $x \in \mathbb{Z}$ modulo 216, 8, 27, respectively. Moreover, we have:

$$\varphi^{-1} : (\mathbb{Z}/8\mathbb{Z}) \times (\mathbb{Z}/27\mathbb{Z}) \to \mathbb{Z}/216\mathbb{Z}$$

$$(\dot{x}, \tilde{y}) \mapsto \overline{81x - 80y}.$$

So our problem can be translated into the ring $\mathbf{R} := (\mathbb{Z}/2^3\mathbb{Z}) \times (\mathbb{Z}/3^3\mathbb{Z})$ in which the considered ideal becomes

$$\begin{aligned} I &= \varphi(J) = \langle f_1 = (0,9), f_2 = (1,1)X + (2,0)Y + (0,1), \\ f_3 &= (2,0)X^2 + (1,0), \\ f_4 &= (0,3)Y^2, f_5 = (0,1)Y^3 + (0,13)Y^2 - (0,12) \rangle. \end{aligned}$$

Let us denote by

$$I_1 := \pi_1(I) = \langle g_1 = X + 2Y, g_2 = 2X^2 + 1 \rangle,$$

and

$$I_2 := \pi_2(I) = \langle h_1 = 9, h_2 = X + 1, h_3 = 3Y^2, h_4 = Y^3 + 13Y^2 - 12 \rangle.$$

Using the algorithm given in Sect. 3.3.2, one finds that $\mathscr{G}_1 = \{1\}$ and $\mathscr{G}_2 = \{h_1, h_2, h_3, h_4\}$ are reduced Gröbner bases for I_1 and I_2 respectively, and thus, $\mathscr{G} = \{(\mathbf{e_1}\mathscr{G}_1, \mathbf{e_1}^{\mathbb{N}}), (\mathbf{e_2}\mathscr{G}_2, \mathbf{e_2}^{\mathbb{N}})\}$ is a dynamical Gröbner basis for I in the ring $\mathbf{R}[X,Y]$, where $\mathbf{e_1} = (1,0)$ and $\mathbf{e_2} = (0,1)$. Going back to the ring $\mathbf{A}[X,Y]$, we conclude that $G = \{81, 72, -80X - 80, -24Y^2, \ 80Y^3 + 40Y^2 - 120\}$ is a Gröbner basis for J in the ring $\mathbf{A}[X,Y]$ (and thus so is $\{9, -80X - 80, -24Y^2, -80Y^3 + 40Y^2 - 120\}$).

Remark 282. It is worth pointing out that if the prime factorization of m is not possible then instead of using the Chinese remainder theorem one has to follow the general theory of dynamical Gröbner bases: do as if $\mathbb{Z}/m\mathbb{Z}$ were a valuation domain, or also, as if m was a power of a prime number. When, during the computations, one meets two integers s and r $(2 \le s, r < m)$ which are not comparable under division modulo m, then compute $d = s \wedge r$, write $s = d\, s'$, $r = d\, r'$ where $s' \wedge r' = 1$ and s', r' are not invertible modulo m, or also, $s' \wedge m \ne 1$ and $r' \wedge m \ne 1$. Now, the ring $\mathbb{Z}/m\mathbb{Z}$ has to be replaced by the rings $\mathbf{A}_1 = (\mathbb{Z}/m\mathbb{Z})[\frac{1}{s' \wedge m}]$ and $\mathbf{A}_2 = (\mathbb{Z}/m\mathbb{Z})[\frac{1}{r' \wedge m}]$. We have:

- $(s' \wedge m) \wedge (r' \wedge m) = 1$ as $s' \wedge r' = 1$,

- s divides r in $\mathbf{A}_1$, and

- r divides s i n $\mathbf{A}_2$.

In fact, this can be rephrased as follows: we found two coprime factors $s' \wedge m$ and $r' \wedge m$ of m which can be used to partially factorize m and then to write $\mathbb{Z}/m\mathbb{Z}$ as a product of simpler rings by the Chinese remainder theorem.

3.3.7   A Parallelisable Algorithm for Computing Gröbner Bases Over $(\mathbb{Z}/p^{\alpha}\mathbb{Z}) \times (\mathbb{Z}/p^{\alpha}\mathbb{Z})$

Many attacks showed that cryptosystems based on Gröbner bases over a field are not secure. The analysis of all known attacks, like for example the linear algebra attack, showed that they use in some step, the solution of a linear system on the underlying field. Hence one solution to avoid such attack (proposed in [144]) is to work with a ring over which linear systems are difficult to solve. Precisely, over a Dedekind ring with many zero-divisors. For this objective, the ring $(\mathbb{Z}/p^{\alpha}\mathbb{Z}) \times (\mathbb{Z}/p^{\alpha}\mathbb{Z})$ (where p is a prime number) may be interesting as, in this ring, the probability that an element is a zero-divisor is equal to $\frac{2p-1}{p^2}$ $(= \frac{3}{4}$ if $p = 2$, see Exercise 386).

The following proposition shows that $(\mathbb{Z}/p^{\alpha}\mathbb{Z}) \times (\mathbb{Z}/p^{\alpha}\mathbb{Z})$ is a Dedekind ring. Moreover, we explicitly code $(\mathbb{Z}/p^{\alpha}\mathbb{Z}) \times (\mathbb{Z}/p^{\alpha}\mathbb{Z})$ as the ring $\mathbb{Z}[t]$ modulo an ideal of $\mathbb{Z}[t]$.

Proposition 283.   (1) *If* $\mathbf{R}$ *and* $\mathbf{T}$ *are Gröbner valuation rings (resp., coherent Noetherian valuation rings), then* $\mathbf{R} \times \mathbf{T}$ *is a Gröbner arithmetical ring (resp., a Dedekind ring).*

(2) $\mathbb{Z}[t]/\langle p^{\alpha}, t^2 - t \rangle \overset{\varphi}{\cong} (\mathbb{Z}/p^{\alpha}\mathbb{Z}) \times (\mathbb{Z}/p^{\alpha}\mathbb{Z})$.

Proof.   (1) Denoting by $\mathbf{e} = (1, 0)$ and $\mathbf{1} = (1, 1)$, we have

$$(\mathbf{R} \times \mathbf{T})[\frac{1}{\mathbf{e}}] \cong (\mathbf{R} \times \mathbf{T})/\langle \mathbf{1} - \mathbf{e} \rangle \cong \mathbf{R} \ \& \ (\mathbf{R} \times \mathbf{T})[\frac{1}{\mathbf{1} - \mathbf{e}}] \cong (\mathbf{R} \times \mathbf{T})/\langle \mathbf{e} \rangle \cong \mathbf{T}.$$

(2) This is very classical, take $\varphi(\bar{f}) = (\overline{f(0)}, \overline{f(1)})$ for $f \in \mathbb{Z}[t]$ (resp., for $c \in \mathbb{Z}$), $\bar{g}$ (resp., for $\bar{c}$) denotes the class of g modulo $\langle p^{\alpha}, t^2 - t \rangle$ (resp., modulo $\langle p^{\alpha} \rangle$).

It is worth pointing out that denoting by $f = \sum_{i=0}^{m} a_i t^i \in \mathbb{Z}[t]$, we

$$\bar{f} = \bar{a}_0 + (\sum_{i=1}^{m} \bar{a}_i)t, \text{ and}$$

$$\varphi(\bar{f}) = (\bar{a}_0, \sum_{i=0}^{m} \bar{a}_i), \quad \varphi^{-1}(\bar{a}, \bar{b}) = \bar{a} + (\bar{b} - \bar{a})t.$$

$\square$

3.3.7.1 How to Compute a Reduced Dynamical Gröbner Basis Over $(\mathbb{Z}/p^{\alpha}\mathbb{Z}) \times (\mathbb{Z}/p^{\alpha}\mathbb{Z})[X_1, \ldots, X_n]$ and $(\mathbb{Z}[t]/\langle p^{\alpha}, t^2 - t \rangle)[X_1, \ldots, X_n]$

Our objective now is to explain how to construct a dynamical Gröbner bases over $(\mathbb{Z}/p^{\alpha}\mathbb{Z}) \times (\mathbb{Z}/p^{\alpha}\mathbb{Z})$. The advantage of working with the ring $(\mathbb{Z}/p^{\alpha}\mathbb{Z}) \times (\mathbb{Z}/p^{\alpha}\mathbb{Z})$ is that we know in advance that the binary tree we will construct when computing dynamically a Gröbner basis of an ideal of $(\mathbb{Z}/p^{\alpha}\mathbb{Z}) \times (\mathbb{Z}/p^{\alpha}\mathbb{Z})[X_1, \ldots, X_n]$ is formed by only two leaves as follows (we denote by $\mathbf{V} = (\mathbb{Z}/p^{\alpha}\mathbb{Z}) \times (\mathbb{Z}/p^{\alpha}\mathbb{Z})$, $\mathbf{e} = (1,0)$ and $\mathbf{1} = (1,1)$):

$$\mathbf{V}$$
$$\mathbf{V}[\tfrac{1}{\mathbf{e}}] \qquad \mathbf{V}[\tfrac{1}{1-\mathbf{e}}]$$

Note that, as $\mathbf{V}[\frac{1}{\mathbf{e}}] \cong \mathbf{V}[\frac{1}{1-\mathbf{e}}] \cong \mathbb{Z}/p^{\alpha}\mathbb{Z}$, in order to compute a dynamical Gröbner basis of an ideal I of $\mathbf{V}[X_1, \ldots, X_n]$, one only has to execute twice Buchberger's Algorithm 269 over $\mathbb{Z}/p^{\alpha}\mathbb{Z}$ (possibly in a parallel way). The first one is with $\pi_1(I)$ and the second one is with $\pi_2(I)$, where π_1 and π_2 are the first and second canonical projection of $(\mathbb{Z}/p^{\alpha}\mathbb{Z}) \times (\mathbb{Z}/p^{\alpha}\mathbb{Z})$ over $\mathbb{Z}/p^{\alpha}\mathbb{Z}$. Moreover, if at each ith leave $(i = 1, 2)$, the computed Gröbner basis is denoted by G_i, then the dynamical Gröbner basis is $G = \{(\mathscr{G}_1, \mathbf{e}^{\mathbb{N}}), (\mathscr{G}_2, (1 - \mathbf{e})^{\mathbb{N}})\}$ and is also reduced, where $\mathscr{G}_1 = \mathbf{e} G_1 = \{\mathbf{e}g : g \in G_1\}$, and $\mathscr{G}_2 = (1 - \mathbf{e}) G_2 = \{(1 - \mathbf{e})g : g \in G_2\}$.

Note that if J is a finitely-generated ideal of $(\mathbb{Z}[t]/\langle p^{\alpha}, t^2 - t \rangle)[X_1, \ldots, X_n]$, denoting by G_i the reduced Gröbner basis for $\pi_i(\phi(J))$ $(i = 1, 2)$, the reduced dynamical Gröbner basis for J in $(\mathbb{Z}[t]/\langle p^{\alpha}, t^2 - t \rangle)[X_1, \ldots, X_n]$ is defined as

$$G = \{(G_1, (1 - t)^{\mathbb{N}}), (G_2, t^{\mathbb{N}})\}.$$

Here ϕ stands for the extension of φ to $(\mathbb{Z}[t]/\langle p^{\alpha}, t^2 - t \rangle)[X_1, \ldots, X_n]$ by setting $\phi(X_j) = X_j$. Of course, as mentioned above, G_1 and G_2 can be computed in a parallel way.

Moreover, for $h \in (\mathbb{Z}[t]/\langle p^{\alpha}, t^2 - t \rangle)[X_1, \ldots, X_n]$, its unique remainder on division by G is

$$\overline{h}^G = \phi^{-1}(\overline{\phi(h)}^{G_1}, \overline{\phi(h)}^{G_2}),$$

and we have:

$$h \in J \Leftrightarrow \overline{h}^G = 0 \Leftrightarrow \overline{\phi(h)}^{G_1} = \overline{\phi(h)}^{G_2} = 0.$$

Example 284. Take $p = 2$ and $\alpha = 3$. Let

$$J = \langle P_1 = (2-t)X + (1+t)Y + 1 - t, P_2 = (1+t)X^2 + 1 \rangle \subseteq (\mathbb{Z}[t]/\langle 8, t^2 - t \rangle)[X,Y].$$

If coded in the ring $(\mathbb{Z}/8\mathbb{Z}) \times (\mathbb{Z}/8\mathbb{Z})[X,Y]$, J becomes

$$I = \langle f_1 = (2,1)X + (1,2)Y + (1,0), f_2 = (1,2)X^2 + (1,1) \rangle.$$

Computing a Reduced Dynamical Gröbner basis

Let us denote by

$$I_1 := \pi_1(I) = \langle g_1 = 2X + Y + 1, g_2 = X^2 + 1 \rangle,$$

and

$$I_2 := \pi_2(I) = \langle h_1 = X + 2Y, h_2 = 2X^2 + 1 \rangle.$$

Let us fix the lexicographic order as monomial order with $X > Y$. We will give all the details of the dynamical computation:

$S(g_1, g_1) = 4g_1 = 4Y + 4 =: g_3, S(g_3, g_3) = 2g_3 = 0, S(g_2, g_2) = 0,$

$S(g_1, g_2) = Xg_1 - 2g_2 = XY + X - 2 := g_4, S(g_4, g_4) = 0,$

$S(g_1, g_3) = 2Yg_1 - Xg_3 = 2Y^2 + 2Y - 4X \xrightarrow{g_1} 2Y^2 - 2 =: g_5, S(g_5, g_5) = 0,$

$S(g_1, g_5) = Y^2 g_1 - Xg_5 = 2X + Y^3 + Y^2 \xrightarrow{g_1} Y^3 + Y^2 - Y - 1 =: g_6, S(g_6, g_6) = 0,$

$S(g_1, g_6) = Y^3 g_1 - 2Xg_6 = -2XY^2 + 2XY + 2X + Y^4 + Y^3 \xrightarrow{g_1} Y^4 + 2Y^3 - 2Y - 1 \xrightarrow{g_1} 0, S(g_2, g_3) = 4Yg_2 - X^2 g_3 = -4X^2 + 4Y \xrightarrow{g_2} 4Y + 4 \xrightarrow{g_2} 0,$

$S(g_2, g_4) = Yg_2 - Xg_4 = -X^2 + 2X + Y \xrightarrow{g_2} 2X + Y + 1 \xrightarrow{g_1} 0,$

$S(g_2, g_5) = 2Y^2 g_2 - X^2 g_5 = 2X^2 + 2Y^2 \xrightarrow{g_2} 2Y^2 - 1 \xrightarrow{g_5} 0,$

$S(g_2, g_6) = Y^3 g_2 - X^2 g_6 = -X^2 Y^2 + X^2 Y + X^2 + Y^3 \xrightarrow{g_1} Y^3 + Y^2 - Y - 1 \xrightarrow{g_6} 0,$

$S(g_3, g_4) = Xg_3 - 4g_4 = 0, S(g_3, g_5) = Yg_3 - 2g_5 = 4Y + 4 \xrightarrow{g_3} 0,$

$S(g_3, g_6) = Y^2 g_3 - 4g_6 = 4Y + 4 \xrightarrow{g_3} 0,$

$S(g_4, g_5) = 2Yg_4 - Xg_5 = 2XY + 2X - 4Y \xrightarrow{g_1} -Y^2 - 6Y - 1 \longrightarrow Y^2 + 6Y + 1 =: g_7,$

$S(g_7, g_7) = 0, g_5, g_6 \xrightarrow{g_7, g_3} 0, S(g_4, g_7) = Yg_4 - Xg_7 = -5XY - X - 2Y \xrightarrow{g_4, g_1, g_3} 0.$

Thus, $\mathcal{G}_1 = \{2X + Y + 1, X^2 + 1, 4Y + 4, XY + X - 2, Y^2 + 6Y + 1\}$ is a reduced Gröbner for I_1.

$S(h_1, h_1) = 0, S(h_2, h_2) = 4h_1 = 4 =: h_3, S(h_3, h_3) = 0, S(h_1, h_2) = 2Xh_1 - h_2 = 4XY - 1 \xrightarrow{h_3} -1 \longrightarrow 1.$

Thus, $\mathcal{G}_2 = \{1\}$ is a reduced Gröbner for I_2.

As a conclusion, a reduced dynamical Gröbner basis for I in the ring $(\mathbb{Z}/8\mathbb{Z}) \times (\mathbb{Z}/8\mathbb{Z})[X,Y]$ is

$$\mathcal{G} = \{((\{(2,0)X + (1,0)Y + (1,0), (1,0)X^2 + (1,0), (4,0)Y + (4,0), (1,0)XY$$
$$+ (1,0)X - (2,0), (1,0)Y^2 + (6,0)Y + (1,0)\}, e^{\mathbb{N}}), (\{(0,1)\}, (1-e)^{\mathbb{N}})\}.$$

In the ring $(\mathbb{Z}[t]/\langle 8, t^2 - t\rangle)[X,Y]$, a reduced dynamical Gröbner basis for J is

$$G=\{(\{2X+Y+1, X^2+1, 4Y+4, XY+X-2, Y^2+6Y+1\}, (1-t)^{\mathbb{N}}), (\{1\}, t^{\mathbb{N}})\}.$$

Denoting by $\mathbf{W} = \mathbb{Z}[t]/\langle 8, t^2 - t\rangle$, this dynamical Gröbner basis corresponds to the binary tree:

$$\mathbf{W}$$
$$\mathbf{W}[\tfrac{1}{1-t}] \qquad \mathbf{W}[\tfrac{1}{t}]$$

Answering the Ideal Membership Problem

$$h = (1+t)X^2 + (4+3t)Y + 5 - 3t \in ?J.$$

If coded in the ring $(\mathbb{Z}/8\mathbb{Z}) \times (\mathbb{Z}/8\mathbb{Z})[X,Y]$, this problem becomes

$$
\begin{aligned}
u &= (u_1, u_2) = (1,2)X^2 + (4,7)Y + (5,2) \in ?I, \text{ or also}\\
u_1 &= X^2 + 4Y + 5 \in ?I_1 \ \& \ u_2 = 2X^2 + 7Y + 2 \in ?I_2.
\end{aligned}
$$

We have $u_1 \in I_1$ as $\overline{u_1}^{\mathscr{G}_1} = 0$. More precisely,

$$
\begin{aligned}
u_1 &= g_2 + g_3 = 4g_1 + g_2, \text{ or also}\\
(1,0)\,u &= (4,0)f_1 + (1,0)f_2. \tag{3.3}
\end{aligned}
$$

On the other hand, of course $u_2 \in I_2 = \langle 1\rangle$, and we have

$$
\begin{aligned}
u_2 &= -2X(2X^2+7Y+2)h_1 + (2X^2+7Y+2)h_2, \text{ or also}\\
(0,1)\,u &= -2X(2X^2+7Y+2)(0,1)f_1 + (2X^2+7Y+2)(0,1)f_2. \tag{3.4}
\end{aligned}
$$

$(3.3)+(3.4) \Rightarrow u = (4, -2X(2X^2+7Y+2))f_1 + (1, 2X^2+7Y+2)f_2 \in I$, and thus,

$$h = (4 - (4X^3 + 14XY + 4X - 4)t)P_1 + (1 + (2X^2 + 7Y + 1)t)P_2 \in J.$$

3.3.8 Dynamical Gröbner Bases Over $\mathbb{F}_2[a,b]/\langle a^2-a, b^2-b\rangle$

One main feature of the use of dynamical Gröbner bases is that it enables to easily resolve the delicate problem caused by zero-divisors appearing as leading coefficients. Cai and Kapur concluded their paper [28] by mentioning the open question of how to generalize Buchberger's algorithm for Boolean rings (see also [85] where Boolean rings are used to model prepositional calculus). As a typical example of a problematical situation, they studied the case where the base ring is $\mathbf{A} = \mathbb{F}_2[a,b]\langle a^2 - a, b^2 - b\rangle$. In that case, the method they proposed in [28] does not work due to the fact that an annihilator of $ab+a+b+1 \in \mathbf{A}$ can be either a or b and thus there may exist noncomparable multiannihilators for an element in $\mathbf{A}$. Dynamical Gröbner bases allow to fairly overcome this difficulty. As a matter of fact, in this precise case, as will be explained below, we will be led to compute a dynamical Gröbner bases made up of four Gröbner bases on localizations of $\mathbf{A}$. We will see that at each leaf of the constructed binary tree, the problem Cai and Kapur pointed disappears completely. Thus, by systemizing the dynamical construction above, it is straightforward that dynamical Gröbner bases over Dedekind rings could be a satisfactory solution to this open problem.

It is folklore that the presence of a nontrivial idempotent $\mathbf{e}$ in a ring $\mathbf{B}$ (i.e., $\mathbf{e}^2 = \mathbf{e}$ and $\mathbf{e} \notin \{0, 1\}$) splits $\mathbf{B}$ into two subrings as follows:
$$\mathbf{B} = \mathbf{e}\mathbf{B} + (1-\mathbf{e})\mathbf{B} \cong \mathbf{e}\mathbf{B} \times (1-\mathbf{e})\mathbf{B} \cong \mathbf{B}/\langle 1-\mathbf{e}\rangle \times \mathbf{B}/\langle \mathbf{e}\rangle \cong \mathbf{B}[\frac{1}{\mathbf{e}}] \times \mathbf{B}[\frac{1}{1-\mathbf{e}}].$$

Let us consider the ring $\mathbf{A} = \mathbb{F}_2[a,b]\langle a^2 - a, b^2 - b\rangle = \mathbb{F}_2 + \mathbb{F}_2 a + \mathbb{F}_2 b + \mathbb{F}_2 ab$ with the relations $a^2 = a$ and $b^2 = b$. When working with the ring $\mathbf{A}$, we know in advance that the binary tree we will construct when computing dynamically a Gröbner basis of an ideal of $\mathbf{A}[X_1, \ldots, X_n]$ is formed by only four leaves as follows:

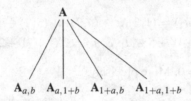

where

$$\mathbf{A}_{a,b} \quad := \quad \mathbf{A}[\frac{1}{a}, \frac{1}{b}] \cong \mathbf{A} \text{ with } \begin{cases} a = 1 \\ b = 1 \end{cases} \cong \mathbb{F}_2,$$

$$\mathbf{A}_{a,1+b} \quad := \quad \mathbf{A}[\frac{1}{a}, \frac{1}{1+b}] \cong \mathbf{A} \text{ with } \begin{cases} a = 1 \\ b = 0 \end{cases} \cong \mathbb{F}_2,$$

$$\mathbf{A}_{1+a,b} \quad := \quad \mathbf{A}[\frac{1}{1+a}, \frac{1}{b}] \cong \mathbf{A} \text{ with } \begin{cases} a = 0 \\ b = 1 \end{cases} \cong \mathbb{F}_2,$$

$$\mathbf{A}_{1+a,1+b} \quad := \quad \mathbf{A}[\frac{1}{1+a}, \frac{1}{1+b}] \cong \mathbf{A} \text{ with } \begin{cases} a = 0 \\ b = 0 \end{cases} \cong \mathbb{F}_2.$$

So, computing a dynamical Gröbner basis over $\mathbf{A}$ amounts to computing four (classical) Gröbner bases over $\mathbb{F}_2$ (possibly, in a parallel way). Moreover, we can define a "reduced dynamical Gröbner basis" as a set $\{(G_1, a^{\mathbb{N}} b^{\mathbb{N}}), (G_2, a^{\mathbb{N}}(1 + b)^{\mathbb{N}}), (G_3, (1+a)^{\mathbb{N}} b^{\mathbb{N}}), (G_4, (1+a)^{\mathbb{N}}(1+b)^{\mathbb{N}})\}$ where each G_i is a reduced Gröbner basis over $\mathbb{F}_2$.

Example 285. Let us consider the ideal $I = \langle f = 1 + (1 + a + b + ab)X\rangle$ of $\mathbf{A}[X]$ where $\mathbf{A} = \mathbb{F}_2[a,b]$ with $a^2 = a$ and $b^2 = b$.
The reduced dynamical Gröbner basis of I is

$$\{(\{1\}, a^{\mathbb{N}} b^{\mathbb{N}}), (\{1\}, a^{\mathbb{N}}(1+b)^{\mathbb{N}}), (\{1\}, (1+a)^{\mathbb{N}} b^{\mathbb{N}}), (\{1+X\}, (1+a)^{\mathbb{N}}(1+b)^{\mathbb{N}})\}.$$

Now, suppose that we want to answer to the ideal membership problem

$$g = 1 + (a + b + ab)X + X^2 \in ? I.$$

Obviously, over the first three localizations, the answer is yes. On the other hand, taking $a = b = 0$, g becomes $1 + X^2 = (1 + X)^2 \in \langle 1 + X \rangle$, and thus the global answer to the ideal membership problem is positive. Moreover, as the relation $g = (1 + X)f$ holds at the four localizations (i.e., by successively taking $a = 1$, $b = 1$; $a = 1$, $b = 0$; $a = 0$, $b = 1$; $a = 0$, $b = 0$), it holds globally, i.e., over $\mathbf{A}$.

We will propose, in the following three subsections, three methods for computing Gröbner bases over the integers.

3.3.9 Dynamical Gröbner Bases Over the Integers

We propose in this subsection to explain how to compute dynamical Gröbner bases over the integers ($\mathbb{Z}$ can be replaced by any Bezout domain with Krull dimension ≤ 1) following the general theory of dynamical Gröbner bases. We start as if $\mathbb{Z}$ were a valuation domain. Suppose that two incomparable (under division) elements a, b in $\mathbb{Z}$ appear as leading coefficients, of f and g, respectively, when computing an $S(f, g)$. A key fact is that writing $a = (a \wedge b)a'$, $b = (a \wedge b)b'$, with $a' \wedge b' = 1$, then a divides b in $\mathbb{Z}[\frac{1}{a'}]$, b divides a in $\mathbb{Z}[\frac{1}{b'}]$, and the two multiplicative subsets $a'^{\mathbb{N}}$ and $b'^{\mathbb{N}}$ are comaximal as $1 \in \langle a', b' \rangle$. Then the ring $\mathbb{Z}$ splits into the rings $\mathbb{Z}[\frac{1}{a'}]$ and $\mathbb{Z}[\frac{1}{b'}]$:

$$
\begin{array}{ccc}
 & \mathbb{Z} & \\
\swarrow & & \searrow \\
\mathbb{Z}[\frac{1}{a'}] & & \mathbb{Z}[\frac{1}{b'}]
\end{array}
$$

and one continues as if the rings $\mathbb{Z}[\frac{1}{a'}]$ and $\mathbb{Z}[\frac{1}{b'}]$ were valuation domains. Denoting by $\mathrm{mdeg}(f) = \alpha$, $\mathrm{mdeg}(g) = \beta$, and $\gamma = (\gamma_1, \ldots, \gamma_n)$, where $\gamma_i = \max(\alpha_i, \beta_i)$ for each i, $S(f, g)$ is computed as follows:

In the ring $\mathbb{Z}[\frac{1}{b'}]$: $S(f, g) = \frac{X^\gamma}{X^\alpha}f - \frac{a'}{b'}\frac{X^\gamma}{X^\beta}g =: S_1$.

In the ring $\mathbb{Z}[\frac{1}{a'}]$: $S(f, g) = \frac{b'}{a'}\frac{X^\gamma}{X^\alpha}f - \frac{X^\gamma}{X^\beta}g =: S_2$.

For $a_1, \ldots, a_n \in \mathbb{Z}$, we denote by

$$\mathscr{M}(a_1, \ldots, a_n) := a_1^{\mathbb{N}} \cdots a_n^{\mathbb{N}}$$

the monoid generated by $a_1, \ldots, a_n$. The localization of $\mathbb{Z}$ at $\mathscr{M}(a_1, \ldots, a_n)$ will be denoted by

$$\mathbb{Z}_{a_1.a_2.\ldots.a_n} := \mathscr{M}(a_1, \ldots, a_n)^{-1}\mathbb{Z} = \mathbb{Z}[\frac{1}{a_1 \cdots a_n}].$$

Example 286. a) Suppose that we want to construct a dynamical Gröbner basis for $I = \langle f_1 = 10XY + 1, f_2 = 6X^2 + 3 \rangle$ in $\mathbb{Z}[X, Y]$.

Let fix the lexicographic order as monomial order with $X > Y$. We will give all the details of the computations only for one leaf.

Since $10 \wedge 6 = 2$, $10 = 2 \times 5$, and $6 = 2 \times 3$, one has to open two branches:

$$
\begin{array}{ccc}
 & \mathbb{Z} & \\
\swarrow & & \searrow \\
\mathbb{Z}_5 & & \mathbb{Z}_3
\end{array}
$$

In $\mathbb{Z}_5$:

$S(f_1,f_2) = \frac{3}{5}Xf_1 - Yf_2 = \frac{3}{5}X - 3Y := f_3$. But, when computing $S(f_1,f_3)$ we see that the leading coefficients of f_1 and f_3 are not comparable under division. Since $10 \wedge \frac{3}{5} = 2 \wedge 3 = 1$, one has to open two new branches:

$$
\begin{array}{ccc}
 & \mathbb{Z}_5 & \\
\swarrow & & \searrow \\
\mathbb{Z}_{5.2} & & \mathbb{Z}_{5.3}
\end{array}
$$

In $\mathbb{Z}_{5.2}$:

$$
\begin{aligned}
S(f_1,f_3) &= \frac{\frac{3}{5}}{10}f_1 - Yf_3 = 3Y^2 + \frac{3}{50} := f_4. \\
S(f_1,f_4) &= \frac{3}{10}Yf_1 - Xf_4 = -\frac{3}{50}X + \frac{3}{10}Y \\
&= -\frac{1}{10}f_3 \xrightarrow{f_3} 0 (\text{reduction modulo} f_3). \\
S(f_2,f_3) &= f_2 - \frac{6 \times 5}{3}Xf_3 = 30XY + 3 = 3f_1 \xrightarrow{f_1} 0. \\
S(f_2,f_4) &= Y^2 f_2 - 2X^2 f_4 = -\frac{3}{25}X^2 + 3Y^2 \xrightarrow{f_2} f_4 \xrightarrow{f_4} 0. \\
S(f_3,f_4) &= Y^2 f_3 - \frac{1}{5}Xf_4 = -\frac{3}{250}X - 3Y^3 \xrightarrow{f_3} -Yf_4 \xrightarrow{f_4} 0.
\end{aligned}
$$

Thus, $G_1 = \{10XY + 1, 6X^2 + 3, \frac{3}{5}X - 3Y, 3Y^2 + \frac{3}{50}\}$ is a Gröbner basis for $\langle 10XY + 1, 6X^2 + 3 \rangle$ at the leaf $\mathcal{M}(5,2)^{-1}\mathbb{Z} = \mathbb{Z}_{5.2}$.

At the leaf $\mathbb{Z}_{5.3}$, we find $G_2 = \{10XY + 1, 6X^2 + 3, \frac{3}{5}X - 3Y, 2Y^2 + \frac{1}{25}, -\frac{3}{25}X^2 + 3Y^2\}$ as a Gröbner basis for $\langle 10XY + 1, 6X^2 + 3 \rangle$.

Let's handle the right subtree:

$$
\begin{array}{ccc}
 & \mathbb{Z}_3 & \\
\swarrow & & \searrow \\
\mathbb{Z}_{3.2} & & \mathbb{Z}_{3.5}
\end{array}
$$

At the leaf $\mathbb{Z}_{3.2}$, we find $G_3 = \{10XY + 1, 6X^2 + 3, X - 5Y, 50Y^2 + 1, 25Y^2 + \frac{1}{2}\}$ as a Gröbner basis for $\langle 10XY + 1, 6X^2 + 3 \rangle$. Of course, at the leaf $\mathbb{Z}_{3.5} = \mathbb{Z}_{5.3}$, G_1 is a Gröbner basis for $\langle 10XY + 1, 6X^2 + 3 \rangle$.

As a conclusion, the dynamical evaluation of the problem of constructing a Gröbner basis for I produces the following evaluation tree:

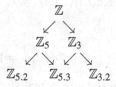

The obtained dynamical Gröbner basis of I is

$$G = \{(G_1, \mathscr{M}(5,2)), (G_2, \mathscr{M}(5,3)), (G_3, \mathscr{M}(3,2))\}.$$

b) Suppose that we have to deal with the ideal membership problem:

$$f = 62X^3Y + 11X^2 + 10XY^2 + 56XY + Y + 8 \in? \langle 10XY + 1, 6X^2 + 3 \rangle \text{ in } \mathbb{Z}[X,Y].$$

The responses to this ideal membership problem in the rings $\mathbb{Z}_{5.2}[X,Y]$, $\mathbb{Z}_{5.3}[X,Y], \mathbb{Z}_{3.2}[X,Y]$ are all positive. One obtains:

$$5f = (31X^2 + 5Y + 28)f_1 + 4f_2, \text{ and}$$

$$6f = (6Y + 15)f_1 + (62XY + 11)f_2.$$

Together with the Bezout identity $6 - 5 = 1$, one obtains:

$$f = (-31X^2 + Y - 13)f_1 + (62XY + 7)f_2, \text{ a complete positive answer.}$$

3.3.10 Gröbner Bases Over the Integers via Prime Factorization

In this section, $\mathbb{Z}$ can be replaced by any Dedekind domain **R** with complete prime factorization of principal ideals.

Let $I = \langle f_1, \ldots, f_s \rangle$ be a finitely-generated ideal of $\mathbb{Z}[X_1, \ldots, X_n]$. Let us fix a monomial order $>$ on $\mathbb{Z}[X_1, \ldots, X_n]$ and consider a normalized Gröbner basis $G_0 = \{g_1, \ldots, g_m\}$ for $J := I \otimes_{\mathbb{Z}} \mathbb{Q}$ in $\mathbb{Q}[X_1, \ldots, X_n]$. Denote by $g_i = \frac{h_i}{d_i}$ where $h_i \in \mathbb{Z}[X_1, \ldots, X_n]$, $d_i = \text{LC}(h_i) \in \mathbb{Z}$, and take $d = \text{lcm}(d_1, \ldots, d_m)$. If $d = \pm 1$ (for a Dedekind domain, if $d \in \mathbf{R}^{\times}$), then there is nothing to do, G_0 is already a Gröbner basis in $\mathbb{Z}[X_1, \ldots, X_n]$. Suppose now that d is not a unit and that its prime factorization is feasible. Write

$$\langle d \rangle = \pm \prod_{i=1}^{\ell} p_i^{n_i},$$

where the p_i's are distinct prime numbers (for a Dedekind domain, one should write $\langle d \rangle = \prod_{i=1}^{\ell} \mathfrak{p}_i^{n_i}$, where the $\mathfrak{p}_i$'s are nonzero distinct prime ideals of **R**), and $n_i \in \mathbb{N} \setminus \{0\}$.

Since for any $t \in d^{\mathbb{N}}$, $t_1 \in \mathbb{Z} \setminus p_1\mathbb{Z}, \ldots, t_\ell \in \mathbb{Z} \setminus p_\ell\mathbb{Z}$, we have

$$\gcd(t, t_1, \ldots, t_\ell) = 1,$$

there exist $z, z_1, \ldots, z_\ell \in \mathbb{Z}$ such that

$$zt + z_1 t_1 + \cdots + z_\ell t_\ell = 1.$$

(for a Dedekind domain $\mathbf{R}$, one should say that since no prime of $\mathbf{R}$ contains the ideal $\langle d, d_1, \ldots, d_\ell \rangle$, we obtain that $1 \in \langle t, t_1, \ldots, t_\ell \rangle$).

By this Bezout identity, we see that the monoids

$$S_0 = d^{\mathbb{N}}, \ S_1 = \mathbb{Z} \setminus p_1\mathbb{Z}, \ldots, \ S_\ell = \mathbb{Z} \setminus p_\ell\mathbb{Z}$$

are comaximal (for a Dedekind domain, the monoids $S_0 = d^{\mathbb{N}}$, $S_1 = \mathbf{R} \setminus \mathfrak{p}_1, \ldots, S_\ell = \mathbf{R} \setminus \mathfrak{p}_\ell$ are comaximal).

Now, one has to compute ℓ Gröbner basis $G_1, \ldots, G_\ell$, over $\mathbb{Z}_{p_1\mathbb{Z}}, \ldots, \mathbb{Z}_{p_\ell\mathbb{Z}}$, respectively, following Sect. 3.3.1 (for a Dedekind domain, one should compute ℓ Gröbner bases $G_1, \ldots, G_\ell$, over $(\mathbf{R} \setminus \mathfrak{p}_1)^{-1}\mathbf{R}, \ldots, (\mathbf{R} \setminus \mathfrak{p}_\ell)^{-1}\mathbf{R}$, respectively).

As a conclusion,

$$G = \{(G_0, S_0), (G_1, S_1), \ldots, (G_\ell, S_\ell)\}$$

is a dynamical Gröbner basis for I in $\mathbb{Z}[X_1, \ldots, X_n]$.

Note that, for $1 \leq i \leq \ell$, denoting $G_i = \{\frac{1}{a_{i,1}} h_{i,1}, \ldots, \frac{1}{a_{i,m_i}} h_{i,m_i}\}$ with $a_{i,j} \in \mathbb{Z} \setminus p_i\mathbb{Z}$ and $h_{i,j} \in \mathbb{Z}[X_1, \ldots, X_n]$, then

$$\mathscr{G} = \{h_1, \ldots, h_m, h_{1,1}, \ldots, h_{1,m_1}, \ldots, h_{\ell,1}, \ldots, h_{\ell,m_\ell}\}$$

is a Gröbner basis for I in $\mathbb{Z}[X_1, \ldots, X_n]$.

Example 287. Let $I = \langle f = 20Y - X, g = 8X^2 \rangle \subseteq \mathbb{Z}[X,Y]$, and fix the lexicographic monomial order $>$ on $\mathbb{Z}[X,Y]$ with $Y > X$. Then:

- *Over $\mathbb{Q}$:*

$$S(f,g) = X^2 f - \frac{5}{2} Y g = -X^3 \xrightarrow{g} 0.$$

  Thus, $G_0 = \{Y - \frac{1}{20}X, X^2\} = \{\frac{20Y - X}{20}, X^2\}$ is a normalized Gröbner basis for $\langle f, g \rangle$ in $\mathbb{Q}[X,Y]$. As $20 = 2^2 \times 5$, one has to compute two more Gröbner bases for $\langle f, g \rangle$ in $\mathbb{Z}_{2\mathbb{Z}}[X,Y]$ and $\mathbb{Z}_{5\mathbb{Z}}[X,Y]$. Thus, we obtain the following three leaves tree:

$$
\begin{array}{ccc}
 & \mathbb{Z} & \\
\swarrow & \downarrow & \searrow \\
\mathbb{Z}[\frac{1}{20}] & \mathbb{Z}_{2\mathbb{Z}} & \mathbb{Z}_{5\mathbb{Z}}
\end{array}
$$

- *Over $\mathbb{Z}_{2\mathbb{Z}}$:*

$$
\begin{aligned}
S(f,g) &= \frac{2}{5}X^2f - Yg = -\frac{2}{5}X^3 \xrightarrow{\text{normalization}} 2X^3 =: h_1, \\
S(f,h_1) &= X^3f - 10Yh_1 = -X^4 \xrightarrow{\text{normalization}} X^4 =: h_2, \\
S(f,h_2) &= X^4f - 20Yh_2 = -X^5 \xrightarrow{h_2} 0, \\
S(g,h_1) &= S(g,h_2) = S(h_1,h_2) = 0.
\end{aligned}
$$

Thus, $G_1 = \{f,g,2X^3,X^4\}$ is a Gröbner basis for $\langle f,g \rangle$ in $\mathbb{Z}_{2\mathbb{Z}}[X,Y]$.

- *Over $\mathbb{Z}_{5\mathbb{Z}}$:*

$$
S(f,g) = X^2f - \frac{5}{2}Yg = -X^3 \xrightarrow{g} 0.
$$

Thus, $G_2 = \{f,g\}$ is a Gröbner basis for $\langle f,g \rangle$ in $\mathbb{Z}_{5\mathbb{Z}}[X,Y]$.

As a conclusion, $G = \{(\{f,g\},20^{\mathbb{N}}),\ (\{f,g,2X^3,X^4\},\mathbb{Z}\setminus 2\mathbb{Z}),\ (\{f,g\},\mathbb{Z}\setminus 5\mathbb{Z})\}$ is a dynamical Gröbner basis for I in $\mathbb{Z}[X,Y]$.

We conclude also that $\mathscr{G} = \{f,g,2X^3,X^4\}$ is a Gröbner basis for I in $\mathbb{Z}[X,Y]$.

3.3.11 A Branching-Free Algorithm for Computing Gröbner Bases Over the Integers

In this section, $\mathbb{Z}$ can be replaced by any Bezout domain of Krull dimension ≤ 1. We propose an important simplification [119] of the dynamical method for the construction of dynamical Gröbner bases over the integers [181]. We will benefit from the fact that our base ring $\mathbb{Z}$ is more than a Dedekind domain, it is a one-dimensional Bezout domain. As explained above, we start as if $\mathbb{Z}$ were a valuation domain. Suppose that two incomparable (under division) elements a,b in $\mathbb{Z}$ appear as leading coefficients, of f and g respectively, when computing an $S(f,g)$. A key fact is that writing $a = (a \wedge b)a'$, $b = (a \wedge b)b'$, with $a' \wedge b' = 1$, then a divides b in $\mathbb{Z}[\frac{1}{a'}]$, b divides a in $\mathbb{Z}[\frac{1}{b'}]$, and the two multiplicative subsets $a'^{\mathbb{N}}$ and $b'^{\mathbb{N}}$ are comaximal as $1 \in \langle a',b' \rangle$. Then $\mathbb{Z}$ splits into $\mathbb{Z}[\frac{1}{a'}]$ and $\mathbb{Z}[\frac{1}{b'}]$, and we can continue as if $\mathbb{Z}$ were a valuation ring. Denoting by $\mathrm{mdeg}(f) = \alpha$, $\mathrm{mdeg}(g) = \beta$, and $\gamma = (\gamma_1,\ldots,\gamma_n)$, where $\gamma_i = \max(\alpha_i,\beta_i)$ for each i, $S(f,g)$ is computed as follows:

In the ring $\mathbb{Z}[\frac{1}{b'}]$: $S(f,g) = \frac{X^\gamma}{X^\alpha}f - \frac{a'}{b'}\frac{X^\gamma}{X^\beta}g =: S_1$.

In the ring $\mathbb{Z}[\frac{1}{a'}]$: $S(f,g) = \frac{b'}{a'}\frac{X^\gamma}{X^\alpha}f - \frac{X^\gamma}{X^\beta}g =: S_2$.

But, denoting by $S := b'\frac{X^\gamma}{X^\alpha}f - a'\frac{X^\gamma}{X^\beta}g$, we have:

$$
S = b'S_1 = a'S_2.
$$

As S is associated (i.e., equal up to a unit) to S_1 in $\mathbb{Z}[\frac{1}{b'}]$ and to S_2 in $\mathbb{Z}[\frac{1}{a'}]$, it can replace both of them, and thus there was no need to open the two branches $\mathbb{Z}[\frac{1}{a'}]$ and $\mathbb{Z}[\frac{1}{b'}]$ (we retrieve the same construction as in [3, 136]).

It is worth pointing out that $\mathbb{Z}$ can be replaced by any Bezout domain with Krull dimension ≤ 1.

3.3.11.1   Division Algorithm Over $\mathbb{Z}$

It is the global version of the (local) case when the base ring is a valuation domain. Let $f, f_1, \ldots, f_s \in \mathbb{Z}[X_1, \ldots, X_n] \setminus \{0\}$, and consider a monomial order $>$ on $\mathbb{Z}[X_1, \ldots, X_n]$. We want to compute a quotient Q and a remainder R of f on division by $[f_1, \ldots, f_s]$.

Initialization: $Q := [0, \ldots, 0]$, $R := 0$, $p := f$ (p is an intermediary variable with the relation $f = Q[1]f_1 + \cdots + Q[s]f_s + p + R$).

WHILE $p \neq 0$ DO
$\quad D := \{i \mid \mathrm{LM}(f_i) \text{ divides } \mathrm{LM}(p)\} \subseteq \{1, \ldots, s\}$; $b_i := \mathrm{LC}(f_i)$;
$\quad d := \wedge_{i \in D} b_i = \sum_{i \in D} \alpha_i b_i \in \mathbb{Z}$ with $\alpha_i \in \mathbb{Z}$, and $\mathrm{LC}(p) = qd + r$
$\quad$ (Euclidean division of $\mathrm{LC}(p)$ by d);
$\quad R := R + r\mathrm{LM}(p)$;
$\quad Q[i] := Q[i] + \alpha_i q \frac{\mathrm{LM}(p)}{\mathrm{LM}(f_i)}$ if $i \in D$;
$\quad p := p - \sum_{i \in D} \alpha_i q \frac{\mathrm{LM}(p)}{\mathrm{LM}(f_i)} f_i - r\mathrm{LM}(p)$;

A remainder of $f \in \mathbb{Z}[X_1, \ldots, X_n]$ on division by a finite set $F = \{f_1, \ldots, f_s\}$ of polynomials in $\mathbb{Z}[X_1, \ldots, X_n]$ will be denoted by $\overline{f}^F$.

3.3.11.2   Computation of S-Polynomials

Let $f, g \in \mathbb{Z}[X_1, \ldots, X_n] \setminus \{0\}$, and consider a monomial order $>$ on $\mathbb{Z}[X_1, \ldots, X_n]$. Denoting by $\mathrm{LT}(f) = aX^\alpha$, $\mathrm{LT}(g) = bX^\beta$, $a = (a \wedge b)a'$, $b = (a \wedge b)b'$, $\gamma = (\gamma_1, \ldots, \gamma_n)$ where $\gamma_i = \max(\alpha_i, \beta_i)$ for each i, the S-polynomial of f and g is the combination:

$$S(f, g) := b' \frac{X^\gamma}{X^\alpha} f - a' \frac{X^\gamma}{X^\beta} g.$$

Algorithm 288. *(Buchberger's Algorithm Over the Integers)*

Input: $g_1, \ldots, g_s \in \mathbb{Z}[X_1, \ldots, X_n]$ and $>$ a monomial order on $\mathbb{Z}[X_1, \ldots, X_n]$
Output: a Gröbner basis G for $\langle g_1, \ldots, g_s \rangle$ with $\{g_1, \ldots, g_s\} \subseteq G$

$\quad G := \{g_1, \ldots, g_s\}$
$\quad$ REPEAT

$G' := G$

For each pair f_i, f_j $(i \neq j)$ in G' DO

$\quad S := \overline{S(f_i, f_j)}^{G'}$

$\quad$ If $S \neq 0$ THEN $G := G' \cup \{S\}$

UNTIL $G = G'$

Example 289. Let us consider the ideal $I = \langle f_1 = 10X + 2Y, f_2 = 6X + 3 \rangle$ of $\mathbb{Z}[X, Y]$ and fix the lexicographic order as monomial order with $X > Y$. Then:

$$S(f_1, f_2) = 3f_1 - 5f_2 = 6Y - 15 =: f_3,$$
$$S(f_1, f_3) = 3Yf_1 - 5Xf_3 = 6Y^2 + 75X \xrightarrow{f_3, f_2} 3X + 3Y - 6 =: f_4,$$
$$S(f_1, f_4) = 3f_1 - 10f_4 = -24Y + 60 \xrightarrow{f_3} 0,$$
$$S(f_2, f_3) = Yf_2 - Xf_3 = 3Y + 15X \xrightarrow{f_4, f_3} 0,$$
$$S(f_2, f_4) = f_2 - 2f_4 = -6Y + 15 \xrightarrow{f_3} 0,$$
$$S(f_3, f_4) = Xf_3 - 2Yf_4 = -15X - 6Y^2 + 12Y \xrightarrow{f_4, f_3} 0.$$

Thus, $G = \{f_1, f_2, f_3, f_4\}$ is a Gröbner basis for I.

3.3.11.3 Reduced Gröbner Basis Over the Integers

Analogously to the case where the base ring is a field, a reduced Gröbner basis G of an ideal I of $\mathbb{Z}[X_1, \ldots, X_n]$ can be fairly defined as a Gröbner basis of I such that $\forall p \in G$ and for any term T of p, the remainder of T on division by the list formed by the $LT(q)$ with $q \in G \setminus \{p\}$ is equal to T. It is not necessarily unique. For uniqueness one need to compute the so-called "strong" Gröbner basis [3].

Example 290. (Example 289 Continued) We transform G into a reduced Gröbner basis as follows:

$$f_1 \xrightarrow{f_2, f_4} X - Y + 3 =: f_1,$$
$$f_2 \xrightarrow{f_1, f_3} 0 =: f_2,$$
$$f_4 \xrightarrow{f_1, f_3} 0 =: f_4.$$

Thus, $\{X - Y + 3, 6Y - 15\}$ is a reduced Gröbner basis of I.

Dynamical Gröbner bases in the noncommutative case have been studied in [118].

3.4 Computing Syzygy Modules with Polynomial Rings Over Gröbner Arithmetical Rings

3.4.1 Computing Syzygy Modules with Polynomial Rings Over Gröbner Valuation Rings

3.4.1.1 Computing Syzygy Modules Over Gröbner Valuation Rings by Direct Computation of Gröbner Bases

The following theorem gives a generating set for syzygies of monomials with coefficients in a valuation ring. It is a generalization of [43, Proposition 8 (page 104)] to valuation rings.

Theorem 291. (Generating set of the syzygy module of a finite set of terms over a Gröbner valuation ring)

Let $\mathbf{V}$ be a valuation ring, $c_1, \ldots, c_s \in \mathbf{V} \setminus \{0\}$, and $M_1, \ldots, M_s$ be monomials in $\mathbf{V}[X_1, \ldots, X_n]$. Denoting $\mathrm{LCM}(M_i, M_j)$ by $M_{i,j}$, and the canonical basis of $\mathbf{V}[X_1, \ldots, X_n]^{s \times 1}$ by $(e_1, \ldots, e_s)$, the syzygy module $\mathrm{Syz}(c_1 M_1, \ldots, c_s M_s)$ is generated by

$$\{S_{i,j} \in \mathbf{V}[X_1, \ldots, X_n]^s \mid 1 \leq i \leq j \leq s\},$$

where for $i \neq j$,

$$S_{i,j} = \begin{cases} \frac{M_{i,j}}{M_i} e_i - \frac{c_i}{c_j} \frac{M_{i,j}}{M_j} e_j & \text{if } c_j \mid c_i \\ \frac{c_j}{c_i} \frac{M_{i,j}}{M_i} e_i - \frac{M_{i,j}}{M_j} e_j & \text{else,} \end{cases}$$

and

$$S_{i,i} = d_i e_i,$$

with d_i a generator of the annihilator of c_i in $\mathbf{V}$ ($S_{i,i}$ is defined up to a unit).

Proof. It is clear that for all $i \leq j$, $S_{i,j}$ is a syzygy of $M = (c_1 M_1, \ldots, c_s M_s)$.

Now, in order to verify that $\{S_{i,j}, 1 \leq i \leq j \leq s\}$ is really a syzygy basis, we need to show that every syzygy H of M can be written as $H = \sum_{1 \leq i \leq j \leq s} u_{i,j} S_{i,j}$ where $u_{i,j} \in \mathbf{V}[X_1, \ldots, X_n]$.

For this, let $H = {}^{\mathrm{t}}(h_1, \ldots, h_s)$ be a syzygy of M, that is, such that $MH = 0$. Letting $\gamma(H) = \max_{1 \leq i \leq s} \mathrm{mdeg}(h_i M_i)$, we have

$$\sum_{\mathrm{mdeg}(h_i M_i) = \gamma(H)} c_i h_i M_i + \sum_{\mathrm{mdeg}(h_i M_i) < \gamma(H)} c_i h_i M_i = 0.$$

Thus,

$$\sum_{\mathrm{mdeg}(h_i M_i) = \gamma(H)} c_i \mathrm{LT}(h_i) M_i + \sum_{\mathrm{mdeg}(h_i M_i) = \gamma(H)} c_i (h_i - \mathrm{LT}(h_i)) M_i$$

$$+ \sum_{\mathrm{mdeg}(h_i M_i) < \gamma(H)} c_i h_i M_i = 0.$$

We can write $H = G + \widetilde{G}$, where $G = (g_1, \ldots, g_s)$ with $g_i = \mathrm{LT}(h_i)$ if $\mathrm{mdeg}(h_i M_i) = \gamma(H)$, 0 else; $\widetilde{G} = (\widetilde{g}_1, \ldots, \widetilde{g}_s)$ with $\widetilde{g}_i = h_i - \mathrm{LT}(h_i)$ if $\mathrm{mdeg}(h_i M_i) = \gamma(H)$, h_i else.

Since $\gamma(\widetilde{G}) < \gamma(H)$, it suffices, by induction on $\gamma(H)$ (the induction is legitimated by Corollary 210), to prove the result for G. In particular, we can assume that $h_i = a_i M'_i$ with $a_i \in \mathbf{V}$ (a_i can be zero). Let $i_1 < i_2 < \ldots < i_t$ be the indices corresponding to the nonzero a_i's, and denote $\gamma(H)$ by γ. The facts that $a_1 M'_1 c_1 M_1 + \cdots + a_s M'_s c_s M_s = 0$ and $a_i M'_i c_i M_i = a_i c_i X^\gamma$ imply that

$$a_{i_1} c_{i_1} + \cdots + a_{i_t} c_{i_t} = 0. \quad (*)$$

It follows that

$$
\begin{aligned}
(h_1, \ldots, h_s) &= (a_1 M'_1, \ldots, a_s M'_s) = a_{i_1} M'_{i_1} e_{i_1} + \ldots + a_{i_t} M'_{i_t} e_{i_t} \\
&= a_{i_1} \frac{X^\gamma}{M_{i_1}} e_{i_1} + \cdots + a_{i_t} \frac{X^\gamma}{M_{i_t}} e_{i_t}.
\end{aligned}
$$

As $\mathbf{V}$ is a valuation ring, there exists an integer $q \in \{1, \ldots, t\}$ such that c_{i_q} divides all the c_{i_j}'s. So, the previous expression can be written as

$$
\begin{aligned}
a_{i_1} \frac{X^\gamma}{M_{i_1}} e_{i_1} + \cdots + a_{i_t} \frac{X^\gamma}{M_{i_t}} e_{i_t} &= \sum_{1 \le j \le q-1} a_{i_j} \frac{X^\gamma}{M_{i_j, i_q}} \left(\frac{M_{i_j, i_q}}{M_{i_j}} e_{i_j} - \frac{c_{i_j}}{c_{i_q}} \frac{M_{i_j, i_q}}{M_{i_q}} e_{i_q} \right) \\
&\quad - \sum_{q+1 \le j \le t} a_{i_j} \frac{X^\gamma}{M_{i_j, i_q}} \left(\frac{c_{i_j}}{c_{i_q}} \frac{M_{i_j, i_q}}{M_{i_q}} e_{i_q} - \frac{M_{i_j, i_q}}{M_{i_j}} e_{i_j} \right) \\
&\quad + \left(\sum_{j \ne q} a_{i_j} \frac{c_{i_j}}{c_{i_q}} + a_{i_q} \right) \frac{X^\gamma}{M_{i_q}} e_{i_q}. \quad (**)
\end{aligned}
$$

Note that we have $\left(\sum_{j \ne q} a_{i_j} \frac{c_{i_j}}{c_{i_q}} + a_{i_q} \right) c_{i_q} = 0$ by virtue of $(*)$, and thus, $\left(\sum_{j \ne q} a_{i_j} \frac{c_{i_j}}{c_{i_q}} + a_{i_q} \right) e_{i_q} \in \langle d_{i_q} \rangle e_{i_q}$. Note also that, for $1 \le j \le q-1$, $\frac{M_{i_j, i_q}}{M_{i_j}} e_{i_j} - \frac{c_{i_j}}{c_{i_q}} \frac{M_{i_j, i_q}}{M_{i_q}} e_{i_q} = S_{i_j, i_q}$, and for $q+1 \le j \le t$,

$$
\frac{c_{i_j}}{c_{i_q}} \frac{M_{i_j, i_q}}{M_{i_q}} e_{i_q} - \frac{M_{i_j, i_q}}{M_{i_j}} e_{i_j} = \begin{cases} \frac{c_{i_j}}{c_{i_q}} S_{i_q, i_j} & \text{if } c_{i_j} \mid c_{i_q} \\ S_{i_q, i_j} & \text{else.} \end{cases}
$$

Thus, $\mathrm{Syz}(c_1 M_1, \ldots, c_s M_s) \subseteq \langle S_{ij}, 1 \le i \le j \le s \rangle$.

$\square$

Example 292. Let $\mathbf{V} = \mathbb{Z}/8\mathbb{Z}$, $f_1 = 4X^2$, $f_2 = 2XY^3$, $f_3 = 6Y$, $f_4 = 5$ in $\mathbf{V}[X, Y]$. With the previous notation, we have

$$S_{1,1} = (2,0,0,0), \; S_{2,2} = (0,4,0,0), \; S_{3,3} = (0,0,4,0), \; S_{4,4} = (0,0,0,0).$$

In addition, since $c_4 \mid c_3 \mid c_2 \mid c_1$, the syzygy module $\mathrm{Syz}(f_1,\ldots,f_4)$ is generated by $\{S_{i,j} = \frac{M_{i,j}}{M_i}e_i - \frac{c_i}{c_j}\frac{M_{i,j}}{M_j}e_j \mid 1 \le i < j \le 4\} \cup \{S_{1,1}, S_{2,2}, S_{3,3}\}$, that is,

$$\mathrm{Syz}(f_1,\ldots,f_4) = \langle\, {}^{\mathrm{t}}(Y^3, 6X, 0, 0), {}^{\mathrm{t}}(Y, 0, 2X^2, 0), {}^{\mathrm{t}}(1, 0, 0, 4X^2), {}^{\mathrm{t}}(0, 1, 5XY^2, 0),$$
$$ {}^{\mathrm{t}}(0, 1, 0, 6XY^3), {}^{\mathrm{t}}(0, 0, 1, 2Y), {}^{\mathrm{t}}(2, 0, 0, 0), {}^{\mathrm{t}}(0, 4, 0, 0), {}^{\mathrm{t}}(0, 0, 4, 0)\,\rangle.$$

Notation 293. Let $\mathbf{V}$ be a coherent valuation ring, $>$ a monomial order on $\mathbf{V}[X_1,\ldots,X_n]$, $f_1,\ldots,f_s \in \mathbf{V}[X_1,\ldots,X_n] \setminus \{0\}$, and $\{g_1,\ldots,g_t\}$ a Gröbner basis for $\langle f_1,\ldots,f_s\rangle$. Denote by $c_i = \mathrm{LC}(g_i)$ and $M_i = \mathrm{LM}(g_i)$. In order to compute the syzygy module $\mathrm{Syz}(f_1,\ldots,f_s)$, we will first compute $\mathrm{Syz}(g_1,\ldots,g_t)$. Recall that for each $1 \le i < j \le t$, the S-polynomial of g_i and g_j is given by

$$S(g_i,g_j) := \begin{cases} \dfrac{M_{i,j}}{M_i}g_i - \dfrac{c_i}{c_j}\dfrac{M_{i,j}}{M_j}g_j & \text{if } c_j \mid c_i \\[2mm] \dfrac{c_j}{c_i}\dfrac{M_{i,j}}{M_i}g_i - \dfrac{M_{i,j}}{M_j}g_j & \text{else.} \end{cases}$$

Moreover, $S(g_i,g_i) := d_i g_i$, where d_i is a generator of the annihilator of c_i in $\mathbf{V}$ (it is defined up to a unit).
For some $h_{i,j,k} \in \mathbf{V}[X_1,\ldots,X_n]$, we have

$$S(g_i,g_j) = \sum_{k=1}^{t} h_{i,j,k}g_k \quad \text{with } \mathrm{mdeg}(S(g_i,g_j)) = \max_{1 \le k \le t} \mathrm{mdeg}(h_{i,j,k}g_k) \qquad (\star).$$

(The polynomials $h_{i,j,k}$ are given by the division algorithm.)
For $1 \le i < j \le t$, let

$$\varepsilon_{i,j} = \begin{cases} \dfrac{M_{i,j}}{M_i}e_i - \dfrac{c_i}{c_j}\dfrac{M_{i,j}}{M_j}e_j & \text{if } c_j \mid c_i \\[2mm] \dfrac{c_j}{c_i}\dfrac{M_{i,j}}{M_i}e_i - \dfrac{M_{i,j}}{M_j}e_j & \text{else,} \end{cases}$$

and $\varepsilon_{i,i} = d_i e_i$, where d_i is a generator of the annihilator of c_i in $\mathbf{V}$. For $1 \le i \le j \le t$, denote by

$$s_{i,j} = \varepsilon_{i,j} - \sum_{k=1}^{t} h_{i,j,k}e_k.$$

Theorem 294. (Syzygy module of a Gröbner basis over a coherent valuation ring) *With the previous notation,*

$$\mathrm{Syz}(g_1,\ldots,g_t) = \langle s_{i,j} \mid 1 \le i \le j \le t\rangle.$$

Proof. "$\supseteq$" Let $G = (g_1,\ldots,g_t)$. For each $1 \le i \le j \le t$, we have $G s_{i,j} = S(g_i,g_j) - \sum_{k=1}^{t} h_{i,j,k}g_k = 0$. Thus, $s_{i,j} \in \mathrm{Syz}(g_1,\ldots,g_t)$.
"$\subseteq$" Let $U = {}^{\mathrm{t}}(u_1,\ldots,u_t) \in \mathrm{Syz}(g_1,\ldots,g_t)$, and set $\gamma(U) = \max_{1 \le i \le t} \{\mathrm{mdeg}(u_i g_i)\}$. We will proceed by induction on $\gamma(U)$ (the induction is legitimated by Corollary 210).

Letting $S = \{i \in \{1, \ldots, t\} \mid \mathrm{mdeg}(u_i g_i) = \gamma(U)\}$, we have

$$\sum_{i \in S} u_i g_i + \sum_{i \notin S} u_i g_i = 0 \ \Rightarrow \ \sum_{i \in S} \mathrm{LT}(u_i) g_i + \sum_{i \in S}(u_i - \mathrm{LT}(u_i)) g_i + \sum_{i \notin S} u_i g_i = 0,$$

and so, $\sum_{i \in S} \mathrm{LT}(u_i)\mathrm{LT}(g_i) = 0$, that is, $(\mathrm{LT}(u_i))_{i \in S} \in \mathrm{Syz}(\mathrm{LT}(g_i))_{i \in S}$. By virtue of Theorem 291, we can write

$$(\mathrm{LT}(u_i))_{i \in S} = \sum_{1 \leq i \leq j \leq t,\, i,j \in S} h_{i,j} \varepsilon_{i,j}. \quad (\star\star)$$

Let $U = W + {}^t(u'_1, \ldots, u'_t)$ with $W = {}^t(w_1, \ldots, w_t)$ and $w_i = \begin{cases} 0 & \text{if } i \notin S \\ \mathrm{LT}(u_i) & \text{if } i \in S, \end{cases}$

in such a way we have

$$U = \sum_{1 \leq i \leq j \leq t,\, i,j \in S} h_{i,j} \varepsilon_{ij} + {}^t(u'_1, \ldots, u'_t).$$

We can write $U = \overline{V} + V$ where

$$\overline{V} = \sum_{1 \leq i \leq j \leq t,\, i,j \in S} h_{i,j} s_{i,j} \quad \text{and} \quad V = \sum_{1 \leq i \leq j \leq t,\, i,j \in S} h_{i,j} \sum_{k=1}^{t} h_{i,j,k} e_k + {}^t(u'_1, \ldots, u'_t).$$

It is clear that $\overline{V} \in \langle s_{ij},\ 1 \leq i \leq j \leq t \rangle$. Denoting by $V = {}^t(v_1, \ldots, v_t)$, we have

$$\begin{aligned}
\mathrm{mdeg}(v_l g_l) &= \mathrm{mdeg}(u'_l g_l + \sum_{1 \leq i \leq j \leq t,\, i,j \in S} h_{i,j} h_{i,j,l} g_l) \\
&\leq \max_{1 \leq i \leq j \leq t,\, i,j \in S} \{\mathrm{mdeg}(u'_l g_l), \mathrm{mdeg}(h_{i,j} h_{i,j,l} g_l)\}.
\end{aligned}$$

By definition of ${}^t(u'_1, \ldots, u'_t)$, we have $\mathrm{mdeg}(u'_l g_l) < \gamma(U)$. Moreover, from $(\star\star)$, we have

$$(\mathrm{LT}(u_i))_{i \in S}{}^t(g_i)_{i \in S} = \sum_{1 \leq i \leq j \leq t,\, i,j \in S} h_{i,j} S(g_i, g_j). \quad (\star\star\star)$$

In the equality $(\star\star\star)$, all the terms $\mathrm{LT}(u_i) g_i$ on the left-hand side are homogeneous with multidegree $\gamma(U)$ since $\mathrm{mdeg}(\mathrm{LT}(u_i)\mathrm{LT}(g_i)) = \gamma(U)\ \forall\, i \in S$. This property must also be satisfied on the right-hand side. Thus, $\mathrm{mdeg}(h \frac{M_{i,j}}{M_i} M_i) \leq \gamma(U)$, $\mathrm{mdeg}(h_{i,j} \frac{c_i}{c_j} \frac{M_{i,j}}{M_j} M_j) \leq \gamma(U)$, and $\mathrm{mdeg}(h_{i,j} M_{i,j}) \leq \gamma(U)$.

On the other hand, $\forall\, 1 \leq k \leq t$, $\mathrm{mdeg}(h_{i,j,k} g_k) \leq \mathrm{mdeg}(S(g_i, g_j))$ since by $(\star)$ we have $\mathrm{mdeg}(S(g_i, g_j)) = \max_{1 \leq k \leq t} \mathrm{mdeg}(h_{i,j,k} g_k)$.

Hence $\mathrm{mdeg}(h_{i,j} h_{i,j,l} g_l) \leq \mathrm{mdeg}(h_{i,j} S(g_i, g_j)) < \mathrm{mdeg}(h_{i,j} M_{i,j}) \leq \gamma(U)$, $\mathrm{mdeg}(v_l g_l) < \gamma(U)\ \forall\, 1 \leq l \leq t$, and finally, $\gamma(V) < \gamma(U)$ as desired. $\qquad \square$

Example 295. Let $\mathbf{V} = \mathbb{Z}/8\mathbb{Z}$ and $g_1 = 2X^3 + 6X^2, g_2 = 6Y^2, g_3 = 5XY - 5Y \in \mathbf{V}[X, Y]$. Let us fix the lexicographic order as monomial order with $X > Y$. Then

$G = \{g_1, g_2, g_3\}$ is a Gröbner basis for $\langle g_1, g_2, g_3 \rangle$ in $(\mathbb{Z}/8\mathbb{Z})[X, Y]$ as

$$S(g_1, g_1) = 4g_1 = 0, \; S(g_2, g_2) = 4g_2 = 0, \; S(g_3, g_3) = 0g_3 = 0,$$
$$S(g_1, g_2) = Y^2 g_1 - 3X^3 g_2 = X^2 g_2 \xrightarrow{g_2} 0,$$
$$S(g_1, g_3) = Y g_1 - 2X^2 g_3 = 0,$$
$$S(g_2, g_3) = X g_2 - 6Y g_3 = g_2 \xrightarrow{g_2} 0.$$

Keeping the previous notation, we have

$$h_{1,2,1} = 0, \; h_{1,2,2} = X^2, h_{123} = 0 \;\Rightarrow\; s_{1,2} = {}^t(Y^2, -3X^3 - X^2, 0).$$

In the same way, we obtain that $s_{1,3} = {}^t(Y, 0, -2X^2)$, $s_{2,3} = {}^t(0, X-1, -6Y)$, $s_{1,1} = {}^t(4, 0, 0)$, $s_{2,2} = {}^t(0, 4, 0)$, and $s_{3,3} = {}^t(0, 0, 0)$. Thus,

$$\mathrm{Syz}(g_1, g_2, g_3) = \langle {}^t(Y^2, 5X^3 + 7X^2, 0), {}^t(Y, 0, 6X^2), {}^t(0, X+7, 2Y), {}^t(4, 0, 0), {}^t(0, 4, 0) \rangle.$$

Keeping Notation 293, and denoting by $F = [f_1 \cdots f_s]$ and $G = [g_1 \cdots g_t]$, there exist two matrices S and T respectively of size $t \times s$ and $s \times t$ such that $F = GS$ and $G = FT$. We can first compute a generator set $\{s_1, \ldots, s_r\}$ of $\mathrm{Syz}(G)$. For each $i \in \{1, \ldots, r\}$, we have $0 = Gs_i = (FT)s_i = F(Ts_i)$. So $\langle Ts_i \mid i \in \{1, \ldots, r\} \rangle \subseteq \mathrm{Syz}(F)$. Also, denoting by $\mathbf{I}_s$ the identity matrix of size s, we have

$$F(\mathbf{I}_s - TS) = F - FTS = F - GS = F - F = 0.$$

This equality shows that the columns $r_1, \ldots, r_s$ of $\mathbf{I}_s - TS$ are also in $\mathrm{Syz}(F)$. The converse holds as stated by the following theorem.

Theorem 296. (Syzygy module over a Gröbner valuation ring: general case) *With the previous notation, we have*

$$\mathrm{Syz}(f_1, \ldots, f_s) = \langle Ts_1, \ldots, Ts_r, r_1, \ldots, r_s \rangle.$$

Proof. Let $s = (a_1, \ldots, a_s) \in \mathrm{Syz}(f_1, \ldots, f_s)$. As $0 = Fs = GSs$, we have $Ss \in \mathrm{Syz}(g_1, \ldots, g_t)$.

By definition of $s_1, \ldots, s_r$, we have $Ss = \sum_{i=1}^{r} h_i s_i$ for $h_i \in V[X_1, \ldots, X_n]$, which implies

that $TSs = \sum_{i=1}^{r} h_i(Ts_i)$. Thus, $s = s - TSs + TSs = (\mathbf{I}_s - TS)s + \sum_{i=1}^{r} h_i(Ts_i) =$

$\sum_{i=1}^{s} a_i r_i + \sum_{i=1}^{r} h_i(Ts_i)$, and $\mathrm{Syz}(f_1, \ldots, f_s) \subseteq \langle Ts_1, \ldots, Ts_r, r_1, \ldots, r_s \rangle$. We conclude that $\mathrm{Syz}(f_1, \ldots, f_s) = \langle Ts_1, \ldots,$ $Ts_r, r_1, \ldots, r_s \rangle$. $\square$

Example 297. Let $f_1 = 2XY, f_2 = Y^3 + 1, f_3 = X^2 + X \in \mathbf{V}[X, Y] = (\mathbb{Z}/4\mathbb{Z})[X, Y]$, and $F = [f_1 \; f_2 \; f_3]$. Computing a Gröbner basis for $\langle f_1, f_2, f_3 \rangle$ using the lexicographic order with $X > Y$ as monomial order, we obtain:

$S(f_1, f_1) = 2f_1 = 0, S(f_2, f_2) = 0f_2 = 0, S(f_3, f_3) = 0f_3 = 0,$

$S(f_1, f_2) = Y^2 f_1 - 2X f_2 = 2X =: f_4, S(f_4, f_4) = 2f_4 \xrightarrow{f_4} 0,$

$S(f_1, f_3) = X f_1 - 2Y f_3 = 2XY \xrightarrow{f_1} 0,$

$S(f_2, f_3) = X^2 f_2 - Y^3 f_3 = 3X^2 + XY^3 \xrightarrow{f_3, f_2} 0,$

$f_1 \xrightarrow{f_4} 0, S(f_2, f_4) = 2X f_2 - Y^3 f_4 = 2X \xrightarrow{f_4} 0,$

$S(f_3, f_4) = 2f_3 - X f_4 = 2X \xrightarrow{f_4} 0.$

Thus, $\{f_2, f_3, f_4\}$ is a Gröbner basis for $\langle f_1, f_2, f_3 \rangle$ in $\mathbf{V}[X, Y]$. Let us denote by $G = [g_1 \ g_2 \ g_3]$ with $g_1 = f_4, g_2 = f_2, g_3 = f_3$.

We have $G = FT$ with $T = \begin{pmatrix} Y^2 & 0 & 0 \\ -2X & 1 & 0 \\ 0 & 0 & 1 \end{pmatrix}$ and $F = GS$ with $S = \begin{pmatrix} Y & 0 & 0 \\ 0 & 1 & 0 \\ 0 & 0 & 1 \end{pmatrix}$.

The nonzero vectors $s_{i,j}$ we found are

$s_{1,1} = {}^t(2, 0, 0), s_{1,2} = {}^t(Y^3 - 1, 2X, 0), s_{1,3} = {}^t(X - 1, 0, 2),$ and

$s_{2,3} = {}^t(0, X^2 + X, -Y^3 - 1).$

And so

$$T s_{1,1} = \begin{pmatrix} 2Y^2 \\ 0 \\ 0 \end{pmatrix}, \ T s_{1,2} = \begin{pmatrix} Y^5 - Y^2 \\ 2XY^3 \\ 0 \end{pmatrix}, \ T s_{1,3} = \begin{pmatrix} XY^2 - Y^2 \\ 2X + 2X^2 \\ 2 \end{pmatrix},$$

$$T s_{2,3} = \begin{pmatrix} 0 \\ X^2 + X \\ -Y^3 - 1 \end{pmatrix}.$$

Moreover, we have $\mathbf{I}_3 - TS = \begin{pmatrix} 1 - Y^3 & 0 & 0 \\ 2XY & 0 & 0 \\ 0 & 0 & 0 \end{pmatrix}$. So, denoting the first column of $\mathbf{I}_3 - TS$ by r_1, we have

$$\begin{aligned} \mathrm{Syz}(F) &= \langle T s_{1,1}, T s_{1,2}, T s_{1,3}, T s_{2,3}, r_1 \rangle \\ &= \langle {}^t(2Y^2, 0, 0), {}^t(Y^5 - Y^2, 2XY^3, 0), {}^t(XY^2 - Y^2, 2X + 2X^2, 2), \\ & \quad {}^t(0, X^2 + X, -Y^3 - 1), {}^t(1 - Y^3, 2XY, 0) \rangle. \end{aligned}$$

3.4.1.2 Computing Syzygy Modules Over Gröbner Domains via Saturation

We will hereafter present a method for computing syzygies over Gröbner domains via saturation (and which is known over $\mathbb{Z}$).

Let $\mathbf{R}$ be a Gröbner domain with quotient field $\mathbf{K}$ and consider $p_1, \ldots, p_m \in \mathbf{R}[X_1, \ldots, X_k]$. Recall that the syzygy module of $(p_1, \ldots, p_m)$ over $\mathbf{A}$ ($\mathbf{A} = \mathbf{R}$ or $\mathbf{K}$) is

$$\mathrm{Syz}_{\mathbf{A}}(p_1, \ldots, p_m) = \{(q_1, \ldots, q_m) \in \mathbf{A}[X_1, \ldots, X_k]^m \mid p_1 q_1 + \cdots + p_m q_m = 0\}.$$

We know that $\mathrm{Syz}_{\mathbf{K}}(p_1,\ldots,p_m)$ is finitely-generated (it is a particular case of Theorem 296, the base ring being a discrete field), say

$$\mathrm{Syz}_{\mathbf{K}}(p_1,\ldots,p_m) = \mathbf{K}[X_1,\ldots,X_k]s_1 + \cdots + \mathbf{K}[X_1,\ldots,X_k]s_n$$

for some $s_i \in \mathbf{R}[X_1,\ldots,X_k]^m$. Now $\mathrm{Syz}_{\mathbf{R}}(p_1,\ldots,p_m)$ is nothing but the $\mathbf{R}$-saturation $\mathrm{Sat}_{\mathbf{R}}(S)$ ($\mathrm{Sat}(S)$ is there is no confusion risk) of $S := \mathbf{R}[X_1,\ldots,X_k]s_1 + \cdots + \mathbf{R}[X_1,\ldots,X_k]s_n$, i.e.,

$$\begin{aligned}
\mathrm{Sat}(S) &:= \{s \in \mathbf{R}[X_1,\ldots,X_k]^m \mid \alpha s \in S \text{ for some } \alpha \in \mathbf{R} \setminus \{0\}\} \\
&= (S \otimes_{\mathbf{R}} \mathbf{K}) \cap \mathbf{R}[X_1,\ldots,X_k]^m.
\end{aligned}$$

The computation of the $\mathbf{R}$-saturation of finitely-generated submodules of $\mathbf{R}[X_1,\ldots,X_k]^m$ can be done as in the case where $\mathbf{R} = \mathbb{Z}$. To simplify, we will compute the $\mathbf{R}$-saturation of finitely-generated ideals of $\mathbf{R}[X_1,\ldots,X_k]$ (the case of submodules of $\mathbf{R}[X_1,\ldots,X_k]^m$ is analogous).

Notation 298. Let $\mathbf{A}$ be a ring, $a \in \mathbf{A}$ and I an ideal of $\mathbf{A}$. We denote by

$$(I : a^{\infty}) = \{x \in \mathbf{A} \mid \exists n \in \mathbb{N} \mid xa^n \in I\}.$$

It is an ideal of $\mathbf{A}$ containing I.

Proposition 299. *Let $\mathbf{R}$ be a domain with quotient field $\mathbf{K}$, take $X_1,\ldots, X_n, Z$ independent variables over $\mathbf{K}$, and consider an ideal $I = \langle f_1,\ldots,f_s \rangle$ of $\mathbf{R}[X_1,\ldots,X_n]$. Let us fix a monomial order $>$ on $\mathbf{R}[X_1,\ldots,X_n]$ and consider a normalized Gröbner basis $G = \{g_1,\ldots,g_m\}$ for $I \otimes_{\mathbf{R}} \mathbf{K}$. Denoting by $g_i = \frac{h_i}{d_i}$ where $h_i \in \mathbf{R}[X_1,\ldots,X_n]$, $d_i = \mathrm{LC}(h_i) \in \mathbf{R}$, $J = \langle h_1,\ldots,h_s \rangle \subseteq \mathbf{R}[X_1,\ldots,X_n]$, and taking $d = \prod_{i=1}^{m} d_i$ (or $d = \mathrm{lcm}(d_1,\ldots,d_m)$ if such a notion exists), we have:*

$$\begin{aligned}
\mathrm{Sat}(I) &= \mathrm{Sat}(J) = (I \otimes_{\mathbf{R}} \mathbf{K}) \cap \mathbf{R}[X_1,\ldots,X_n] = (J : d^{\infty}) \\
&= \langle h_1,\ldots,h_s,dZ-1 \rangle \cap \mathbf{R}[X_1,\ldots,X_n].
\end{aligned}$$

Proof. The fact that $\mathrm{Sat}(I) = \mathrm{Sat}(J) = (I \otimes_{\mathbf{R}} \mathbf{K}) \cap \mathbf{R}[X_1,\ldots,X_n]$ is straightforward. The fact that $(J : d^{\infty}) = \langle h_1,\ldots,h_s,dZ-1 \rangle \cap \mathbf{R}[X_1,\ldots,X_n]$ follows from Lemma 260. It remains to prove that $(I \otimes_{\mathbf{R}} \mathbf{K}) \cap \mathbf{R}[X_1,\ldots,X_n] = (J : d^{\infty})$. For this, let $f \in (I \otimes_{\mathbf{R}} \mathbf{K}) \cap \mathbf{R}[X_1,\ldots,X_n]$. As $G = \{g_1,\ldots,g_m\}$ is a Gröbner basis for $(I \otimes_{\mathbf{R}} \mathbf{K})$, by the Division Algorithm 211, one can find $q_1,\ldots,q_m \in \mathbf{K}[X_1,\ldots,X_n]$ such that $f = q_1 g_1 + \cdots + q_m g_m$. But as all the g_i's are monic (i.e., with 1 as leading coefficient), when dividing f by G all the denominators which may appear are powers of d, that is, $q_1,\ldots,q_m \in \frac{1}{d^r}\mathbf{R}[X_1,\ldots,X_n]$ for some $r \in \mathbb{N}$. As a matter of fact, as $g_i = \frac{1}{d_i}h_i \in \frac{1}{d}\mathbf{R}[X_1,\ldots,X_n]$ and $f \in \mathbf{R}[X_1,\ldots,X_n]$, the coefficient of the first quotient term appearing when starting the division of f by one of the g_i's is in $\mathbf{R}$, the second is in $\frac{1}{d}\mathbf{R}$, the third in $\frac{1}{d^2}\mathbf{R}$, and so on. It follows that

$$d^{r+1}f = \sum_{i=1}^{m} d^r q_i \frac{d}{d_i} h_i \in J, \text{ and thus, } f \in (J : d^{\infty}).$$

Conversely, if $f \in (J : d^\infty)$, then $d^k f \in J \subseteq (I \otimes_{\mathbf{R}} \mathbf{K})$ for some $k \in \mathbb{N}$, and hence $f \in (I \otimes_{\mathbf{R}} \mathbf{K}) \cap \mathbf{R}[X_1, \ldots, X_n]$. □

Corollary 300. *Let $\mathbf{R}$ be a Gröbner domain. Denote its quotient field by $\mathbf{K}$, and consider an ideal $I = \langle f_1, \ldots, f_s \rangle$ of $\mathbf{R}[X_1, \ldots, X_n]$. Let us fix a monomial order $>$ on $\mathbf{R}[X_1, \ldots, X_n]$ and consider a Gröbner basis $\{h_1, \ldots, h_m\}$ for I in $\mathbf{R}[X_1, \ldots, X_n]$. Denoting by $\delta = \prod_{i=1}^m \mathrm{LC}(h_i)$ (or $\delta = \mathrm{lcm}(\mathrm{LC}(h_1), \ldots, \mathrm{LC}(h_m))$ if such a notion exists), we have*

$$\mathrm{Sat}(I) = (I : \delta^\infty).$$

Proof. It suffices to use Proposition 299 and the fact that $\{\frac{1}{\mathrm{LC}(h_1)} h_1, \ldots, \frac{1}{\mathrm{LC}(h_m)} h_m\}$ is a normalized Gröbner basis for $I \otimes_{\mathbf{R}} \mathbf{K}$. □

From Corollary 300 ensues the following algorithm for computing the saturation of finitely-generated ideals of $\mathbf{R}[X_1, \ldots, X_n]$, where $\mathbf{R}$ is a Gröbner domain.

Algorithm 301. (An Algorithm for Computing the Saturation of Finitely-Generated Ideals of $\mathbf{R}[X_1, \ldots, X_n]$, Where $\mathbf{R}$ Is a Gröbner Domain)

Input: $f_1, \ldots, f_s \in \mathbf{R}[X_1, \ldots, X_n]$, where $\mathbf{R}$ be a Gröbner domain.
Output: a finite generating set $G \subseteq \mathbf{R}[X_1, \ldots, X_n]$ of $\mathrm{Sat}(\langle f_1, \ldots, f_s \rangle)$.

1. Compute a Gröbner basis $\{h_1, \ldots, h_m\}$ for $\langle f_1, \ldots, f_s \rangle$ in $\mathbf{R}[X_1, \ldots, X_n]$ with respect to the lexicographic monomial order $>$ on $\mathbf{R}[X_1, \ldots, X_n]$ with $X_n > \cdots > X_1$.

2. $\delta := \prod_{i=1}^m \mathrm{LC}(h_i)$ (or $\delta := \mathrm{lcm}(\mathrm{LC}(h_1), \ldots, \mathrm{LC}(h_m)) = \mathrm{LC}(h_1) \vee \cdots \vee \mathrm{LC}(h_m))$ if such a notion exists, for example if $\mathbf{R} = \mathbb{Z}$).

3. Compute a Gröbner basis $\{g_1, \ldots, g_r\}$ for $\langle f_1, \ldots, f_s, \delta Z - 1 \rangle$ in $\mathbf{R}[X_1, \ldots, X_n, Z]$ with respect to the lexicographic monomial order $>$ on $\mathbf{R}[X_1, \ldots, X_n, Z]$ with $Z > X_n > \cdots > X_1$.

4. $G := \{g_1, \ldots, g_r\} \cap \mathbf{R}[X_1, \ldots, X_n]$.

Example 302. (Example 290 Continued) Let us compute a finite generating set for $\mathrm{Sat}(I)$, where $I = \langle f_1 = 10X + 2Y, f_2 = 6X + 3 \rangle$ is an ideal of $\mathbb{Z}[X, Y]$. We know, via Example 290, that $\{h_1 = X - Y + 3, h_2 = 6Y - 15\}$ is a reduced Gröbner basis of I with respect to the lexicographic order with $X > Y$. We have $\delta := \mathrm{LC}(h_1) \vee \mathrm{LC}(h_2)) = 1 \vee 6 = 6$. We obtain $\{X - Y + 3, 2Y - 5, Z - Y + 2\}$ as a Gröbner basis of $\langle h_1, h_2, 6Z - 1 \rangle$ with respect to the lexicographic order with $Z > X > Y$. Thus, $G = \{X - Y + 3, 2Y - 5\}$ is a generating set of $\mathrm{Sat}(I)$.

Proposition 303. *Let $I = \langle f_1, \ldots, f_s \rangle$ be a finitely-generated ideal of $\mathbb{Z}[X_1, \ldots, X_n]$. Let us fix a monomial order $>$ on $\mathbb{Z}[X_1, \ldots, X_n]$, and consider a reduced Gröbner basis $\{h_1, \ldots, h_m\}$ for I in $\mathbb{Z}[X_1, \ldots, X_n]$ (computed with Algorithm 288 for example). Denoting by $\delta = \mathrm{lcm}(\mathrm{LC}(h_1), \ldots, \mathrm{LC}(h_m))$ and suppose that we can compute the set*

$\{p_1,\ldots,p_r\}$ *of the prime numbers dividing* δ *(these primes are called the essential primes of* I *[82]). Then:*

(i) $\mathrm{Sat}_{\mathbb{Z}}(I) = S^{-1}I \cap \mathbb{Z}[X_1,\ldots,X_n]$, *where* S *is the monoid* $p_1^{\mathbb{N}}\cdots p_r^{\mathbb{N}}$.

(ii) I *is* $\mathbb{Z}$-*saturated if and only if* $I\mathbb{Z}_{p_i\mathbb{Z}}[X_1,\ldots,X_n]$ *is* $\mathbb{Z}_{p_i\mathbb{Z}}$-*saturated for all* $\leq i \leq r$.

(iii) *If* , *for* $1 \leq i \leq r$, $\{\frac{f_{i,1}}{a_{i,1}},\ldots,\frac{f_{i,\ell_i}}{a_{i,\ell_i}}\}$ *is a generating set for* $\mathrm{Sat}_{\mathbb{Z}_{p_i\mathbb{Z}}}(I\mathbb{Z}_{p_i\mathbb{Z}}$ $[X_1,\ldots,X_n])$, *with* $f_{i,j} \in \mathbb{Z}[X_1,\ldots,X_n]$ *and* $a_{i,j} \in \mathbb{Z}\setminus p_i\mathbb{Z}$, *then*

$$\mathrm{Sat}_{\mathbb{Z}}(I) = \langle f_1,\ldots,f_s,f_{1,1},\ldots,f_{1,\ell_1},\ldots,f_{r,1},\ldots,f_{r,\ell_r}\rangle.$$

Proof.   (i) By virtue of Lemma 260 and Corollary 300, we have

$$\mathrm{Sat}(I) = (I\mathbb{Z}[\tfrac{1}{\delta}][X_1,\ldots,X_n]) \cap \mathbb{Z}[X_1,\ldots,X_n].$$

But $\mathbb{Z}[\frac{1}{\delta}]$ is nothing but $S^{-1}\mathbb{Z}$.

(ii) "$\Rightarrow$" Let $g = \frac{h}{a}$, with $a \in \mathbb{Z}\setminus p_i\mathbb{Z}$ and $h \in \mathbb{Z}[X_1,\ldots,X_n]$, such that $\alpha g \in I\mathbb{Z}_{p_i\mathbb{Z}}[X_1,\ldots,X_n]$ for some $\alpha \in \mathbb{Z}_{p_i\mathbb{Z}}\setminus\{0\}$. It follows that there exists $b \in \mathbb{Z}\setminus\{0\}$ such that $bh \in I$. But, as I is $\mathbb{Z}$-saturated, we infer that $h \in I$, and, thus, $g \in I\mathbb{Z}_{p_i\mathbb{Z}}[X_1,\ldots,X_n]$.

Note that the above proof applies for any monoid of $\mathbb{Z}$, not only for $\mathbb{Z}\setminus p_i\mathbb{Z}$.

"$\Leftarrow$" Let $f \in \mathrm{Sat}_{\mathbb{Z}}(I)$. By virtue of (i), there exists $m \in \mathbb{N}$ such that

$$\delta^m f \in I. \qquad (0)$$

In addition, for all $\leq i \leq r$, as $I\mathbb{Z}_{p_i\mathbb{Z}}[X_1,\ldots,X_n]$ is $\mathbb{Z}_{p_i\mathbb{Z}}$-saturated, there exists $\alpha_i \in \mathbb{Z}\setminus p_i\mathbb{Z}$ such that

$$\alpha_i f \in I. \qquad (i)$$

As $\gcd(\delta^m,\alpha_1,\ldots,\alpha_r) = 1$, there exist $\beta,\beta_1,\ldots,\beta_r \in \mathbb{Z}$ such that $\beta\delta^m + \beta_1\alpha_1 + \cdots + \beta_r\alpha_r = 1$ ($\mathbb{Z}$ being a Bezout domain), and, thus, combining (0), (1),…,(r), one gets $f \in I$, as desired.

(iii) It is clear that $\mathrm{Sat}_{\mathbb{Z}}(I) \supseteq \langle f_1,\ldots,f_s,f_{1,1},\ldots,f_{1,\ell_1},\ldots,f_{r,1},\ldots,f_{r,\ell_r}\rangle =: J$. Conversely, let $f \in \mathrm{Sat}_{\mathbb{Z}}(I)$. By virtue of (i), there exists $m \in \mathbb{N}$ such that

$$\delta^m f \in \langle f_1,\ldots,f_s\rangle. \qquad (0)$$

In addition, for all $\leq i \leq r$, there exists $\alpha_i \in \mathbb{Z} \setminus p_i\mathbb{Z}$ such that

$$\alpha_i f \in \langle f_{i,1}, \ldots, f_{i,\ell_i} \rangle. \qquad (i)$$

As $\gcd(\delta^m, \alpha_1, \ldots, \alpha_r) = 1$, there exist $\beta, \beta_1, \ldots, \beta_r \in \mathbb{Z}$ such that $\beta\delta^m + \beta_1\alpha_1 + \cdots + \beta_r\alpha_r = 1$, and, thus, combining (0), (1),...,(r), one gets $f \in J$, as desired.

$\square$

From Proposition 303 ensues an obvious algorithm for computing the $\mathbb{Z}$-saturation of a finitely-generated ideal $I = \langle f_1, \ldots, f_s \rangle$ of $\mathbb{Z}[X_1, \ldots, X_n]$. This algorithm (Algorithm 301, too, when used with $\mathbf{R} = \mathbb{Z}$) requires an initial Gröbner basis computation over $\mathbb{Z}$ (see Sect. 3.3.11 for such computation) as well saturation computation over the rings $\mathbb{Z}_{p_1\mathbb{Z}}, \ldots, \mathbb{Z}_{p_r\mathbb{Z}}$, where $p_1, \ldots, p_r$ are the essentials primes of I. Unfortunately, this initial computation is often the bottleneck of the whole algorithm, since a Gröbner basis calculation over the integers can be several orders of magnitude slower than a Gröbner basis calculation over a field. We propose hereafter two solutions to this problem:

- The use of the dynamical method (see Sect. 3.4.2).

- The use Algorithm 305 below following the philosophy of Sect. 3.3.10 (Gröbner bases over the integers via prime factorization).

We need first to the following result.

Proposition 304. *Consider an ideal* $I = \langle f_1, \ldots, f_s \rangle$ *of* $\mathbb{Z}[X_1, \ldots, X_n]$. *Let us fix a monomial order* $>$ *on* $\mathbb{Z}[X_1, \ldots, X_n]$ *and consider a normalized Gröbner basis* $G = \{g_1, \ldots, g_m\}$ *for* $J := I \otimes_{\mathbb{Z}} \mathbb{Q}$ *in* $\mathbb{Q}[X_1, \ldots, X_n]$. *Denote by* $g_i = \frac{h_i}{d_i}$ *where* $h_i \in \mathbb{Z}[X_1, \ldots, X_n]$, $J = \langle h_1, \ldots, h_m \rangle$, $d_i = \mathrm{LC}(h_i) \in \mathbb{Z}$, *take* $d = \mathrm{lcm}(d_1, \ldots, d_m)$, *and denote by* $\{p_1, \ldots, p_r\}$ *the set of the prime numbers dividing* d. *Then:*

(i) $\mathrm{Sat}_{\mathbb{Z}}(I) = \mathrm{Sat}_{\mathbb{Z}}(J) = S^{-1}J \cap \mathbb{Z}[X_1, \ldots, X_n]$, *where* S *is the monoid* $p_1^{\mathbb{N}} \cdots p_r^{\mathbb{N}}$.

(ii) J *is* $\mathbb{Z}$-*saturated if and only if* $J\mathbb{Z}_{p_i\mathbb{Z}}[X_1, \ldots, X_n]$ *is* $\mathbb{Z}_{p_i\mathbb{Z}}$-*saturated for all* $1 \leq i \leq r$.

(iii) *If , for* $1 \leq i \leq r$, $\{\frac{f_{i,1}}{a_{i,1}}, \ldots, \frac{f_{i,\ell_i}}{a_{i,\ell_i}}\}$ *is a generating set for* $\mathrm{Sat}_{\mathbb{Z}_{p_i\mathbb{Z}}}(I\mathbb{Z}_{p_i\mathbb{Z}}$ $[X_1, \ldots, X_n])$, *with* $f_{i,j} \in \mathbb{Z}[X_1, \ldots, X_n]$ *and* $a_{i,j} \in \mathbb{Z} \setminus p_i\mathbb{Z}$, *then*

$$\mathrm{Sat}_{\mathbb{Z}}(I) = \langle h_1, \ldots, h_m, f_{1,1}, \ldots, f_{1,\ell_1}, \ldots, f_{r,1}, \ldots, f_{r,\ell_r} \rangle.$$

Proof. (i) This is Proposition 299.

(ii) Do as in the proof of item (ii) in Proposition 303.

(iii) Do as in the proof of item (iii) in Proposition 303 and use the fact that $\text{Sat}_{\mathbb{Z}}(I) = \text{Sat}_{\mathbb{Z}}(J)$ and $\text{Sat}_{\mathbb{Z}_{p_i\mathbb{Z}}}(I\mathbb{Z}_{p_i\mathbb{Z}}[X_1,\dots,X_n]) = \text{Sat}_{\mathbb{Z}_{p_i\mathbb{Z}}}(J\mathbb{Z}_{p_i\mathbb{Z}}[X_1,\dots,X_n])$.

□

For an analogue to Proposition 304 for finitely-generated sub-$\mathbb{Z}$-modules of $\mathbb{Z}[X_1,\dots,X_k]^m$, see Proposition 329.

Algorithm 305. (An algorithm for Computing the Saturation of Finitely-Generated Ideals of $\mathbb{Z}[X_1,\dots,X_n]$ via Prime Factorization)

Input: $f_1,\dots,f_s \in \mathbb{Z}[X_1,\dots,X_n]$.
Output: a finite generating set $H \subseteq \mathbb{Z}[X_1,\dots,X_n]$ of $\text{Sat}(\langle f_1,\dots,f_s\rangle)$.

1. Fix a monomial order $>$ on $\mathbb{Z}[X_1,\dots,X_n]$, and compute a normalized Gröbner basis $G = \{g_1,\dots,g_m\}$ for $J := I \otimes_{\mathbb{Z}} \mathbb{Q}$ in $\mathbb{Q}[X_1,\dots,X_n]$ (use Algorithm 235 with **R** a discrete field). Denote by $g_i = \frac{h_i}{d_i}$ where $h_i \in \mathbb{Z}[X_1,\dots,X_n]$, $d_i = \text{LC}(h_i) \in \mathbb{Z}$, take $d = \text{lcm}(d_1,\dots,d_m)$, and compute the set $\{p_1,\dots,p_r\}$ of the prime numbers dividing d.

2. For $\leq i \leq r$, compute a finite generating set $\{\frac{f_{i,1}}{a_{i,1}},\dots,\frac{f_{i,\ell_i}}{a_{i,\ell_i}}\}$ for $\text{Sat}_{\mathbb{Z}_{p_i\mathbb{Z}}}(I\mathbb{Z}_{p_i\mathbb{Z}}[X_1,\dots,X_n])$, with $f_{i,j} \in \mathbb{Z}[X_1,\dots,X_n]$ and $a_{i,j} \in \mathbb{Z} \setminus p_i\mathbb{Z}$ (for example, with Algorithm 301).

3. $H := \{h_1,\dots,h_m,f_{1,1},\dots,f_{1,\ell_1},\dots,f_{r,1},\dots,f_{r,\ell_r}\}$.

For an analogue to Algorithm 305 for finitely-generated sub-$\mathbb{Z}$-modules of $\mathbb{Z}[X_1,\dots,X_k]^m$, see Algorithm 330.

3.4.2 Computing Dynamically a Generating Set for Syzygies of Polynomials Over Gröbner Arithmetical Rings

Let **R** be a Gröbner arithmetical ring and consider $f_1,\dots,f_s \in \mathbf{R}[X_1,\dots,X_n] \setminus \{0\}$. Our goal is to compute a generating set for $\text{Syz}(f_1,\dots,f_s)$. We have first to compute a dynamical Gröbner basis $G = \{(S_1,G_1),\dots,(S_k,G_k)\}$ for the ideal $\langle f_1,\dots,f_s\rangle$ of $\mathbf{R}[X_1,\dots,X_n]$. Denoting by $H_j = \{h_{j,1},\dots,h_{j,p_j}\}$ a generating set for $\text{Syz}(f_1,\dots,f_s)$ over $(S_j^{-1}\mathbf{R})[X_1,\dots,X_n]$, $1 \leq j \leq k$, for each $1 \leq i \leq p_j$, there exists $d_{j,i} \in S_j$ such that $d_{j,i}h_{j,i} \in \mathbf{R}[X_1,\dots,X_n]$. Under these hypotheses, we have:

Theorem 306. *(Syzygies Over Gröbner Arithmetical Rings) As an* $\mathbf{R}[X_1,\dots,X_n]$*-module,*

$$\text{Syz}(f_1,\dots,f_s) = \langle d_{1,1}h_{1,1},\dots,d_{1,p_1}h_{1,p_1},\dots,d_{k,1}h_{k,1},\dots,d_{k,p_k}h_{k,p_k}\rangle.$$

Proof. It is clear that $\langle d_{1,1}h_{1,1},\ldots,d_{1,p_1}h_{1,p_1},\ldots,d_{k,1}h_{k,1},\ldots,d_{k,p_k}h_{k,p_k}\rangle \subseteq$ $\mathrm{Syz}(f_1,\ldots,f_s)$. For the converse, let $h \in \mathrm{Syz}(f_1,\ldots,f_s)$ over $\mathbf{R}[X_1,\ldots,X_n]$. It is also a syzygy for $(f_1,\ldots,f_s)$ over $(S_j^{-1}\mathbf{R})[X_1,\ldots,X_n]$ for each $1 \le j \le k$. Hence, for some $d_j \in S_j$, $d_j h \in \langle d_{j,1}h_{j,1},\ldots,d_{j,p_j}h_{j,p_j}\rangle$ over $\mathbf{R}[X_1,\ldots,X_n]$. On the other hand, as $S_1,\ldots,S_k$ are comaximal multiplicative subsets of $\mathbf{R}$, there exist $\alpha_1,\ldots,\alpha_k \in \mathbf{R}$ such that $\sum_{j=1}^{k}\alpha_j d_j = 1$. From the fact that $h = \sum_{j=1}^{k}\alpha_j d_j h$, we infer that

$$h \in \langle d_{1,1}h_{1,1},\ldots,d_{1,p_1}h_{1,p_1},\ldots,d_{k,1}h_{k,1},\ldots,d_{k,p_k}h_{k,p_k}\rangle$$

over $\mathbf{R}[X_1,\ldots,X_n]$. □

3.4.2.1 A Dynamical Method for Computing the Syzygy Module for Polynomials Over a Gröbner Arithmetical Ring

Let $\mathbf{R}$ be a Gröbner arithmetical ring and consider $f_1,\ldots,f_s \in \mathbf{R}[X_1,\ldots,X_n] \setminus \{0\}$. Our goal is to give a dynamical way of computing a generating set for $\mathrm{Syz}(f_1,\ldots,f_s)$. This method works like the case where the base ring is a Gröbner valuation ring (Sect. 3.4.1). The only difference is when one has to handle two incomparable (under division) elements a,b in $\mathbf{R}$. In that situation, one should first compute $u,v,w \in \mathbf{R}$ such that

$$\begin{cases} ub = va \\ wb = (1-u)a. \end{cases}$$

Now, one opens two branches: the computations are pursued in $\mathbf{R}_u$ and $\mathbf{R}_{1-u}$.

3.4.2.2 An Example of Dynamical Computation

Example 307. Let $I = \langle f_1 = 3XY + 1, \ f_2 = (4+2\theta)Y + 9 \rangle$ in $\mathbb{Z}[\theta][X,Y]$ where $\theta = \sqrt{-5}$. The ring $\mathbb{Z}[\theta]$ is a Dedekind domain which is not principal.

Let us fix the lexicographic order with $X > Y$ as monomial order.

a) Computing a dynamical Gröbner basis and the syzygy module:

We will first compute a dynamical Gröbner basis for I in $\mathbb{Z}[\theta][X,Y]$. We will give all the details of the computations only for one leaf. Since $x_1 := 3$ and $x_2 := 4 + 2\theta$ are not comparable, we have to find $u,v,w \in \mathbb{Z}[\theta]$ such that:

$$\begin{cases} ux_2 = vx_1 \\ wx_2 = (1-u)x_1. \end{cases}$$

Note that as the ring $\mathbb{Z}[\theta]$ has a $\mathbb{Z}$-basis (it is a rank 2 free $\mathbb{Z}$-module), u,v,w can be computed by solving an underdetermined linear system over the integers. A solution is given by: $u = 5 + 2\theta$, $v = 6\theta$, $w = -3$. Then we can open two branches:

$$\mathbb{Z}[\theta]$$
$$\swarrow \quad \searrow$$
$$\mathbb{Z}[\theta]_{4+2\theta} \quad \mathbb{Z}[\theta]_{5+2\theta}$$

Recall that $\mathbb{Z}[\theta]_\alpha := \mathbb{Z}[\theta][\frac{1}{\alpha}]$ for $\alpha \in \mathbb{Z}[\theta]$.

In $\mathbb{Z}[\theta]_{5+2\theta}$:

$$S(f_1, f_2) = \frac{6\theta}{5+2\theta} f_1 - X f_2 = -9X + \frac{6\theta}{5+2\theta} =: f_3,$$
$$S(f_1, f_3) = -3f_1 - Y f_3 = -\frac{6\theta}{5+2\theta} Y - 3 =: f_4,$$
$$S(f_1, f_4) = -\frac{2\theta}{5+2\theta} f_1 - X f_4 = 3X - \frac{2\theta}{5+2\theta} =: f_5,$$
$$f_2 \xrightarrow{f_4} 0, \ f_3 \xrightarrow{f_5} 0,$$
$$S(f_1, f_5) = f_1 - Y f_5 = \frac{2\theta}{5+2\theta} Y + 1 =: f_6,$$
$$f_4 \xrightarrow{f_6} 0, \ S(f_2, f_5) = X f_2 - \frac{6\theta}{5+2\theta} Y f_5 \xrightarrow{f_5, f_6} 0.$$

As 2 and 3 are not comparable under division in $\mathbb{Z}[\theta]_{5+2\theta}$, we open two news branches:

$$\mathbb{Z}[\theta]_{5+2\theta}$$
$$\swarrow \quad \searrow$$
$$\mathbb{Z}[\theta]_{(5+2\theta).3} \quad \mathbb{Z}[\theta]_{(5+2\theta).2}$$

Recall that $\mathbb{Z}[\theta]_{\alpha.\beta} := \mathbb{Z}[\theta][\frac{1}{\alpha}][\frac{1}{\beta}] = \mathbb{Z}[\theta][\frac{1}{\alpha\beta}]$ for $\alpha, \beta \in \mathbb{Z}[\theta]$.

In $\mathbb{Z}[\theta]_{(5+2\theta).3}$:

$$S(f_1, f_6) = \frac{2\theta}{3(5+2\theta)} f_1 - X f_6 = -\frac{1}{3} f_5 \xrightarrow{f_5} 0,$$
$$S(f_5, f_6) = \frac{2\theta}{3(5+2\theta)} Y f_5 - X f_6 = \frac{20}{3(5+2\theta)^2} Y - X \xrightarrow{f_5} \frac{20}{3(5+2\theta)^2} Y - \frac{2\theta}{3(5+2\theta)} \xrightarrow{f_6} 0.$$

Thus, $G_1 = \{3XY + 1, 3X - \frac{2\theta}{5+2\theta}, \frac{2\theta}{5+2\theta} Y + 1\}$ is a Gröbner basis for $\langle 3XY + 1, (4 + 2\theta)Y + 9 \rangle$ in $\mathscr{M}(5 + 2\theta, 3)^{-1}\mathbb{Z}[\theta] = \mathbb{Z}[\theta]_{(5+2\theta).3}$.

Denoting by $F = [f_1 \ f_2]$ and $G = [g_1 \ g_2 \ g_3]$ with $g_1 = 3XY + 1$, $g_2 = 3X - \frac{2\theta}{5+2\theta}$, $g_3 = \frac{2\theta}{5+2\theta} Y + 1$, we have $G = FT$ with

$$T = \begin{pmatrix} 1 & 3X - \frac{2\theta}{5+2\theta} + \frac{6\theta}{5+2\theta} XY & -3XY + \frac{2\theta}{5+2\theta} Y - \frac{6\theta}{5+2\theta} XY^2 + 1 \\ 0 & -X^2 Y & X^2 Y^2 \end{pmatrix}, \text{ and } F = GS$$

with $S = \begin{pmatrix} 1 & 0 \\ 0 & 0 \\ 0 & 9 \end{pmatrix}$.

$$\mathbf{I}_2 - TS = \begin{pmatrix} 0 & 27XY - 9 - (4+2\theta)Y + 3(4+2\theta)XY^2 \\ 0 & 1 - 9X^2 Y^2 \end{pmatrix},$$

$$r_1 = \begin{pmatrix} 27XY - 9 - (4+2\theta)Y + 3(4+2\theta)XY^2 \\ 1 - 9X^2 Y^2 \end{pmatrix} \in \text{Syz}(F),$$

$$s_{12} = {}^t(1, -Y, -1), s_{13} = {}^t(\tfrac{2\theta}{3(5+2\theta)}, \tfrac{1}{3}, -X), s_{23} = {}^t(0, \tfrac{2\theta}{3(5+2\theta)} Y + \tfrac{1}{3}, -X + \tfrac{2\theta}{3(5+2\theta)}),$$

$Ts_{12}=\begin{pmatrix} 0 \\ 0 \end{pmatrix}$, $Ts_{13}=\begin{pmatrix} 3X^2Y+\frac{4+2\theta}{3}X^2Y^2 \\ \frac{-1}{3}X^2Y-X^3Y^2 \end{pmatrix}$, and $Ts_{23}=Ts_{13}$. Thus, over $\mathbb{Z}[\theta]_{(5+2\theta).3}$ $[X,Y]$,

$$\mathrm{Syz}(F)=\langle \begin{pmatrix} 3X^2Y+\frac{4+2\theta}{3}X^2Y^2 \\ \frac{-1}{3}X^2Y-X^3Y^2 \end{pmatrix}, \begin{pmatrix} 27XY-9-(4+2\theta)Y+3(4+2\theta)XY^2 \\ 1-9X^2Y^2 \end{pmatrix} \rangle.$$

In $\mathbb{Z}[\theta]_{(5+2\theta).2}$:

$G_2=\{3XY+1, 3X-\frac{2\theta}{5+2\theta}, \frac{2\theta}{5+2\theta}Y+1\}$ is a Gröbner basis for $\langle 3XY+1, (4+2\theta)Y+9\rangle$. Thus, over $\mathbb{Z}[\theta]_{(5+2\theta).2}[X,Y]$,

$$\mathrm{Syz}(F)=\langle \begin{pmatrix} \frac{9X^2Y(5+2\theta+2\theta Y)}{2\theta} \\ \frac{-(5+2\theta)(3X^3Y^2+X^2Y)}{2\theta} \end{pmatrix}, \begin{pmatrix} 27XY-9-(4+2\theta)Y+3(4+2\theta)XY^2 \\ 1-9X^2Y^2 \end{pmatrix}\rangle.$$

In $\mathbb{Z}[\theta]_{(4+2\theta)}$:

$G_3=\{3XY+1, (4+2\theta)Y+9, \frac{-27}{4+2\theta}X+1\}$ is a Gröbner basis for $\langle 3XY+1, (4+2\theta)Y+9\rangle$. Over $\mathbb{Z}[\theta]_{(4+2\theta)}[X,Y]$, we have

$$\mathrm{Syz}(F)=\langle \begin{pmatrix} -\frac{9}{4+2\theta}-Y \\ \frac{1}{4+2\theta}+\frac{3XY}{4+2\theta} \end{pmatrix}\rangle.$$

Finally, in $\mathbb{Z}[\theta]$: Over $\mathbb{Z}[\theta][X,Y]$, we have

$$\mathrm{Syz}(F)=\langle \begin{pmatrix} -(4+2\theta)Y-9 \\ 3XY+1 \end{pmatrix}, \begin{pmatrix} 27XY-9-(4+2\theta)Y+3(4+2\theta)XY^2 \\ 1-9X^2Y^2 \end{pmatrix}\rangle$$

$$=\langle \begin{pmatrix} -(4+2\theta)Y-9 \\ 3XY+1 \end{pmatrix}\rangle.$$

As a conclusion, the dynamical evaluation of the problem of constructing a Gröbner basis for I produces the following evaluation tree:

$$\mathbb{Z}[\theta]$$
$$\swarrow \qquad \searrow$$
$$\mathbb{Z}[\theta]_{4+2\theta} \qquad \mathbb{Z}[\theta]_{5+2\theta}$$
$$\swarrow \qquad \searrow$$
$$\mathbb{Z}[\theta]_{(5+2\theta).3} \quad \mathbb{Z}[\theta]_{(5+2\theta).2}$$

The obtained dynamical Gröbner basis of I is

$$G=\{(\mathcal{M}(5+2\theta),G_1), (\mathcal{M}(4+2\theta),G_2)\}.$$

b) The ideal membership problem: Suppose that we have to deal with the ideal membership problem:

$$f=(4\theta-1)X^2Y+6\theta XY^2+9\theta X^2+3X-4Y-9 \in? I$$

Let us first execute the dynamical division algorithm of f by $G_1 = \{f_1 = 3XY + 1, f_5 = -3X + \frac{2\theta}{5+2\theta}, f_6 = \frac{2\theta}{5+2\theta}Y + 1\}$ in the ring $\mathbb{Z}[\theta]_{(5+2\theta).3}[X,Y]$. With the same notation as in the Division Algorithm 211, one obtains

q_1	q_5	q_6	p
$\frac{4\theta-1}{3}X$	0	0	$6\theta XY^2 + 9\theta X^2 + \frac{10-4\theta}{3}X - 4Y - 9$
$\frac{4\theta-1}{3}X + 2\theta Y$	0	0	$9\theta X^2 + \frac{10-4\theta}{3}X - (4+2\theta)Y - 9$
$\frac{4\theta-1}{3}X + 2\theta Y$	$-3\theta X$	0	$-(4+2\theta)Y - 9$
$\frac{4\theta-1}{3}X + 2\theta Y$	$-3\theta X$	-9	0

Thus, the answer to this ideal membership problem in the ring $\mathbb{Z}[\theta]_{(5+2\theta).3}[X,Y]$ is positive and one obtains

$$f = (\tfrac{4\theta-1}{3}X + 2\theta Y)f_1 - 3\theta X f_5 - 9f_6.$$

But since

$$f_5 = (\frac{-6\theta}{5+2\theta}XY - 3X + \frac{2\theta}{5+2\theta})f_1 - X^2Y f_2, \text{ and}$$

$$f_6 = (\frac{-6\theta}{5+2\theta}XY^2 - 3XY + \frac{2\theta}{5+2\theta}Y + 1)f_1 - X^2Y^2 f_2, \text{ one infers that}$$

$$f = [\frac{-90}{5+2\theta}X^2Y + 9\theta X^2 + \frac{54\theta}{5+2\theta}XY^2 + 27XY + \frac{6\theta+15}{5+2\theta}X - 4Y - 9]f_1$$
$$+ [3\theta X^3Y + 9X^2Y^2]f_2.$$

Seeing that 3 does not appear in the denominators of the relation above, we can say that we have a positive answer to our ideal membership problem in the ring $\mathbb{Z}[\theta]_{5+2\theta}[X,Y]$ without dealing with the leaf $\mathbb{Z}[\theta]_{(5+2\theta).2}$. Clearing the denominators, we get:

$$(5+2\theta)f = [-90X^2Y + 45(\theta-2)X^2 + 54\theta XY^2 + 27(5+2\theta)XY + (6\theta+15)X$$
$$-4(5+2\theta)Y - 9(5+2\theta)]f_1 + [15(\theta-2)X^3Y + 9(5+2\theta)X^2Y^2]f_2. \quad \text{(A)}$$

It remains to execute the dynamical division algorithm of f by $G_2 = \{f_1 = 3XY + 1, f_7 = -\frac{27}{4+2\theta}X + 1, f_8 = Y + \frac{9}{4+2\theta}\}$ in the ring $\mathbb{Z}[\theta]_{4+2\theta}[X,Y]$. The division is as follows

q_1	q_7	q_8	p
0	0	$(4\theta-1)X^2$	$6\theta XY^2 - \frac{81}{4+2\theta}X^2 + 3X - 4Y - 9$
$2\theta Y$	0	$(4\theta-1)X^2$	$\frac{-81}{4+2\theta}X^2 + 3X - (4+2\theta)Y - 9$
$2\theta Y$	$3X$	$(4\theta-1)X^2$	$-(4+2\theta)Y - 9$
$2\theta Y$	$3X$	$(4\theta-1)X^2 - (4+2\theta)$	0

Thus, the answer to this ideal membership problem in the ring $\mathbb{Z}[\theta]_{4+2\theta}[X,Y]$ is positive and one obtains:

$$f = 2\theta Y f_1 + 3X f_7 + ((4\theta - 1)X^2 - (4 + 2\theta))f_8.$$

But since

$$
\begin{aligned}
f_7 &= f_1 - \frac{3}{4+2\theta}X f_2, \text{ and} \\
f_8 &= (Y + \frac{9}{4+2\theta})f_1 - \frac{3}{4+2\theta}XY f_2, \text{ one infers that} \\
(4+2\theta)f &= [(14\theta - 44)X^2Y + 9(4\theta - 1)X^2 - 4(4+2\theta)Y + 3(4+2\theta)X \\
&\quad -9(4+2\theta)]f_1 + [-9X^2 - 3(4\theta - 1)X^3Y + 3(4+2\theta)XY]f_2. \quad \text{(B)}
\end{aligned}
$$

Using the Bezout identity $(5+2\theta) - (4+2\theta) = 1$, (A) $-$ (B) $\Rightarrow$

$$
\begin{aligned}
f &= [(46 - 14\theta)X^2Y + 9(\theta - 9)X^2 + 54\theta XY^2 + 27(5+2\theta)XY + 3X - 4Y - 9]f_1 \\
&\quad + [3(9\theta - 11)X^3Y + 9(5+2\theta)X^2Y^2 + 9X^2 - 3(4+2\theta)XX]f_2,
\end{aligned}
$$

a complete positive answer.

Chapter 4

Syzygies in Polynomial Rings Over Valuation Domains

It is folklore (see for example Theorem 7.3.3 in [68]) that if V is a valuation domain, then $V[X_1,\ldots,X_k]$ ($k \in \mathbb{N}$) is coherent: that is, syzygy modules of finitely-generated ideals of $V[X_1,\ldots,X_k]$ are finitely-generated. The proof in the above-mentioned reference relies on a profound and difficult result published in a huge paper by Gruson and Raynaud [75]. There is nevertheless no known general algorithm for this remarkable result, and it seems difficult to compute the syzygy module even for small polynomials. An exception is the case where the valuation domain is Gröbner (or, equivalently, a valuation domain of Krull dimension ≤ 1, see Corollary 257). In that case the syzygy module of polynomials in $V[X_1,\ldots,X_k]$ can be computed via Gröbner bases (see Sect. 3.4.1.1).

The main objective of this chapter is to give a general algorithm [50] for computing a finite generating set for the syzygies of any finitely-generated ideal of $V[X_1,\ldots,X_k]$ (V a valuation domain with explicit divisibility) which neither relies on Noetherianity nor on Krull dimension. We will in fact give an algorithm for computing a finite generating set for the V-saturation of any finitely-generated submodule of $V[X_1,\ldots,X_k]^n$. This algorithm is based on a notion of "echelon form" which ensures its correctness. The proposed algorithm terminates when two (Hilbert) series on the quotient field of V and the residue field of V coincide. Computing syzygies over $V[X_1,\ldots,X_k]$ is one important application of the saturation algorithm we give. In the univariate case [113, 48], we will give precise complexity bounds.

© Springer International Publishing Switzerland 2015

I. Yengui, *Constructive Commutative Algebra*, Lecture Notes in Mathematics 2138,

DOI 10.1007/978-3-319-19494-3_4

4.1   Preliminary Tools

It may be useful to recall the following terminology.

Terminology: Recall that, for an arbitrary ring $\mathbf{R}$, we denote by $\mathbf{R}^{\times}$ its group of units. The ring $\mathbf{R}$ is said to be *discrete* if there is an algorithm deciding if $x = 0$ or $x \neq 0$ for an arbitrary element of $\mathbf{R}$. A ring $\mathbf{R}$ is said to be *local* if we have explicitly the implication

$$\forall x, y \in \mathbf{R}, x + y \in \mathbf{R}^{\times} \implies (x \in \mathbf{R}^{\times} \vee y \in \mathbf{R}^{\times}).$$

Recall that a local ring $\mathbf{R}$ has as unique maximal ideal its *Jacobson radical* $\mathrm{Rad}(\mathbf{R}) = \{x \in \mathbf{R} \mid 1 + x\mathbf{R} \subseteq \mathbf{R}^{\times}\}$. The quotient ring $\mathbf{k} = \mathbf{R}/\mathrm{Rad}(\mathbf{R})$ is a field, called *residual field* of $\mathbf{R}$. The local ring $\mathbf{R}$ is said to be residually discrete if we have explicitly the disjunction $\forall x \in \mathbf{R}, (x \in \mathbf{R}^{\times} \vee x \in \mathrm{Rad}(\mathbf{R}))$. In that case, the residual field is discrete. We have an algorithm deciding the disjunction "$x = 0$ or x is invertible" for all $x \in \mathbf{k}$.

Note that it may happen that the residually discrete local ring $\mathbf{R}$ is the result of a construction in a proof while it is unknown whether $\mathbf{R}$ is trivial or not. In the case where $\mathbf{R}$ is trivial, the "residual field" is also the trivial ring and satisfies "any element is zero or a unit".

We need the following series of definitions.

Definition 308. Let $\mathbf{R}$ be a nontrivial residually discrete local ring, and fix a monomial order $>$ on $\mathbf{R}[X_1, \ldots, X_k]$.

(1) A polynomial $f \in \mathbf{R}[X_1, \ldots, X_k]$ is said to be *primitive* if it has an invertible coefficient.

(2) A vector $u = (u_1, \ldots, u_m) \in \mathbf{R}[X_1, \ldots, X_k]^m$ is said to be *primitive* if it has a primitive component. The position i of the last (from left to right) primitive u_i will be denoted by $\mathrm{index}(u)$, the last monomial (i.e., the one with highest multi-degree) of $u_{\mathrm{index}(u)}$ which has an invertible coefficient will be denoted by $\mathrm{PrimMon}(u)$, and the coefficient of this monomial will be denoted by $\mathrm{PrimCoeff}(u)$. For example, if $\mathbf{R} = \mathbb{Z}_{2\mathbb{Z}} = \{\frac{a}{b} \in \mathbb{Q} \mid (a, b) \in \mathbb{Z} \times \mathbb{Z}$ and b is odd$\}$ and $u = (-4 + 2XY, 1 - 6X^3, 5X^7 - 3X^4Y + 6XY^4)$, fixing the lexicographic order with $X < Y$ as monomial order, we have $\mathrm{index}(u) = 3$, $\mathrm{PrimMon}(u) = X^4Y$, and $\mathrm{PrimCoeff}(u) = -3$.

(3) We suppose that $\mathbf{R}$ is a valuation ring (recall that a valuation ring $\mathbf{R}$ is a ring in which for all $a, b \in \mathbf{R}$, either a divides b or b divides a, with an explicit "or"). For a vector $u = (u_1, \ldots, u_m) \in \mathbf{R}[X_1, \ldots, X_k]^m \setminus \{0\}$, as all the coefficients of the u_i's are comparable under division, denoting by a the right-most coefficient (components from left to right and then accordingly to increasing multi-degrees) of u dividing all the others, the primitive vector $\frac{1}{a}u$ is called *the primitive version* of u and denoted by $\mathrm{Prim}(u)$. For example, if $\mathbf{R} = \mathbb{Z}_{2\mathbb{Z}}$ and $u = (8X - 4, 14 - 18X^2, 2 - 6X + 4X^2)$, then $\mathrm{Prim}(u) = -\frac{1}{6}u$.

(4) For a primitive vector $u = (u_1,\ldots,u_m) \in \mathbf{R}[X_1,\ldots,X_k]^m$, we denote by $\mathscr{I}(u)$ the couple $(\mathrm{index}(u), \mathrm{mdeg}(\mathrm{PrimMon}(u))) \in [\![1,m]\!] \times \mathbb{N}^k$. It will be called the *height* of u.

(5) Let u, v be two primitive vectors in $\mathbf{R}[X_1,\ldots,X_k]^m$. By the result of *the reduction of v by u* we mean the vector $w = v + \alpha u$ where $\alpha \in \mathbf{R}$ is chosen such that the term of multi-degree $\mathrm{mdeg}(\mathrm{PrimMon}(u))$ and in position $\mathrm{index}(u)$ does not appear in w (we denote $v \xrightarrow{u} w$). Let us take again the example seen in item 2 above with $\mathrm{index}(u) = 3$ and $\mathrm{PrimMon}(u) = X^4Y$ and consider the vector $v = (0, X + X^2Y, 2XY^2 + 5X^4Y)$. The term $5X^4Y$ in v disappears with $\alpha = \frac{5}{3}$ as follows

$$v \xrightarrow{u} v + \frac{5}{3}u = (-\frac{20}{3} + \frac{10}{3}XY, \frac{5}{3} - 10X^3, \frac{25}{3}X^7 + 10XY^4 + 10XY^2).$$

(6) For a list $S = [s_1, s_2,\ldots]$ of vectors in $\mathbf{R}[X_1,\ldots,X_k]^m$, by MS, where M is a monomial, we mean the list $[Ms_1, Ms_2,\ldots]$. Moreover, if $\mathscr{N}$ is a totally ordered set of monomials $N_0 < N_1 < \cdots$ in $\mathbf{R}[X_1,\ldots,X_k]$, by $\mathscr{N}S$ we mean the list $[N_0S, N_1S,\cdots]$.

(7) The total degree of a vector $u \in \mathbf{R}[X_1,\ldots,X_k]^m$ is the maximum of the total degrees of its entries. It will be denoted by $\mathrm{tdeg}(u)$.

Definition 309. Let $\mathbf{R}$ be a residually discrete local ring, $L := [L_1, L_2,\ldots]$ a list of vectors in $\mathbf{R}[X_1,\ldots,X_k]^m$ ($m \geq 1$), and fix a monomial order on $\mathbf{R}[X_1,\ldots,X_k]$.

(1) We say that L is *primitive triangular* if all the L_i's are primitive vectors and for each $i \geq 1$, denoting by $\mathrm{PrimMon}(L_i) = M_i$ and $\mathrm{index}(L_i) = n_i$, M_i does not appear (i.e., has coefficient 0) in the n_ith component of L_j for all $j > i$.

(2) We say that L is *in an echelon form* if all the L_i's are primitive vectors and the $\mathscr{I}(L_i)$'s are pairwise different (we disregard the L_i's which are null). Of course, if L is primitive triangular then it is in an echelon form.

(3) Let $S := [s_1, s_2,\ldots]$ a list of vectors in $\mathbf{R}[X_1,\ldots,X_k]^m$ which is in an echelon form. We say that L is *in an echelon form with respect to S* if for all $i, j \geq 1$, denoting by $\mathrm{PrimMon}(s_i) = N_i$ and $\mathrm{index}(s_i) = n_i$, N_i does not appear (i.e., has coefficient 0) in the n_ith component of L_j.

Definition 310. Let $\mathbf{R}$ be a residually discrete local ring, fix a monomial order on $\mathbf{R}[X_1,\ldots,X_k]$, and consider $u, v \in \mathbf{R}[X_1,\ldots,X_k]^m$ ($m \geq 1$).

(1) By an operation of type 1, we mean an operation of type

$$v \leftarrow v + \alpha u,$$

(reduction of v by $u \neq (0,\ldots,0)$, where $\alpha \in \mathbf{R}$ is such that the term of multi-degree $\mathrm{mdeg}(\mathrm{PrimMon}(u))$ and in position $\mathrm{index}(u)$ disappears from v).

(2) By an operation of type 2, we mean an operation of type

$$v \leftarrow \mathrm{Prim}(v), \text{ where } v \neq (0,\dots,0).$$

The following simple but precious three lemmas will be at the heart of the saturation algorithms we will give in this chapter.

Lemma 311. *Let* $\mathbf{R} \subseteq \mathbf{T}$ *be an extension of rings where* $\mathbf{R}$ *is a residually discrete local ring. Fix a monomial order on* $\mathbf{R}[X_1,\dots,X_k]$, *and consider a primitive triangular list* $S = [u_1, u_2, \dots]$ *of vectors in* $\mathbf{R}[X_1,\dots,X_k]^m$ $(m \geq 1)$. *Then,*

$$\left(\sum_{i \geq 1} \mathbf{T} u_i\right) \cap \mathbf{R}[X_1,\dots,X_k]^m = \sum_{i \geq 1} \mathbf{R} u_i.$$

In particular, if S *generates a finitely-generated submodule* M *of* $\mathbf{T}[X_1,\dots,X_k]^m$ *as a* $\mathbf{T}$*-module then it also generates* $M \cap \mathbf{R}[X_1,\dots,X_k]^m$ *as an* $\mathbf{R}$*-module.*

Proof. Let $u \in M \cap \mathbf{R}[X_1,\dots,X_k]^m$. There exist $a_1 \dots, a_s \in \mathbf{T}$ such that

$$u = a_1 u_1 + \cdots + a_s u_s.$$

For each $1 \leq i \leq s$, by identifying the coefficient in component number $\mathrm{index}(u_i)$ and of multidegree $\mathrm{mdeg}(\mathrm{PrimMon}(u_i))$ and denoting $c_i = \mathrm{PrimCoeff}(u_i)$, we obtain:

$$\begin{cases} c_1 a_1 \in \mathbf{R} \\ b_{2,1} a_1 + c_2 a_2 \in \mathbf{R} \\ \vdots \\ b_{s,1} a_1 + b_{s,2} a_2 + \cdots + b_{s,s-1} a_{s-1} + c_s a_s \in \mathbf{R} \end{cases}$$

with $b_{i,j} \in \mathbf{R}$. As $c_1,\dots,c_s \in \mathbf{R}^{\times}$, this triangular system yields to $a_1,\dots,a_s \in \mathbf{R}$, as desired. $\qquad\square$

Lemma 312. *Let* $\mathbf{R}$ *be a residually discrete local ring, fix a monomial order on* $\mathbf{R}[X_1,\dots,X_k]$, *and consider two primitive vectors* u, v *in* $\mathbf{R}[X_1,\dots,X_k]^m$ $(m \geq 1)$ *such that* $\mathscr{I}(u) \neq \mathscr{I}(v)$. *Then the result* w *of the reduction of* v *by* u *is primitive and* $\mathscr{I}(w) = \mathscr{I}(v)$.

Proof. First, suppose that $\mathrm{index}(u) \neq \mathrm{index}(v)$. It suffices to deal with the following two subcases:

– $\mathrm{index}(u) = 1$ and $\mathrm{index}(v) = 2$: write $u = (u_1, u_2, \dots)$ and $v = (v_1, v_2, \dots)$ where u_1, v_2 are primitive polynomials and $v_1 \in \mathrm{Rad}(\mathbf{R})[X_1,\dots,X_k]$. We have $w = (v_1 + \alpha u_1, v_2 + \alpha u_2, \dots)$ for some $\alpha \in \mathrm{Rad}(\mathbf{R})$ and the result clearly follows.

– $\mathrm{index}(u) = 2$ and $\mathrm{index}(v) = 1$: write $u = (u_1, u_2, \dots)$ and $v = (v_1, v_2, \dots)$ where u_2, v_1 are primitive polynomials and $u_1 \in \mathrm{Rad}(\mathbf{R})[X_1,\dots,X_k]$. We have $w = (v_1 + \beta u_1, v_2 + \beta u_2, \dots)$ for some $\beta \in \mathbf{R}$ and the result clearly follows.

The case index(u) = index(v) and mdeg(PrimMon(u)) $\neq$ mdeg(PrimMon(v)) is analogous. $\qquad\square$

Lemma 313. *Let* **R** *be a residually discrete local ring, fix a monomial order on* **R**$[X_1,\ldots,X_k]$*, and consider a list* $L = [u_1, u_2,\ldots]$ *of primitive vectors in* **R**$[X_1,\ldots,X_k]^m$ ($m \geq 1$) *which is in an echelon form. Then we can (theoretically, i.e., as far as we want) transform L into a primitive triangular list* $L' = [u'_1, u'_2,\ldots]$ *only by means of operations of type 1.*

Proof. As in the gaussian algorithm, this can be done with operations of type 1 and 2. But Lemma 312 guaranties that all the vectors computed when reducing L to L' are primitive and so there is no need of operations of type 2.

$\qquad\square$

Definition 314. Let **R** be a domain with quotient field **K** and consider vectors $s_1, s_2,\ldots \in$ **R**$[X_1,\ldots,X_k]^m$ ($m \geq 1$). By the **R**-saturation of $M := \sum_{i=1}^{\infty}$ **R** s_i we mean

$$\text{Sat}(M) := \{s \in \mathbf{R}[X_1,\ldots,X_k]^m \mid \alpha s \in M \text{ for some } \alpha \in \mathbf{R} \setminus \{0\}\}$$
$$= (M \otimes_{\mathbf{R}} \mathbf{K}) \cap \mathbf{R}[X_1,\ldots,X_k]^m.$$

If Sat(M) = M, we say that M is **R**-saturated.

Now we reach the main result the proposed saturation algorithms will be based on.

Proposition 315. *Let* **R** *be a residually discrete local domain, fix a monomial order on* **R**$[X_1,\ldots,X_k]$*, and consider a list* $L = [u_1, u_2,\ldots]$ *of primitive vectors in* **R**$[X_1,\ldots,X_k]^m$ ($m \geq 1$). *If L is in an echelon form then* $\sum_{i=1}^{\infty}$ **R** u_i *is* **R**-saturated.

Proof. On the one hand, by virtue of Lemma 313, we can (theoretically) transform the list L into a primitive triangular list $L' = [u'_1, u'_2,\ldots]$ only with the help of operations of type 1, and thus $\sum_{i=1}^{\infty}$ **R** $u_i = \sum_{i=1}^{\infty}$ **R** u'_i. On the other hand, using Lemma 311 with $S = L'$ and **T** being the quotient field of **R**, we infer that $\sum_{i=1}^{\infty}$ **R** u'_i is **R**-saturated. The result clearly follows. $\qquad\square$

Example 316. Let **R** be a residually discrete local domain, fix a monomial order $>$ on **R**$[X_1,\ldots,X_k]$, denote by $1 = N_0 < N_1 < N_2 < \cdots$ the monomials at $X_1,\ldots,X_k$, and consider a primitive vector u in **R**$[X_1,\ldots,X_k]^m$ ($m \geq 1$). Then, obviously, the list $[N_0 u, N_1 u, N_2 u,\ldots]$ is in an echelon form, and thus, the **R**$[X_1,\ldots,X_k]$-module $\langle u \rangle = \sum_{i=0}^{\infty}$ **R** $N_i u$ is **R**-saturated.

4.2 Saturation of Finitely-Generated Sub-V-Modules of V$[X_1,\ldots,X_k]^m$

The following algorithm will be the cornerstone of the saturation algorithm for finitely-generated sub-V-modules of **V**$[X_1,\ldots,X_k]^m$.

Algorithm 317. (Algorithm for reduction modulo a list in an echelon form)

Input: A primitive vector $s \in \mathbf{V}[X_1, \dots, X_k]^m$, a monomial order on $\mathbf{V}[X_1, \dots, X_k]$, and a finite list $S = [s_1, \dots, s_n]$ of vectors in $\mathbf{V}[X_1, \dots, X_k]^m$ in an echelon form, where $\mathbf{V}$ is a valuation domain and $m \geq 1$.

Output: A reduction $u = \mathrm{PrimRed}(s; S) = \mathrm{PrimRed}(s; s_1, \dots, s_n)$ of s modulo S so that $[S, u]$ becomes in an echelon form and $\mathrm{Sat}(\langle s_1, \dots, s_n, s \rangle) = \mathrm{Sat}(\langle s_1, \dots, s_n, u \rangle)$.

$u := \mathrm{Prim}(s)$

FOR i FROM 1 TO n DO

$u := \mathrm{Prim}(u - \mathrm{PrimCoeff}(s_i)^{-1}\mathrm{Coeff}(u, \mathrm{index}(s_i), \deg(\mathrm{PrimMon}(s_i)))s_i)$

(reduction of u by s_i so that the term of degree $\deg(\mathrm{PrimMon}(s_i))$ and in position $\mathrm{index}(s_i)$ disappears from u)

Algorithm 318. (Algorithm for putting a list in an echelon form with respect to a list already in an echelon form)

Input: A list $U = [u_1, \dots, u_r]$ of vectors in $\mathbf{V}[X_1, \dots, X_k]^m$, a monomial order on $\mathbf{V}[X_1, \dots, X_k]$, and a finite list $S = [s_1, \dots, s_n]$ of vectors in $\mathbf{V}[X_1, \dots, X_k]^m$ in an echelon form, where $\mathbf{V}$ is a valuation domain and $m \geq 1$.

Output: A list $L' = \mathrm{EchelResp}(L; S)$ obtained from L such that $\langle L' \rangle_{\mathbf{V}} = \langle L \rangle_{\mathbf{V}}$ and L' is in an echelon form with respect to S.

FOR i FROM 1 TO r DO

$u_i := \mathrm{PrimRed}(u_i; S)$

$L' := [u_1, \dots, u_r]$

Now we are in position to give the following saturation algorithm for finitely-generated sub-$\mathbf{V}$-modules of $\mathbf{V}[X_1, \dots, X_k]^m$.

Algorithm 319. (Saturation algorithm for a finitely-generated sub-$\mathbf{V}$-module of $\mathbf{V}[X_1, \dots, X_k]^m$)

Input: A finite list $S = [s_1, \dots, s_n]$ of vectors in $\mathbf{V}[X_1, \dots, X_k]^m$, where $\mathbf{V}$ is a valuation domain and $m \geq 1$.

Output: A generating list $U = [u_1, \dots, u_n] = \mathrm{Echel}(S)$ for $\mathrm{Sat}(\mathbf{V}s_1 + \cdots + \mathbf{V}s_n)$.

Fix a monomial order on $\mathbf{V}[X_1, \dots, X_k]$.

$u_1 := \mathrm{Prim}(s_1)$

IF $n \geq 2$ THEN FOR i FROM 2 TO n DO

$u_i := \mathrm{PrimRed}(\mathrm{Prim}(s_i); u_1, \dots, u_{i-1})$ (use Algorithm 317)

Example 320. Consider the list $[s_1 = (8 + 2XY, 8 + 8X^2Y^2), s_2 = (XY, 2 - 2X^2Y^2)]$ of vectors in $\mathbb{Z}_{2\mathbb{Z}}[X, Y]^2$. Executing Algorithm 319, we have

$$\mathrm{Sat}(\mathbb{Z}_{2\mathbb{Z}}s_1 + \mathbb{Z}_{2\mathbb{Z}}s_2) = (\mathbb{Q}s_1 + \mathbb{Q}s_2) \cap \mathbb{Z}_{2\mathbb{Z}}[X, Y]^2 = \mathbb{Z}_{2\mathbb{Z}}u_1 + \mathbb{Z}_{2\mathbb{Z}}u_2, \text{ with}$$

$$u_1 \ := \ \mathrm{Prim}(s_1) = \frac{1}{2}s_1 = (4+XY, 4+4X^2Y^2), \ \mathrm{PrimMon}(u_1) = XY \ \text{and index}(u_1) = 1,$$

$$
\begin{aligned}
u_2 \ :=& \ \mathrm{PrimRed}(\mathrm{Prim}(s_2); u_1) = \mathrm{PrimRed}(s_2; u_1) \\
=& \ \mathrm{Prim}(s_2 - u_1) = \mathrm{Prim}((-4, -2 - 6X^2Y^2)) \\
=& \ -\frac{1}{6}(-4, -2 - 6X^2Y^2) = (\frac{2}{3}, \frac{1}{3} + X^2Y^2) = \frac{1}{3}(2, 1 + 3X^2Y^2).
\end{aligned}
$$

4.2.0.3 Case of Bezout Domains

As for dynamical Gröbner bases, one can obtain a dynamical version of Algorithm 319 for Prüfer domains. In the case where the base ring is $\mathbb{Z}$, the situation is easier as it is a Bezout domain.

Begin by fixing a monomial order $>$ on $\mathbb{Z}[X_1,\ldots,X_k]$. For $v \in \mathbb{Z}[X_1,\ldots,X_k]^m \setminus \{(0,\ldots,0)\}$, we denote by $\gcd(v)$ the gcd of all the nonzero coefficients of v and $\mathrm{Prim}(v) := \frac{1}{\gcd(v)} v$ (we convene that $\mathrm{Prim}((0,\ldots,0)) = (0,\ldots,0)$). To explain the dynamical version of Algorithm 319 over the integers, it suffices to give the dynamical versions of type 1 and type 2 operations. This can be done as follows:

(i) By a (dynamical) operation of type 2, we mean an operation of type

$$v \ \leftarrow \mathrm{Prim}(v), \text{ where } v \in \mathbb{Z}[X_1,\ldots,X_k]^m \setminus \{(0,\ldots,0)\}.$$

Denoting by $d_1,\ldots,d_s$ the nonzero coefficients of $\mathrm{Prim}(v)$, we have gcd $(d_1,\ldots,d_s) = 1$. Note that $\mathrm{Prim}(v)$ is a primitive vector (i.e., with one invertible coefficient) over each localization $\mathbb{Z}[\frac{1}{d_i}]$. A Bezout identity between the d_i's guarantees that the monoids $d_1^{\mathbb{N}},\ldots,d_s^{\mathbb{N}}$ are comaximal. Such a vector will be simply called a *primitive* vector (a vector whose coefficients generate the whole ring).

(ii) Let $u, v \in \mathbb{Z}[X_1,\ldots,X_k]^m$ and suppose that u is primitive. To avoid redundancies, as the gcd of the elements of the set E formed by nonzero coefficients of u is 1, one can consider a minimal (for inclusion) subset $\{c_1,\ldots,c_r\}$ of E such that $\gcd(c_1,\ldots,c_r) = 1$. It is worth mentioning that the LLL method [95] provides an effective algorithm for finding a short basis of a given lattice and can be used for finding $c_1,\ldots,c_r$ [78].

Suppose that each c_i is the coefficient of a monomial M_i (recall that M_i has the form $X^\alpha e_j$, where $\alpha \in \mathbb{N}^k$, $1 \leq j \leq m$, and $(e_1,\ldots,e_m)$ stands for the canonical basis of $\mathbb{Z}[X_1,\ldots,X_k]^m$), with $M_r \succ \cdots \succ M_1$. Recall that $\succ$ is a POT order associated to $>$. For more details, see Remark 201.

Also, for $1 \leq i \leq r$, denote by d_i the coefficient of the monomial M_i in v.

By an operations of type 1, we mean an operation of type

$$v \ \leftarrow [c_1 v - d_1 u, \ldots, c_r v - d_r u].$$

At each localization $\mathbb{Z}[\frac{1}{c_i}]$, the vector $c_i v - d_i u = c_i(v - \frac{d_i}{c_i}u)$ is, up to a unit, equal to the reduction of v by u so that the term of multidegree $\mathrm{mdeg}(\mathrm{PrimMon}(u))$ and in position $\mathrm{index}(u)$ disappears from v. Contrary to the local case, a type 1 reduction may produce more that one vector if one wants to do all the reductions at once (i.e., globally over $\mathbb{Z}$) instead of doing the job at each localization $\mathbb{Z}[\frac{1}{c_i}]$ separately.

In the particular case of two vectors, we infer the following result.

Proposition 321. *Let $S = [u, v]$ be a list formed by two vectors in $\mathbb{Z}[X_1, \ldots, X_k]^m$ with $u \neq (0, \ldots, 0)$. Denoting by $c_1, \ldots, c_r$ the nonzero coefficients of u, and by $d_1, \ldots, d_r$ the corresponding coefficients in v, we have*

$$\mathrm{Sat}_{\mathbb{Z}}(\langle u, v \rangle_{\mathbb{Z}}) = \langle \mathrm{Prim}(u), \mathrm{Prim}(c_1 v - d_1 u), \ldots, \mathrm{Prim}(c_r v - d_r u) \rangle_{\mathbb{Z}}.$$

Proof. Let us put "dynamically" the list $[u, v]$ in an echelon form by executing the dynamical version of Algorithm 319 explained above. One has first to replace u by its primitive version $\tilde{u} = \mathrm{Prim}(u) = \frac{1}{c}u$, where $c = \gcd(u)$. The obtained new list is then

$$[\tilde{u}, \mathrm{Prim}(\frac{c_1}{c}v - d_1\tilde{u}), \ldots, \mathrm{Prim}(\frac{c_r}{c}v - d_r\tilde{u})],$$

with $\mathrm{Prim}(\frac{c_i}{c}v - d_i\tilde{u}) = \mathrm{Prim}(\frac{1}{c}(c_i v - d_i u)) = \frac{1}{\gcd(\frac{1}{c}(c_i v - d_i u))}(\frac{1}{c}(c_i v - d_i u)) = \mathrm{Prim}(c_i v - d_i u)$. $\qquad\square$

Example 322. Consider the list

$$S = [u,\ v] = [(6, 6 + 4X),\ (4 + 3X, 3)]$$

of vectors in $\mathbb{Z}[X]^2$. By Proposition 321, we obtain:

$$\begin{aligned}
\mathrm{Sat}_{\mathbb{Z}}(\langle u, v \rangle_{\mathbb{Z}}) &= \langle \mathrm{Prim}(u), \mathrm{Prim}(6v - 4u), \mathrm{Prim}(4v - 0u) \rangle_{\mathbb{Z}} \\
&= \langle \mathrm{Prim}(6, 6 + 4X), \mathrm{Prim}(9X, -3 - 8X), \mathrm{Prim}(4 + 3X, 3) \rangle_{\mathbb{Z}} \\
&= \langle (3, 3 + 2X), (9X, -3 - 8X), (4 + 3X, 3) \rangle_{\mathbb{Z}}.
\end{aligned}$$

Notation 323. Let $\mathbf{V}$ be a residually discrete valuation domain of quotient field $\mathbf{K}$ and residue field $\mathbf{k}$. Consider a list $L = [u_1, \ldots, u_s]$ $(s \geq 1)$ of vectors in $\mathbf{V}[X_1, \ldots, X_k]^m$ $(m \geq 1)$. We denote by $\langle L \rangle_{\mathbf{K}}$ (resp. $\langle L \rangle_{\mathbf{k}}$) the $\mathbf{K}$-vector space (resp. the $\mathbf{k}$-vector space) generated by $u_1, \ldots, u_s$ (resp. by the classes $\bar{u}_1, \ldots, \bar{u}_s$ of $u_1, \ldots, u_s$ modulo $\mathrm{Rad}(\mathbf{V})[X_1, \ldots, X_k]^m$), and

$$\dim_{\mathbf{K}} L := \dim_{\mathbf{K}} \langle L \rangle_{\mathbf{K}} \quad \text{and} \quad \dim_{\mathbf{k}} L := \dim_{\mathbf{k}} \langle L \rangle_{\mathbf{k}}.$$

We also denote by $\langle L \rangle_{\mathbf{V}}$ the $\mathbf{V}$-module generated by $u_1, \ldots, u_s$, and by $\langle L \rangle_{\mathbf{V}[X_1, \ldots, X_k]}$ (resp. $\langle L \rangle_{\mathbf{K}[X_1, \ldots, X_k]}$) the $\mathbf{V}[X_1, \ldots, X_k]$-submodule of $\mathbf{V}[X_1, \ldots, X_k]^m$ (resp. the $\mathbf{K}[X_1, \ldots, X_k]$-submodule of $\mathbf{K}[X_1, \ldots, X_k]^m$) generated by $u_1, \ldots, u_s$.

Lemma 324. *Let $\mathbf{V}$ be a residually discrete valuation domain of quotient field $\mathbf{K}$ and residue field $\mathbf{k}$. If L is a finite list of vectors in $\mathbf{V}[X_1,\dots,X_k]^m$ then $\dim_{\mathbf{K}} L \geq \dim_{\mathbf{k}} L$.*

Proof. Denote by $L = [u_1,\dots,u_s]$, $d = \dim_{\mathbf{k}} L$ and suppose that $\bar{u}_1,\dots,\bar{u}_d$ are $\mathbf{k}$-linearly independent. Then, necessarily, $u_1,\dots,u_d$ are $\mathbf{K}$-linearly independent. To see this, let $\alpha_1,\dots,\alpha_d \in \mathbf{V}$ such that $\alpha_1 u_1 + \dots + \alpha_d u_d = 0$. As $\mathbf{V}$ is a valuation domain, there exists $1 \leq i_0 \leq d$ such that α_{i_0} divides all the α_i's. Necessarily $\alpha_{i_0} = 0$ because, otherwise, we would have $\bar{u}_{i_0} \in \sum_{1 \leq i \leq d;\, i \neq i_0} \mathbf{k}\bar{u}_i$, and, thus, $\alpha_1 = \dots = \alpha_d = 0$. We conclude that $\dim_{\mathbf{K}} L \geq d = \dim_{\mathbf{k}} L$. $\qquad\square$

Now we give a necessary and sufficient condition for a finitely-generated sub-$\mathbf{V}$-module of $\mathbf{V}[X_1,\dots,X_k]^m$ to be $\mathbf{V}$-saturated using its corresponding dimensions as $\mathbf{K}$-vector space and $\mathbf{k}$-vector space.

Theorem 325. *Let L be a finite list of vectors in $\mathbf{V}[X_1,\dots,X_k]^m$, where $\mathbf{V}$ is a residually discrete valuation domain of quotient field $\mathbf{K}$ and residue field $\mathbf{k}$. Then, $\langle L \rangle_{\mathbf{V}}$ is $\mathbf{V}$-saturated if and only if $\dim_{\mathbf{K}} L = \dim_{\mathbf{k}} L$.*

Proof. Denote by $L = [u_1,\dots,u_s]$.

"$\Leftarrow$" We proceed by induction on s.
For $s = 1$, two cases may arise:

- Case 1: $\dim_{\mathbf{K}} L = \dim_{\mathbf{k}} L = 0$. In this case, we have $L = [0]$ and of course $\{0\}$ is $\mathbf{V}$-saturated as $\mathbf{V}$ is a domain.

- Case 2: $\dim_{\mathbf{K}} L = \dim_{\mathbf{k}} L = 1$. Necessarily, u_1 is primitive and, thus, $\mathbf{V}u_1$ is $\mathbf{V}$-saturated.

Suppose now that $s > 1$. Two cases may arise:

- Case 1: u_1 is not primitive (i.e., belongs to $\mathrm{Rad}(\mathbf{V})[X_1,\dots,X_k]^m$). Let us denote by $L' = [u_2,\dots,u_s]$. Necessarily, $u_1 \in \langle L' \rangle_{\mathbf{K}}$ as otherwise we would have
$$\dim_{\mathbf{k}} L' = \dim_{\mathbf{k}} L = \dim_{\mathbf{K}} L = 1 + \dim_{\mathbf{K}} L' \geq 1 + \dim_{\mathbf{k}} L'.$$

  As $\dim_{\mathbf{K}} L = \dim_{\mathbf{K}} L'$, $\dim_{\mathbf{k}} L = \dim_{\mathbf{k}} L'$, and $\dim_{\mathbf{K}} L = \dim_{\mathbf{k}} L$, we infer that $\dim_{\mathbf{K}} L' = \dim_{\mathbf{k}} L'$. The $\mathbf{V}$-module $\langle L' \rangle_{\mathbf{V}}$ is $\mathbf{V}$-saturated by the induction hypothesis. Now, since $u_1 \in \langle L' \rangle_{\mathbf{K}}$, there exist $\beta_1 \in \mathbf{V} \setminus \{0\}$ and $\beta_2,\dots,\beta_s \in \mathbf{V}$ such that $\beta_1 u_1 = \beta_2 u_2 + \dots + \beta_s u_s$, and hence $u_1 \in \langle L' \rangle_{\mathbf{V}}$. It follows that $\langle L \rangle_{\mathbf{V}} = \langle L' \rangle_{\mathbf{V}}$, and, thus, $\langle L \rangle_{\mathbf{V}}$ is $\mathbf{V}$-saturated as desired.

- Case 2: u_1 is primitive. For $2 \leq i \leq s$, we set $v_i := u_i + \alpha_i u_1$, where $\alpha_i \in \mathbf{V}$ is such that the term of degree $\deg(\mathrm{PrimMon}(u_1))$ and in position $\mathrm{index}(u_1)$ does not appear in v_i. Denoting by $S := [v_2,\dots,v_s]$, we have $\langle L \rangle_{\mathbf{V}} = \mathbf{V}u_1 \oplus \langle S \rangle_{\mathbf{V}}$, $\dim_{\mathbf{K}} L = \dim_{\mathbf{K}} S + 1$, and $\dim_{\mathbf{k}} L = \dim_{\mathbf{k}} S + 1$. As $\dim_{\mathbf{K}} L = \dim_{\mathbf{k}} L$, we infer that $\dim_{\mathbf{K}} S = \dim_{\mathbf{k}} S$. As $\langle S \rangle_{\mathbf{V}}$ is $\mathbf{V}$-saturated (by the induction hypothesis) and so is $\mathbf{V}u_1$ (by virtue of the case $s = 1$), the desired conclusion follows.

"$\Rightarrow$" Fixing a monomial order on $\mathbf{V}[X_1,\ldots,X_k]$, we can put L in an echelon form by means of operations of types 1 and 2. Of course, an operation of type 1 does not affect the $\mathbf{V}$-module generated by the current list. Also, so does an operation of type 2 as $\langle L \rangle_{\mathbf{V}}$ is $\mathbf{V}$-saturated. Denoting by U the new list obtained after putting L in an echelon form, we get

$$\dim_{\mathbf{K}} L = \dim_{\mathbf{K}} U = \dim_{\mathbf{k}} U = \dim_{\mathbf{k}} L.$$

$\square$

Example 326. Consider the list $[s_1 = (1+2X, 2Y),\ s_2 = (1+2Y, 2X)]$ of vectors in $\mathbb{Z}_{2\mathbb{Z}}[X,Y]^2$. Clearly, we have

$$\dim_{\mathbb{Q}}(\mathbb{Q}s_1 + \mathbb{Q}s_2) = 2 > \dim_{\mathbb{F}_2}(\mathbb{F}_2\bar{s}_1 + \mathbb{F}_2\bar{s}_2) = \dim_{\mathbb{F}_2}(\mathbb{F}_2(\bar{1},\bar{0}))) = 1.$$

So, by Theorem 325, we know that the $\mathbb{Z}_{2\mathbb{Z}}$-module $\mathbb{Z}_{2\mathbb{Z}}s_1 + \mathbb{Z}_{2\mathbb{Z}}s_2$ is not $\mathbb{Z}_{2\mathbb{Z}}$-saturated. This amounts to saying that $\{s_1, s_2\}$ is not a generating set for $\mathrm{Sat}(\mathbb{Z}_{2\mathbb{Z}}s_1 + \mathbb{Z}_{2\mathbb{Z}}s_2)$. Executing Algorithm 319 (suppose that $X > Y$), we find:

$$
\begin{aligned}
\mathrm{Sat}(\mathbb{Z}_{2\mathbb{Z}}s_1 + \mathbb{Z}_{2\mathbb{Z}}s_2) \ &= \ \mathbb{Z}_{2\mathbb{Z}}u_1 + \mathbb{Z}_{2\mathbb{Z}}u_2, \text{ with} \\
u_1 \ &:= \ \mathrm{Prim}(s_1) = s_1,\ \mathrm{PrimMon}(u_1) = 1 \text{ and } \mathrm{index}(u_1) = 1, \\
u_2 \ &:= \ \mathrm{PrimRed}(\mathrm{Prim}(s_2); u_1) = \mathrm{PrimRed}(s_2; u_1) = \mathrm{Prim}(s_2 - s_1) \\
&= \ \frac{1}{2}(s_2 - s_1) = (Y - X, X - Y).
\end{aligned}
$$

And, of course,

$$\dim_{\mathbb{Q}}(\mathbb{Q}u_1 + \mathbb{Q}u_2) = 2 = \dim_{\mathbb{F}_2}(\mathbb{F}_2\bar{u}_1 + \mathbb{F}_2\bar{u}_2).$$

The following Proposition 328 is the analogue of Corollary 300 for finitely-generated submodules of $\mathbf{R}[X_1,\ldots,X_k]^m$. First, recall the following notation.

Notation 327. Let $\mathbf{A}$ be a ring, $a \in \mathbf{A}$, and M a submodule of a free $\mathbf{A}$-module F (possibly, with an infinite basis). We denote by

$$(M : a) = \{u \in F \mid au \in M\}.$$

It is a submodule of F containing M.

Proposition 328. Let $\mathbf{R}$ be a domain with quotient field $\mathbf{K}$, and consider a finite list $S = [v_1,\ldots,v_s]$ of nonzero vectors in $\mathbf{R}[X_1,\ldots,X_k]^m$ ($s, m \geq 1$). Fix a monomial order on $\mathbf{R}[X_1,\ldots,X_k]$, and denote by $U = \mathrm{Echel}(S) = [u_1,\ldots,u_r]$ ($r = \dim_{\mathbf{K}}(\mathbf{K}v_1 + \cdots + \mathbf{K}v_s) \leq s$) the list obtained after transforming S into a primitive triangular list over the quotient field $\mathbf{K}$ with Algorithm 319. Denoting by $u_i = \frac{w_i}{\delta_i}$ where $w_i \in \mathbf{R}[X_1,\ldots,X_k]^m$, $\delta_i \in \mathbf{R} \setminus \{0\}$, and taking $\delta = \prod_{i=1}^{r} \delta_i$ (or $\delta = \mathrm{lcm}(\delta_1,\ldots,\delta_r)$ if such a notion exists), we have:

$$\mathrm{Sat}(\mathbf{R}v_1 + \cdots + \mathbf{R}v_s) = ((\mathbf{R}w_1 + \cdots + \mathbf{R}w_r) : \delta^r) = ((\mathbf{R}w_1 + \cdots + \mathbf{R}w_r) : \delta^s).$$

Proof. Let $u \in (\mathbf{K}v_1 + \cdots + \mathbf{K}v_s) \cap \mathbf{R}[X_1,\ldots,X_k]^m = (\mathbf{K}u_1 + \cdots + \mathbf{K}u_r) \cap \mathbf{R}[X_1,\ldots,X_k]^m$. There exist $a_1 \ldots, a_r \in \mathbf{K}$ such that

$$u = a_1 u_1 + \cdots + a_r u_r.$$

For each $1 \leq i \leq r$, by identifying the coefficient in component number $\text{index}(u_i)$ and of multidegree $\text{mdeg}(\text{PrimMon}(u_i))$ (note that $\text{PrimCoeff}(u_i) = 1$), we obtain:

$$\begin{cases} a_1 \in \mathbf{R} \\ b_{2,1}a_1 + a_2 \in \mathbf{R} \\ \vdots \\ b_{r,1}a_1 + b_{r,2}a_2 + \cdots + b_{r,r-1}a_{r-1} + a_r \in \mathbf{R} \end{cases}$$

with $b_{i,j} \in \frac{1}{\delta}\mathbf{R}$. It follows that $a_1 \in \mathbf{R}$, $a_2 \in \frac{1}{\delta}\mathbf{R}$, $a_3 \in \frac{1}{\delta^2}\mathbf{R}, \ldots, a_r \in \frac{1}{\delta^{r-1}}\mathbf{R}$, and, thus, $u \in \frac{1}{\delta^{r-1}}(\mathbf{R}u_1 + \cdots + \mathbf{R}u_r) \subseteq \frac{1}{\delta^r}(\mathbf{R}w_1 + \cdots + \mathbf{R}w_r)$. We deduce that $\text{Sat}(\mathbf{R}v_1 + \cdots + \mathbf{R}v_s) = ((\mathbf{R}w_1 + \cdots + \mathbf{R}w_r) : \delta^r)$. As

$$\begin{aligned} \text{Sat}(\mathbf{R}v_1 + \cdots + \mathbf{R}v_s) &= ((\mathbf{R}w_1 + \cdots + \mathbf{R}w_r) : \delta^r) \subseteq ((\mathbf{R}w_1 + \cdots + \mathbf{R}w_r) : \delta^s) \\ &\subseteq \text{Sat}(\mathbf{R}v_1 + \cdots + \mathbf{R}v_s), \end{aligned}$$

the desired second equality follows. $\qquad\square$

The following result is the analogue of Proposition 303 for finitely-generated submodules of $\mathbb{Z}[X_1,\ldots,X_k]^m$.

Proposition 329. *Consider a finite list $S = [v_1,\ldots,v_s]$ of nonzero vectors in $\mathbb{Z}[X_1,\ldots,X_k]^m$ ($s, m \geq 1$). Fix a monomial order on $\mathbb{Z}[X_1,\ldots,X_k]$, and denote by $U = \text{Echel}(S) = [u_1,\ldots,u_r]$ ($r = \dim_{\mathbb{Q}}(\mathbb{Q}v_1 + \cdots + \mathbb{Q}v_s) \leq s$) the list obtained after transforming S into a primitive triangular list over $\mathbb{Q}$ with Algorithm 319. Denote by $u_i = \frac{w_i}{\delta_i}$, where $w_i \in \mathbb{Z}[X_1,\ldots,X_k]^m$, $\delta_i \in \mathbb{Z} \setminus \{0\}$, take $\delta = \text{lcm}(\delta_1,\ldots,\delta_r)$, and suppose that we can compute the set $\{p_1,\ldots,p_t\}$ of the prime numbers dividing δ. Then:*

(1) $\text{Sat}_{\mathbb{Z}}(\mathbb{Z}v_1 + \cdots + \mathbb{Z}v_s) = \text{Sat}_{\mathbb{Z}}(\mathbb{Z}w_1 + \cdots + \mathbb{Z}w_r) = ((\mathbb{Z}w_1 + \cdots + \mathbb{Z}w_r) : \delta^r) = ((\mathbb{Z}w_1 + \cdots + \mathbb{Z}w_r) : \delta^s)$.

(2) *The following assertions are equivalent:*

    (i) $\mathbb{Z}w_1 + \cdots + \mathbb{Z}w_r$ *is $\mathbb{Z}$-saturated.*

    (ii) *For all $1 \leq i \leq t$, $\mathbb{Z}_{p_i\mathbb{Z}}w_1 + \cdots + \mathbb{Z}_{p_i\mathbb{Z}}w_r$ is $\mathbb{Z}_{p_i\mathbb{Z}}$-saturated.*

    (iii) $\dim_{\mathbb{Q}} W = \dim_{\mathbb{F}_{p_1}} W = \cdots = \dim_{\mathbb{F}_{p_t}} W$,
        where $W = [w_1,\ldots,w_r]$, $\dim_{\mathbb{Q}} W$ denotes the dimension of $\mathbb{Q}w_1 + \cdots + \mathbb{Q}w_r$ as $\mathbb{Q}$-vector space, and $\dim_{\mathbb{F}_{p_i}} W$ denotes the dimension of $\mathbb{F}_{p_i}\bar{w}_1 + \cdots + \mathbb{F}_{p_i}\bar{w}_r$ as $\mathbb{F}_{p_i}$-vector space, $\bar{w}_j$ denoting the class of w_j modulo $(p_i\mathbb{Z})[X_1,\ldots,X_k]^m$.

(3) *If, for $1 \leq i \leq t$, $\{\frac{1}{a_{i,1}} v_{i,1}, \ldots, \frac{1}{a_{i,\ell_i}} v_{i,\ell_i}\}$ is a generating set for $\mathrm{Sat}_{\mathbb{Z}_{p_i\mathbb{Z}}}(\mathbb{Z}_{p_i\mathbb{Z}} v_1 +$*
$\cdots + \mathbb{Z}_{p_i\mathbb{Z}} v_s)$ (computed, for example, with Algorithm 319), with $v_{i,j} \in$
$\mathbb{Z}[X_1, \ldots, X_n]^m$ and $a_{i,j} \in \mathbb{Z} \setminus p_i\mathbb{Z}$, then

$$\mathrm{Sat}_{\mathbb{Z}}(\mathbb{Z} v_1 + \cdots + \mathbb{Z} v_s) = \langle w_1, \ldots, w_r, v_{1,1}, \ldots, v_{1,\ell_1}, \ldots, v_{t,1}, \ldots, v_{t,\ell_t} \rangle.$$

Proof.

(1) This is Proposition 328. The fact that $\mathrm{Sat}_{\mathbb{Z}}(\mathbb{Z} v_1 + \cdots + \mathbb{Z} v_s) = \mathrm{Sat}_{\mathbb{Z}}(\mathbb{Z} w_1 + \cdots + \mathbb{Z} w_r)$ is clear.

(2) (ii) $\Leftrightarrow$ (iii) by Theorem 325.

   "(i) $\Rightarrow$ (ii)" Let $v = \frac{1}{a} u$, with $a \in \mathbb{Z} \setminus p_i\mathbb{Z}$ and $u \in \mathbb{Z}[X_1, \ldots, X_k]^m$, such that $\alpha v \in \mathbb{Z}_{p_i\mathbb{Z}} w_1 + \cdots + \mathbb{Z}_{p_i\mathbb{Z}} w_r$ for some $\alpha \in \mathbb{Z}_{p_i\mathbb{Z}} \setminus \{0\}$. It follows that there exists $b \in \mathbb{Z} \setminus \{0\}$ such that $bu \in \mathbb{Z} w_1 + \cdots + \mathbb{Z} w_r$. But, as $\mathbb{Z} w_1 + \cdots + \mathbb{Z} w_r$ is $\mathbb{Z}$-saturated, we infer that $h \in \mathbb{Z} w_1 + \cdots + \mathbb{Z} w_r$, and, thus, $g \in \mathbb{Z}_{p_i\mathbb{Z}} w_1 + \cdots + \mathbb{Z}_{p_i\mathbb{Z}} w_r$.

   Note that the above proof applies for any monoid of $\mathbb{Z}$, not only for $\mathbb{Z} \setminus p_i\mathbb{Z}$.

   "(ii) $\Leftarrow$ (i)" Let $v \in \mathrm{Sat}_{\mathbb{Z}}(\mathbb{Z} w_1 + \cdots + \mathbb{Z} w_r)$. By virtue of (1), we know that

   $$\delta^r v \in \mathbb{Z} w_1 + \cdots + \mathbb{Z} w_r. \qquad (0)$$

   In addition, for all $1 \leq i \leq t$, as $\mathbb{Z}_{p_i\mathbb{Z}} w_1 + \cdots + \mathbb{Z}_{p_i\mathbb{Z}} w_r$ is $\mathbb{Z}_{p_i\mathbb{Z}}$-saturated, there exists $\alpha_i \in \mathbb{Z} \setminus p_i\mathbb{Z}$ such that

   $$\alpha_i v \in \mathbb{Z} w_1 + \cdots + \mathbb{Z} w_r. \qquad (i)$$

   Since $\gcd(\delta^r, \alpha_1, \ldots, \alpha_t) = 1$, there exist $\beta, \beta_1, \ldots, \beta_t \in \mathbb{Z}$ such that $\beta \delta^r + \beta_1 \alpha_1 + \cdots + \beta_t \alpha_t = 1$ ($\mathbb{Z}$ being a Bezout domain), and, thus, combining (0), (1),…,(t), one gets $v \in \mathbb{Z} w_1 + \cdots + \mathbb{Z} w_r$, as desired.

(3) We have $\mathrm{Sat}_{\mathbf{A}}(\mathbf{A} v_1 + \cdots + \mathbf{A} v_s) = \mathrm{Sat}_{\mathbf{A}}(\mathbf{A} w_1 + \cdots + \mathbf{A} w_r)$ with $\mathbf{A} = \mathbb{Z}$ or $\mathbb{Z}_{p_i\mathbb{Z}}$.

   It is clear that $\mathrm{Sat}_{\mathbb{Z}}(\mathbb{Z} v_1 + \cdots + \mathbb{Z} v_s) \supseteq \langle w_1, \ldots, w_r, v_{1,1}, \ldots, v_{1,\ell_1}, \ldots, v_{t,1}, \ldots, v_{t,\ell_t} \rangle =: M$. Conversely, let $v \in \mathrm{Sat}_{\mathbb{Z}}(\mathbb{Z} v_1 + \cdots + \mathbb{Z} v_s)$. By virtue of (1), we have

   $$\delta^r v \in \langle w_1, \ldots, w_r \rangle. \qquad (0)$$

   In addition, for all $1 \leq i \leq t$, there exists $\alpha_i \in \mathbb{Z} \setminus p_i\mathbb{Z}$ such that

   $$\alpha_i v \in \langle v_{i,1}, \ldots, v_{i,\ell_i} \rangle. \qquad (i)$$

   As $\gcd(\delta^r, \alpha_1, \ldots, \alpha_t) = 1$, there exist $\beta, \beta_1, \ldots, \beta_t \in \mathbb{Z}$ such that $\beta \delta^r + \beta_1 \alpha_1 + \cdots + \beta_t \alpha_t = 1$, and, thus, combining (0), (1),…,(t), one gets $v \in M$, as desired.

$\square$

From Corollary 329 ensues the following algorithm for computing the $\mathbb{Z}$-saturation of a finitely-generated sub-$\mathbb{Z}$-module of $\mathbb{Z}[X_1, \ldots, X_k]^m$.

Algorithm 330. (An algorithm for computing the $\mathbb{Z}$-saturation of a finitely-generated sub-$\mathbb{Z}$-module of $\mathbb{Z}[X_1, \ldots, X_k]^m$ via prime factorization)
Input: A finite list $S = [v_1, \ldots, v_s]$ of nonzero vectors in $\mathbb{Z}[X_1, \ldots, X_k]^m$ ($s, m \geq 1$).
Output: A finite generating set $H \subseteq \mathbb{Z}[X_1, \ldots, X_k]^m$ of $\mathrm{Sat}_{\mathbb{Z}}(\mathbb{Z}v_1 + \cdots + \mathbb{Z}v_s)$.

1. Fix a monomial order on $\mathbb{Z}[X_1, \ldots, X_k]$, and compute a finite list $U = \mathrm{Echel}(S) = [u_1, \ldots, u_r]$ by transforming S into a primitive triangular list over $\mathbb{Q}$ with Algorithm 319. Denote by $u_i = \frac{w_i}{\delta_i}$, where $w_i \in \mathbb{Z}[X_1, \ldots, X_k]^m$, $\delta_i \in \mathbb{Z} \setminus \{0\}$, take $\delta = \mathrm{lcm}(\delta_1, \ldots, \delta_r)$, and compute the set $\{p_1, \ldots, p_t\}$ of the prime numbers dividing δ.

2. For $1 \leq i \leq t$, using with Algorithm 319, compute a finite generating set

$$\{\frac{1}{a_{i,1}} v_{i,1}, \ldots, \frac{1}{a_{i,\ell_i}} v_{i,\ell_i}\}$$

for $\mathrm{Sat}_{\mathbb{Z}_{p_i\mathbb{Z}}}(\mathbb{Z}_{p_i\mathbb{Z}} v_1 + \cdots + \mathbb{Z}_{p_i\mathbb{Z}} v_s)$, with $v_{i,j} \in \mathbb{Z}[X_1, \ldots, X_n]^m$ and $a_{i,j} \in \mathbb{Z} \setminus p_i\mathbb{Z}$.

3. $H := \{w_1, \ldots, w_r, v_{1,1}, \ldots, v_{1,\ell_1}, \ldots, v_{t,1}, \ldots, v_{t,\ell_t}\}$.

Example 331. (Example 322 Revisited) Keep notation of Algorithm 330, and consider the list

$$S = [v_1, v_2] = [(6, 6 + 4X), (4 + 3X, 3)]$$

of vectors in $\mathbb{Z}[X]^2$. As, when using Algorithm 319 we will use different rings, by $\mathrm{Echel}_{\mathbf{R}}(U)$ we mean that we consider the vectors in the list U as elements of $\mathbf{R}[X]^2$. The first call of Algorithm 319 is with $\mathbf{R} = \mathbb{Q}$. We obtain:

$$\begin{aligned}
\mathrm{Echel}_{\mathbb{Q}}(S) &= [\frac{1}{4} v_1, \frac{1}{3} v_2] = [(\frac{3}{2}, \frac{3}{2} + X), (\frac{4}{3} + X, 1)] \\
&= [\frac{1}{2}(3, 3 + 2X), \frac{1}{3}(4 + 3X, 3)], \\
\delta &= \mathrm{lcm}(\delta_1, \delta_2) = 2 \vee 3 = 6, \\
W &= [w_1, w_2] = [(3, 3 + 2X), (4 + 3X, 3)], \\
\mathrm{Echel}_{\mathbb{Z}_{2\mathbb{Z}}}(W) &= [\frac{1}{3} w_1, w_2 - w_1] \Rightarrow \mathrm{Sat}_{\mathbb{Z}_{2\mathbb{Z}}}(\mathbb{Z}_{2\mathbb{Z}} w_1 + \mathbb{Z}_{2\mathbb{Z}} w_2) \\
&= \mathbb{Z}_{2\mathbb{Z}} w_1 + \mathbb{Z}_{2\mathbb{Z}} w_2, \text{ and} \\
\mathrm{Echel}_{\mathbb{Z}_{3\mathbb{Z}}}(W) &= [\frac{1}{2} w_1, \frac{1}{4} w_2] \Rightarrow \mathrm{Sat}_{\mathbb{Z}_{3\mathbb{Z}}}(\mathbb{Z}_{3\mathbb{Z}} w_1 + \mathbb{Z}_{3\mathbb{Z}} w_2) = \mathbb{Z}_{3\mathbb{Z}} w_1 + \mathbb{Z}_{3\mathbb{Z}} w_2.
\end{aligned}$$

We conclude that

$$\mathrm{Sat}_{\mathbb{Z}}(\mathbb{Z}v_1 + \mathbb{Z}v_2) = \mathbb{Z}w_1 + \mathbb{Z}w_2 = \mathbb{Z}(3, 3 + 2X) + \mathbb{Z}(4 + 3X, 3).$$

4.3   Saturation of a Finitely-Generated V[X]-Module, with V a Valuation Domain

This section is based on the papers [48, 113]. The univariate question is dealt with separately as it is easier than the multivariate case and is completely resolved (it has a clear combinatorial termination proof, precise complexity bounds, and has been implemented with MAGMA [113, 116]).

First, we need to introduce a notion of (saturation) defect of a finite list of vectors in $V[X]^m$, where V is a valuation domain.

Definition 332. Let V be a residually discrete valuation domain.

(1) Let $s_1, \ldots, s_n \in V[X]^m$ ($n \geq 1$). We say that $s_1, \ldots, s_n$ are in *good position* if $s_1, \ldots, s_n$ are primitive or zero and $\mathrm{index}(s_1), \ldots, \mathrm{index}(s_n)$ (only those of nonzero vectors) are pairwise different.

(2) Let $s_1, \ldots, s_n$ be n primitive vectors in $V[X]^m$ such that $\mathscr{I}(s_1), \ldots, \mathscr{I}(s_n)$ are pairwise different. We define the *defect* of the list $S = [s_1, \ldots, s_n]$ to be

$$\sharp\{i \mid 1 \leq i \leq n \ \& \ \exists\, 1 \leq j \leq n \mid \mathrm{index}(s_i) = \mathrm{index}(s_j)$$

$$\& \ \deg(\mathrm{PrimMon}(s_j)) < \deg(\mathrm{PrimMon}(s_i))\}.$$

The defect is denoted by $\delta(S)$. Of course, $\delta(S) \in [\![1, m]\!]$ and $s_1, \ldots, s_n$ are in good position if and only if $\delta(S) = 0$.

If one of the s_i's is not primitive or if $\mathscr{I}(s_1), \ldots, \mathscr{I}(s_n)$ are not pairwise different, we convene that $\delta(S) = n$.

The same definition holds for a list formed by primitive or zero vectors by disregarding those which are zero.

(3) A primitive vector $u \in S$ is said to be *internal* if there exists a primitive vector $v \in S$ such that $\mathrm{index}(u) = \mathrm{index}(v)$ and $\deg(\mathrm{PrimMon}(v)) > \deg(\mathrm{PrimMon}(u))$. A primitive vector in S which is not internal is said to be *external*. The defect of S is nothing but the number of its internal primitive vectors.

Example 333. Let $S = [s_1, \ldots, s_6] = [(2, 4 + 2X, 0, 0, 1), (0, 0, 8, 4X, 2 - X^2), (4, 0, 0, 2 + X, 2X^2), (8, 4X, 0, X^2, 2 - 4X), (4X^2, 2 + X, 0, 0, 0), (X^3, 0, 0, 4X^4, 8)]$ be a list of vectors in $\mathbb{Z}_{2\mathbb{Z}}[X]^n$ with $n = 5$. It is already in an echelon form. The corresponding heights $\mathscr{I}(s_1), \ldots, \mathscr{I}(s_6)$ are represented in Fig. 4.1 by white circles for heights of external vectors and double white circle for heights of internal vectors. The broken black line joins the points representing the heights of the external vectors. One can see that $\delta(S) = 2$.

Now we reach the main algorithm [48, 113] of this section. This algorithm has been implemented in MAGMA [116] by Claude Quitté [113].

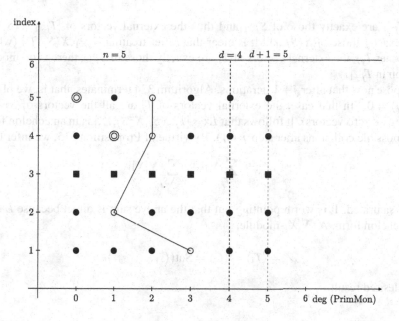

Figure 4.1: $\delta(S) = 2$

Algorithm 334. (Saturation algorithm in case of one variable)

Input: A finite list $S = [s_1, \ldots, s_n]$ of vectors in $\mathbf{V}[X]^m$ with degrees $\leq d$, where $\mathbf{V}$ is a valuation domain and $m \geq 1$.

Output: A finite list $G = T_j$ of vectors in $\mathbf{V}[X]^m$ generating $\mathrm{Sat}(\langle s_1, \ldots, s_n \rangle)$ as a $\mathbf{V}[X]$-module.

Initialization: $j := 0$; $S_0 := \mathrm{Echel}(S)$; $T_0 := S_0$ (see Algorithm 319)

WHILE $\delta(S_j) \neq 0$ DO

$\quad j := j+1$; $S_j := \mathrm{Echel}(\mathrm{EchelResp}(X S_{j-1}; T_{j-1}))$ (use Algorithms 318 and 319)

$\quad T_j = [S_0, \ldots, S_j]$

Remark 335. In Algorithm 334, we are looking for a finite generating set for $\mathrm{Sat}(\langle s_1, \ldots, s_n \rangle)$ as a $\mathbf{V}[X]$-module and not as a $\mathbf{V}$-module. So, in order to avoid superfluous generators, instead of taking $G = T_j$ one can initialize G at $\mathrm{Echel}(S)$ and then add to G only primitive versions of nonprimitive vectors encountered during the while loop.

Proposition 336. *Algorithm 334 is correct if it terminates.*

Proof. We keep the above notation. We will first prove by induction on j that the external vectors of T_j are exactly those of S_j. For $j = 0$ this is clear as $T_0 = S_0$. We suppose now that the result is true for $j - 1$. The list S_j is obtained by putting $X(S_{j-1})$ in an echelon form with respect to T_{j-1} and then by putting it in an echelon form. Thus at step j (i.e., when we want to put $[T_{j-1}, X(S_{j-1})]$ in an echelon form) the vectors of T_{j-1} are not modified. By induction hypothesis, the external vectors of T_{j-1} are exactly those of S_{j-1} and thus the external vectors of $[T_{j-1}, X(S_{j-1})]$ are exactly those of $X(S_{j-1})$. It is clear that, after treating $[T_{j-1}, X(S_{j-1})]$ (which becomes T_j), the external vectors of T_j are exactly those of S_j as there is no modification in T_{j-1}.

Suppose now that after $j+1$ iterations, Algorithm 334 terminates, that is, we obtain $\delta(S_j) = 0$. In that case, the external vectors of T_j are all the vectors of S_j (we disregard zero vectors). It follows that $L := [T_j, XS_j, X^2S_j, \ldots]$ is in an echelon form (no possible collisions after step $j+1$). By virtue of Proposition 315, we infer that

$$\mathscr{S} := \sum_{u \in T_j} \mathbf{V}u \oplus \sum_{v \in XS_j} \mathbf{V}[X]v$$

is $\mathbf{V}$-saturated. It is worth pointing out that the above sum is direct because L is in an echelon form. As $\mathbf{V}[X]$-module,

$$\mathscr{S} = \langle T_j \rangle = \langle G \rangle = \mathrm{Sat}(\langle s_1, \ldots, s_n \rangle),$$

the desired result.

□

Example 337. (Example 333 Continued) We keep the notation of Algorithm 334. Let

$$\begin{aligned}
S = S_0 = [s_1, \ldots, s_6] \;=\; & [(2, 4+2X, 0, 0, 1), (0, 0, 8, 4X, 2-X^2), \\
& (4, 0, 0, 2+X, 2X^2), (8, 4X, 0, X^2, 2-4X), \\
& (4X^2, 2+X, 0, 0, 0), (X^3, 0, 0, 4X^4, 8))]
\end{aligned}$$

be a list of vectors in $\mathbb{Z}_{2\mathbb{Z}}[X]^n$ in an echelon form with $n = 5$. We can take $d = 4$ as a bound on the degrees of the s_i's. The corresponding heights $\mathscr{I}(s_1), \ldots, \mathscr{I}(s_6)$ (recall that $\mathscr{I}(s_i) = (\mathrm{index}(u), \mathrm{mdeg}(\mathrm{PrimMon}(u)))$) are represented in Fig. 4.1 by white circles for heights of external vectors and double white circle for heights of internal vectors. The broken black line joins the points representing the heights of the external vectors. One can see that $\delta(S_0) = 2$. If we only knew the bound $d = 4$ on the degrees of the s_i's, then, obviously, the only information we would have is that the heights of the elements of S_0 belong to the rectangle $\mathscr{R}_0$, where $\mathscr{R}_i = [\![1, m]\!] \times [\![0, d+i]\!]$. Note that this will be the situation when putting XS_0 in an echelon form with respect to S_0 and then in an echelon form (it becomes S_1) as we only know the bound $d + 1 = 5$ for the degrees of the elements of XS_0. The points of $\mathscr{R}_0$ which remain unoccupied by the heights of the elements of S_0 will be represented by black circles or black squares (these latter are used for points of $\mathscr{R}_0$ whose indexes are still unoccupied, i.e., when a whole row is black we put squares to further emphasize).

When "reducing" XS_0, that is, when putting XS_0 in an echelon form with respect to S_0 and then in an echelon form (it becomes S_1), every height of a vector in S_0 will be shifted right one position except where this position is already occupied by the height of a vector in S_0, and then a "collision" occurs. Seeing Fig. 4.2, when

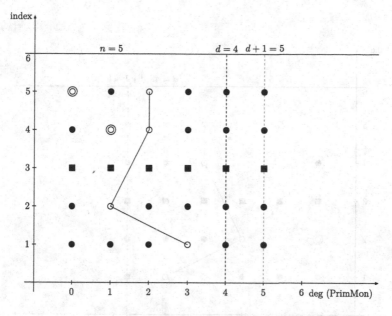

Figure 4.2: $\delta(S_0) = 2$, $\delta(S_1) = ?$

reducing XS_0, only one collision occurs in position $(4,2)$. Then, three cases may arise:

Case 1. The collision in $(4,2)$ produces a point of $\mathscr{R}_1$ located at an occupied row (i.e., is has the same index as a vector in S_0), say $(2,0)$ (Fig. 4.3) or $(2,5)$ (Fig. 4.4) for illustration. In that case, we have $\delta(S_1) = \delta(S_0) = 2$.

Case 2. The collision in $(4,2)$ produces nothing (i.e., the result of the reduction with a type 1 operation is a null vector).

Case 3. The collision in $(4,2)$ produces a point of $\mathscr{R}_1$ located at an unoccupied row, say $(3,3)$ for illustration (see Fig. 4.5). In that case, have $\delta(S_1) = 1 < \delta(S_0) = 2$.

Proposition 338. *Algorithm 334 terminates after at most* $\min(n-1,m)d+1$ *iterations.*

In other words, Algorithm 334 computes after at most $\min(n-1,m)d+1$ *iterations a finite list of vectors in* $\mathbf{V}[X]^m$ *of degrees* $\leq (\min(n-1,m)+1)d$ *generating* $\mathrm{Sat}(\langle s_1,\ldots,s_n\rangle)$ *as a* $\mathbf{V}[X]$*-module.*

Or also, computing $\mathrm{Sat}(\langle s_1,\ldots,s_n\rangle)$ *amounts to performing gaussian elimination on a matrix of size* $n(\min(n-1,m)d+1) \times m$ *and with entries in* $\mathbf{V}[X]$ *of degrees* $\leq (\min(n-1,m)+1)d$.

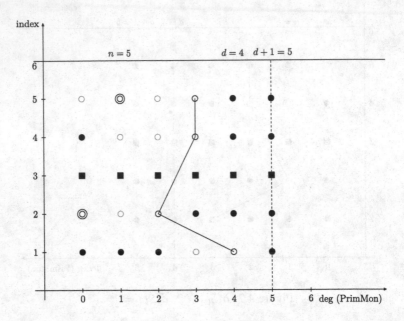

Figure 4.3: Case 1

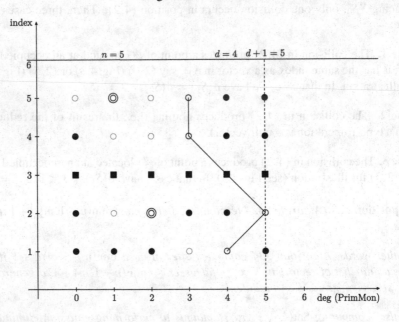

Figure 4.4: Case 1, bis

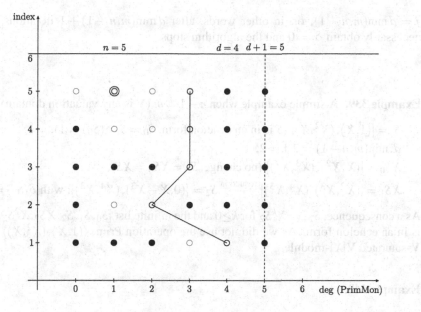

Figure 4.5: Case 2

Proof. We keep the above notation. For $i \leq j$, we denote by e_i the number of external elements of S_i. Thus, the number of (nonzero) elements of S_i is equal to $e_i + \delta_i$, where $\delta_i := \delta(S_i)$. It is worth pointing out the following two remarks:

1. The sequence $(e_i)_{i \leq j}$ is nondecreasing and bounded by m.

2. The sequence $(e_i + \delta_i)_{i \leq j}$ is nonincreasing and bounded by n.

After $j + 1$ iterations, counting the pairwise different primitive monomials of $S_0, \ldots, S_j$ (the left-hand quantity in the inequality below) and their corresponding possible indexes (e_j possible indexes) and degrees ($\leq d + j$), we obtain the following inequality:

$$\sum_{i=0}^{j}(e_i + \delta_i) \leq (1 + d + j)e_j.$$

By Remark 2 above, we can give a lower bound on the terms in the sum above and then obtain:

$$(j+1)(e_j + \delta_j) \leq (1 + d + j)e_j,$$

or also,

$$(j+1)\delta_j \leq de_j.$$

By Remark 1, we have $e_j \leq m$. Moreover, by Remark 2, we have $e_j \leq n - \delta_j$. Thus, if $\delta_j \geq 1$ then $e_j \leq \min(m, n - \delta_j) \leq \min(m, n - 1)$, and $j + 1 \leq (j + 1)\delta_j \leq d \min(m, n - 1)$, and finally $j < d \min(m, n - 1)$. As a conclusion, when

$j = d\min(m, n-1)$, or, in other words, after $d\min(m, n-1) + 1$ iterations, we necessarily obtain $\delta_j = 0$ and the algorithm stops.

$\square$

Example 339. A simple example when $n - 1 < m$ (**V** is any valuation domain).

$S := [(1, X), (X^2, X)]$ (S is in an echelon form, $S_0 = S$, $\delta(S_0) = 1$),

$d.\min(m, n-1) = 2.1 = 2$,

$X S_0 = [(X, X^2), (X^3, X^2)]$ (no change, $S_1 = X S_0 = XS$),

$X S_1 = [(X^2, X^3), (X^4, X^3)] \xrightarrow{\text{reduction}} S_2 = [(0, X - X^3), (X^4, X^3)]$, with $\delta(S_2) = 0$.

As a consequence, $S_{2+k} = X^k S_2$ for $k \geq 0$ and the infinite list $[S_0, S_1, S_2, X S_2, X^2 S_2, \ldots]$ is in an echelon form. As we did not use the operation Prim, $\langle (1, X), (X^2, X) \rangle$ is a **V**-saturated **V**[X]-module.

Example 340.

$$\mathbf{V} \quad = \quad \mathbb{Z}_{2\mathbb{Z}}, S := [(5, 4, -2X^2 - 6X + 12), (2X - 1, 0, -2X^2 + 6X - 4)],$$

$$S \quad \xrightarrow{\text{reduction}} \quad [(5, 4, -2X^2 - 6X + 12), (2X, \frac{4}{5}, -\frac{12}{5}X^2 + \frac{24}{5}X - \frac{8}{5})]$$

$$\xrightarrow{\text{PrimMon}} \quad S_0 = [(5, 4, -2X^2 - 6X + 12),$$

$$(X, \frac{2}{5}, -\frac{6}{5}X^2 + \frac{12}{5}X - \frac{4}{5})], \text{ with } \delta(S_0) = 1,$$

$$X S_0 \quad = \quad [(5X, 4X, -2X^3 - 6X^2 + 12X), (X^2, \frac{2}{5}X, -\frac{6}{5}X^3 + \frac{12}{5}X^2 - \frac{4}{5}X)]$$

$$\xrightarrow{\text{reduction}} \quad [(5X, 4X, -2X^3 - 6X^2 + 12X), (0, 4X - 2, -2X^3 + 4)]$$

$$\xrightarrow{\text{PrimMon}} \quad S_1 = [(5X, 4X, -2X^3 - 6X^2 + 12X),$$

$$(0, 2X - 1, -X^3 + 2)], \text{ with } \delta(S_1) = 0.$$

As a consequence, $S_{1+k} = X^k S_1$ for $k \geq 0$ and the infinite list $[S_0, S_1, X S_1, X^2 S_1, \ldots]$ is in an echelon form. As a conclusion

$$\text{Sat}(\langle (5, 4, -2X^2 - 6X + 12), (2X - 1, 0, -2X^2 + 6X - 4) \rangle)$$

$$= \langle (5, 4, -2X^2 - 6X + 12), (X, \frac{2}{5}, -\frac{6}{5}X^2 + \frac{12}{5}X - \frac{4}{5}), (0, 2X - 1, -X^3 + 2) \rangle.$$

Example 341. (C. Quitté) This example was computed with MAGMA [116]. Keeping notation of Algorithm 334, the base ring is $\mathbf{V} = \mathbb{Z}_{5\mathbb{Z}}$, $n = 2$, $m = 1$, and $d = 7$. The reader can see that the defect becomes zero after exactly $\min(n-1, m)d + 1 = 8$ iterations which is the bound given in Proposition 338. In the computations below, the primitive monomial of a primitive polynomial is computed as the least monomial which has an invertible coefficient.

$Loading"Echelon.magma"S = [[-5*X^7 + 5*X^6 + 5*X^5 + 5*X^4 + 5*X^3 + 5*X^2 + 5*X + 5], [5*X^7 + 5*X^6 + 5*X^5 + 5*X^4 + 5*X^3 + 5*X^2 + 5*X + 5]]$
$> \sharp G; 15$
$> G; [[-X^7 + X^6 + X^5 + X^4 + X^3 + X^2 + X + 1], [2*X^7], [-X^8 + X^6 + X^5 + X^4 + X^3 + X^2 + X], [2*X^8], [-X^9 + X^6 + X^5 + X^4 + X^3 + X^2], [2*X^9], [-X^{10} + X^6 + X^5 + X^4 + X^3], [2*X^{10}], [-X^{11} + X^6 + X^5 + X^4], [2*X^{11}], [-X^{12} + X^6 + X^5], [2*X^{12}], [-X^{13} + X^6], [2*X^{13}], [-X^{14}]]$
$defects : 11111110$

Theorem 342. (Complexity Bounds) *Let* $\mathbf{V}$ *be a valuation domain and consider* n *vectors* $s_1, \ldots, s_n \in \mathbf{V}[X]^m$ ($m \geq 1$) *of degrees* $\leq d$. *Then, with the Saturation Algorithm 334, one can compute a generating set for the* $\mathbf{V}$-*saturation of* $\langle s_1, \ldots, s_n \rangle$ *as a* $\mathbf{V}[X]$-*module formed by at most* $N = n(\min(n-1, m)d + 1)$ *vectors of degrees* $\leq (\min(n-1, m) + 1)d = O(md)$, *and the sequential complexity of this algorithm amounts to* $\frac{N(N-1)}{2} m (\min(n-1, m) + 1) d$ *additions or divisions in* $\mathbf{V}$.

Proof. This follows from Propositions 336 and 338.   □

Corollary 343. *Let* $s_1, \ldots, s_n \in \mathbf{R}[X]^m$ ($m \geq 1$) *where* $\mathbf{R}$ *is a residually discrete local ring. If* $s_1, \ldots, s_n$ *are in good position then the* $\mathbf{R}[X]$-*module* $\langle s_1, \ldots, s_n \rangle$ *is* $\mathbf{R}$-*saturated.*

Proof. It is true that we did not suppose that $\mathbf{R}$ is a valuation domain but the same algorithm works as there is no need of comparability under division.   □

Our goal now is, given a list $[u, s_1, \ldots, s_n]$ of vectors in $\mathbf{V}[X]^m$ where $\mathbf{V}$ is a valuation domain, to present an algorithm to test if $u \in \mathrm{Sat}(\langle s_1, \ldots, s_n \rangle)$, and, in case of a positive answer, to express u as a linear combination of the generators of $\mathrm{Sat}(\langle s_1, \ldots, s_n \rangle)$ computed with Algorithm 334.

Algorithm 344. (Saturation membership test, the univariate case)
Input: A finite list $[u, s_1, \ldots, s_n]$ of vectors in $\mathbf{V}[X]^m$, where $\mathbf{V}$ is a valuation domain of quotient field $\mathbf{K}$ and $m \geq 1$.
Output: An answer to the question $u \in \mathrm{Sat}(\langle s_1, \ldots, s_n \rangle)$? and, in case of positive answer, a finite list $[g_1, \ldots, g_r]$ of vectors in $\mathbf{V}[X]^m$ generating $\mathrm{Sat}(\langle s_1, \ldots, s_n \rangle)$ as a $\mathbf{V}[X]$-module and a list $[u_1, \ldots, u_r]$ of elements in $\mathbf{V}[X]$ such that $u = u_1 g_1 + \cdots + u_r g_r$.

1. Test if $u \in \mathbf{K}[X]s_1 + \ldots + \mathbf{K}[X]s_n$ (with Gröbner bases techniques for example, see Proposition 228). If the answer is NO then return NO. Else, continue.

2. Write u as a $\mathbf{K}$-linear combination of the $X^j s_i$, $j \in \mathbb{N}$, $1 \leq i \leq n$.

3. Use Algorithm 334 to compute a finite list $[g_1,\ldots,g_r]$ of vectors in $\mathbf{V}[X]^m$ generating $\mathrm{Sat}(\langle s_1,\ldots,s_n\rangle)$ as a $\mathbf{V}[X]$-module.

4. Write each $X^j s_i$ in the expression of u as a $\mathbf{K}$-linear combination of the $X^r g_t$, $r \in \mathbb{N}$, $1 \le t \le r$ (by tracing the computations done with Algorithm 334).

5. Write u as a $\mathbf{K}$-linear combination of the $X^r g_t$'s, $r \in \mathbb{N}$, $1 \le t \le r$ (using 2. and 4.). Note that, by virtue of Lemma 311, the obtained $\mathbf{K}$-linear combination is a $\mathbf{V}$-linear combination if and only if $u \in \mathrm{Sat}(\langle s_1,\ldots,s_n\rangle)$.

6. Collect the $\mathbf{V}$-linear combination of the $X^r g_t$'s found in 5. into a $\mathbf{V}[X]$-linear combination $u = u_1 g_1 + \cdots + u_r g_r$ of the g_t's.

4.4 Computing Syzygies Over $\mathbf{R}[X]$, with $\mathbf{R}$ a Prüfer Domain

4.4.1 The Case of a Valuation Domain

Let $\mathbf{V}$ be a valuation domain with quotient field $\mathbf{K}$ and consider m ($m \ge 2$) polynomials $p_1,\ldots,p_m \in \mathbf{V}[X]$. Denote by S and S' the syzygy modules

$$S := \mathrm{Syz}_{\mathbf{V}[X]}(p_1,\ldots,p_m) := \{(u_1,\ldots,u_m) \in \mathbf{V}[X]^m \mid u_1 p_1 + \cdots + u_m p_m = 0\},$$

$$S' := \mathrm{Syz}_{\mathbf{K}[X]}(p_1,\ldots,p_m) := \{(u_1,\ldots,u_m) \in \mathbf{K}[X]^m \mid u_1 p_1 + \cdots + u_m p_m = 0\}.$$

It is straightforward that if $\{s_1,\ldots,s_\ell\}$ is a generating set for S' formed by vectors in $\mathbf{V}[X]^n$, then S is nothing but the $\mathbf{V}$-saturation of $\langle s_1,\ldots,s_\ell\rangle$ as a $\mathbf{V}[X]$-module. So, in order to compute S, one has only to compute $s_1,\ldots,s_\ell$ (with Gröbner bases techniques, see Sect. 3.4) and then to use the Saturation Algorithm 334.

The following is folklore. For example, it is a particular case of Corollary 147.

Proposition 345. *Let $\mathbf{K}$ be a discrete field and consider m polynomials $p_1,\ldots,p_m \in \mathbf{K}[X]$ with $m \ge 2$ and $p_1 \cdots p_m \ne 0$. Then the syzygy module $S' = \mathrm{Syz}_{\mathbf{K}[X]}(p_1,\ldots,p_m)$ is generated as a $\mathbf{K}[X]$-module by the $\binom{m}{2} = \frac{m(m-1)}{2}$ syzygies $q_j e_i - q_i e_j$, $1 \le i < j \le m$ (the obvious syzygies of $(q_1,\ldots,q_m)$), where $(e_1,\ldots,e_m)$ stands for the canonical basis of $\mathbf{K}[X]^m$, $\Delta = \gcd(p_1,\ldots,p_m)$, and $q_i = p_i/\Delta$.*

Note that the generating set given by Proposition 345 is far from being minimal as we know that $\mathrm{Syz}_{\mathbf{K}[X]}(p_1,\ldots,p_m) = \mathrm{Syz}_{\mathbf{K}[X]}(p_1/\Delta,\ldots,p_m/\Delta)$ with $\gcd(p_1/\Delta,\ldots,p_m/\Delta) = 1$. As $\mathbf{K}[X]$ is a Bezout domain, we know that $\mathrm{Syz}_{\mathbf{K}[X]}(p_1,\ldots,p_m)$ is a free $\mathbf{K}[X]$-module of rank $m-1$ (see, for example, [111] page 260).

Algorithm 346. (Computing Syzygies over valuation domains)

Input: $p_1,\ldots,p_m \in \mathbf{V}[X]$ where $m \ge 2$ and $\mathbf{V}$ is a valuation domain.

Output: A list L of vectors in $\mathbf{V}[X]^m$ generating $\text{Syz}_{\mathbf{V}[X]}(p_1, \ldots, p_m)$ as a $\mathbf{V}[X]$-module.

Step 1: Compute a generating set $\{s_1, \ldots, s_\ell\}$ for $\text{Syz}_{\mathbf{K}[X]}(p_1, \ldots, p_m)$ with $s_i \in \mathbf{V}[X]^m$ (by clearing the denominators). One can use Proposition 345 to compute $s_1, \ldots, s_\ell$.

Step 2: Use the Saturation Algorithm 334 to compute a finite list L of vectors in $\mathbf{V}[X]^m$ generating the $\mathbf{V}$-saturation of $\langle s_1, \ldots, s_\ell \rangle$ as a $\mathbf{V}[X]$-module.

Theorem 347. (Complexity Bounds) *Let* $\mathbf{V}$ *be a valuation domain and consider* m *polynomials* $p_1, \ldots, p_m \in \mathbf{V}[X]$ *of degrees* $\leq d$. *Then, with Algorithm 346, one can compute a generating set for* $\text{Syz}_{\mathbf{V}[X]}(p_1, \ldots, p_m)$ *as a* $\mathbf{V}[X]$-*module formed by* $N = \frac{m(m-1)}{2}(md+1) = O(\frac{m^3 d}{2})$ *vectors of degrees* $\leq (m+1)d = O(md)$ *and the sequential complexity of this algorithm amounts to* $\frac{N(N-1)}{2}m(m+1)d = O(\frac{m^8 d^3}{8})$ *additions or divisions in* $\mathbf{V}$.

Example 348. Let $p_1 = 2X + 4$, $p_2 = 2X^2$, $p_3 = 2 + X^2 \in \mathbb{Z}_{2\mathbb{Z}}[X]$, and suppose that we want to compute a generating set for $\text{Syz}_{\mathbb{Z}_{2\mathbb{Z}}[X]}(p_1, p_2, p_3)$.

Using Proposition 345, we have

$$\text{Syz}_{\mathbb{Q}[X]}(p_1, p_2, p_3)$$
$$= \langle s_1 = (2X^2, -2X - 4, 0), \ s_2 = (2 + X^2, 0, -2X - 4), \ s_3 = (0, 2 + X^2, -2X^2) \rangle.$$

Let us denote by $S = S_0 = [s_1, s_2, s_3]$. With the Saturation Algorithm 334, we obtain

$$S \overset{\text{reduction}}{\longrightarrow} [(X^2, -X - 2, 0), (2 + X^2, 0, -2X - 4), (0, 2 + X^2, -2X^2)], \text{ with } \delta(S_0) = 1,$$

$$XS_0 \overset{\text{reduction}}{\longrightarrow} \quad [(X^3 + 4, 6, -2X^2 - 2X - 4), \ (X - 2, -3, -X + 2),$$
$$(-12 + 4X, X^3 - 16, -2X^3 + 16)] = S_1,$$

with $\delta(S_1) = 0$. Thus, ignoring superfluous generators (see Remark 335),

$$\text{Syz}_{\mathbb{Z}_{2\mathbb{Z}}[X]}(p_1, p_2, p_3) = \langle s_1, \ s_2, \ s_3, \ (X - 2, -3, -X + 2) \rangle.$$

Let $\mathbf{V}$ be a valuation domain of Krull dimension ≥ 2 with a corresponding valuation v and group G. The fact that the Krull dimension of $\mathbf{V}$ is ≥ 2 is equivalent to the fact that G is not archimedean (see Exercise 372), that is there exist $a, b \in \mathbf{V}$ such that:

$$v(a) > 0, \text{ and } \forall\, n \in \mathbb{N}^*, \ v(b) > nv(a).$$

It is proven in Proposition 255 (the contrapositive of "(2) $\Rightarrow$ (3)"), that the ideal $\text{LT}(\langle aX + 1, b \rangle)$ of $\mathbf{V}[X]$ is not finitely-generated. In the example below, we will compute $\text{Syz}_{\mathbf{V}[X]}(aX + 1, b)$.

Example 349. Let $\mathbf{R}$ be a strongly discrete domain with quotient field $\mathbf{K}$ and consider $a, b \in \mathbf{R} \setminus \{0\}$. By virtue of Proposition 147, $\mathrm{Syz}_{\mathbf{K}[X]}(aX + 1, b) = \langle s_1 = (-b, aX + 1) \rangle$. As s_1 is obviously in good position, by Corollary 343, we infer that $\mathrm{Syz}_{\mathbf{R}[X]}(aX + 1, b) = \langle (-b, aX + 1) \rangle$ (in fact, this is true with the hypothesis $\mathbf{R}$ is local but it is obvious that this is also true globally).

4.4.2   The Case of a Prüfer Domain

Theorem 350. *Let $\mathbf{R}$ be a Prüfer domain and $k, m \geq 1$. Then the $\mathbf{R}$-saturation of any finitely-generated submodule of $\mathbf{R}[X]^m$ is finitely-generated.*

Proof. Let $s_1, \ldots, s_n \in \mathbf{R}[X]^m$. The proof (algorithm) works in the same way as the case where the base ring is a valuation domain. The only difference occurs when one has to handle two incomparable (under division) elements a, b in $\mathbf{T} = \mathbf{R}_{(I;U)}$ (T is the current ring with the initialization $\mathbf{R}_{(0;1)} = \mathbf{R}$, see Definition and Notation 32). In that situation, one should first compute $u, v, w \in \mathbf{T}$ such that

$$\begin{cases} ub = va \\ wb = (1-u)a. \end{cases}$$

More precisely, one has to open two branches $\mathbf{R}_{(I;u,U)}$ and $\mathbf{R}_{(I;(1-u),U)}$. In the first, a divides b, and in the second b divides a. In the latter branch (the same holds for the first branch), on has to open two sub-branches $\mathbf{R}_{(I;(1-u),w,U)}$ and $\mathbf{R}_{(I,w;(1-u),U)}$. In the first, a and b are associated and a divides b, while in the second b divides strictly a (i.e., $\frac{a}{b}$ is in the radical), and this situation will be preserved in the whole opened sub-branches. $\square$

Corollary 351. *If $\mathbf{R}$ is a Prüfer domain then $\mathbf{R}[X]$ is coherent.*

4.5   The Multivariate Case

4.5.1   Hilbert Series

The goal of this subsection is to provide a rapid overview of Hilbert series and their main properties. For more details, the reader can refer to Kemper's nice book [87] (see also [170]). For sake of simplicity, we will deal with the ideal case.

Definition and notation 352. Let $I = \langle f_1, \ldots, f_s \rangle$ be an ideal in $\mathbf{K}[X_1, \ldots, X_k]$, where $\mathbf{K}$ is a discrete field, and denote by $\mathbf{A} := \mathbf{K}[X_1, \ldots, X_k]/I$.

1. For $i \in \mathbb{N}$, we denote by E_i the $\mathbf{K}$-vector subspace of $\mathbf{A}$ generated by the $\bar{M}$'s, where M is a monomial at $X_1, \ldots, X_k$ of total degree at most i and $\bar{M}$ denotes its class modulo I. Note that E_i is a finite-dimensional $\mathbf{K}$-vector space of dimension at most $\binom{k+i}{i}$. Also, set $F_0 = E_0$, and for $i \geq 1$, decompose $E_i = E_{i-1} \oplus F_i$.

The function $h_I : \mathbb{N} \to \mathbb{N}$ defined by $h_I(i) := \dim_{\mathbf{K}} E_i$ is called the *Hilbert function* of I.

2. The formal power series

$$\mathrm{HS}_I(t) := \sum_{i \geq 0} (\dim_{\mathbf{K}} E_i) t^i,$$

is called the *Hilbert series* of I. Note that, denoting by

$$\mathrm{H}_I(t) := \sum_{i \geq 0} (\dim_{\mathbf{K}} F_i) t^i,$$

we obviously have the relation

$$(1-t)\,\mathrm{HS}_I(t) = \mathrm{H}_I(t).$$

For example,

$$\mathrm{HS}_{\langle 0 \rangle}(t) = \sum_{i \geq 0} \binom{k+i}{i} t^i = \frac{1}{(1-t)^{k+1}},$$

$$\mathrm{HS}_{\langle X_1,\dots,X_k \rangle}(t) = \sum_{i \geq 0} t^i = \frac{1}{1-t},$$

$$\mathrm{HS}_{\langle X_r X_\ell;\, 1 \leq r < \ell \leq k \rangle}(t) = 1 + \sum_{i \geq 1} k t^i = 1 + \frac{kt}{(1-t)} = \frac{1+(k-1)t}{(1-t)}, \text{ with } k \geq 2.$$

Proposition 353. (Computing Hilbert Series with Gröbner Bases [45, 87, 170])
Let $\mathbf{K}$ be a discrete field, and consider a total order $>$ on $\mathbf{K}[X_1,\dots,X_k]$, i.e., a monomial order on $\mathbf{K}[X_1,\dots,X_k]$ such that $M > N$ whenever $\mathrm{tdeg}(M) > \mathrm{tdeg}(N)$. The following holds:

(1) *If $I = \langle f_1,\dots,f_s \rangle$ is an ideal in $\mathbf{K}[X_1,\dots,X_k]$, then*

$$\mathrm{HS}_I(t) = \mathrm{HS}_{\mathrm{LT}(I)}(t).$$

(2) *If I ad J are two finitely-generated homogeneous ideals in $\mathbf{K}[X_1,\dots,X_k]$, then*

$$\mathrm{HS}_{I+J}(t) + \mathrm{HS}_{I \cap J}(t) = \mathrm{HS}_I(t) + \mathrm{HS}_J(t).$$

(3) *If $I = \langle f_1,\dots,f_s \rangle$ is an ideal in $\mathbf{K}[X_1,\dots,X_k]$, then $\mathrm{HS}_I(t)$ can be computed in a finite number of steps with the following algorithm:*
Input: An ideal $I = \langle f_1,\dots,f_s \rangle$ in $\mathbf{K}[X_1,\dots,X_k]$.
Output: The Hilbert series $\mathrm{HS}_I(t)$.

(i) Fix a total order $>$ on $\mathbf{K}[X_1,\ldots,X_k]$, compute a Gröbner basis G of I with respect to $>$, and denote by $M_1,\ldots,M_r$ the leading monomials of the nonzero elements of G.

(ii) If $r = 0$, then return $\mathrm{HS}_I(t) = \frac{1}{(1-t)^{k+1}}$.

(iii) Set $J := (M_2,\ldots,M_r)$ and $\tilde{J} := (\mathrm{lcm}(M_1,M_2),\ldots,\mathrm{lcm}(M_1,M_r))$.

(iv) Compute the Hilbert series $\mathrm{HS}_J(t)$ and $\mathrm{HS}_{\tilde{J}}(t)$ by a recursive call of the algorithm

(v) Return $\mathrm{HS}_I(t) := \frac{1-t^{\mathrm{tdeg}(m_1)}}{(1-t)^{k+1}} + \mathrm{HS}_J(t) - \mathrm{HS}_{\tilde{J}}(t)$.

(4) *If $I = \langle f_1,\ldots,f_s \rangle$ is an ideal in $\mathbf{K}[X_1,\ldots,X_k]$, then, as a consequence of the algorithm given in (3), $\mathrm{HS}_I(t)$ has the form*

$$\mathrm{HS}_I(t) = \frac{a_0 + a_1 t + \cdots + a_n t^n}{(1-t)^{k+1}},$$

with $n \in \mathbb{N}$ and $a_i \in \mathbb{Z}$. Moreover, the Hilbert function is ultimately polynomial. More precisely, the polynomial (called the Hilbert polynomial *of I)*

$$P_I(X) := \sum_{j=0}^{n} a_j \binom{X+k-j}{k} \in \mathbb{Q}[X],$$

satisfies $h_I(i) = P_I(i)$ for sufficiently large integer i.

Moreover, we have

$$\deg(p_I) = \mathrm{Kdim}(\mathbf{K}[X_1,\ldots,X_k]/I),$$

and, therefore, $\mathrm{Kdim}(\mathbf{K}[X_1,\ldots,X_k]/I) = \mathrm{Kdim}(\mathbf{K}[X_1,\ldots,X_k]/\mathrm{LT}(I))$. From this fact ensues the following algorithm for computing $\mathrm{Kdim}(\mathbf{K}[X_1,\ldots,X_k]/I)$.

Input: An ideal $I = \langle f_1,\ldots,f_s \rangle$ in $\mathbf{K}[X_1,\ldots,X_k]$.

Output: $\mathrm{Kdim}(\mathbf{A})$ with $\mathbf{A} = \mathbf{K}[X_1,\ldots,X_k]/I$.

(i) Fix a total order $>$ on $\mathbf{K}[X_1,\ldots,X_k]$, compute a Gröbner basis G of I with respect to $>$, and denote by $M_1,\ldots,M_r$ the leading monomials of the nonzero elements of G.

(ii) If $M_j = 1$ for some j, return $\mathrm{Kdim}(\mathbf{A}) = -1$.

(iii) By an exhaustive search, find a set $E \subseteq \{X_1,\ldots,X_k\}$ of minimal size such that every M_j involves at least one indeterminate from E.

(iv) Return $\mathrm{Kdim}(\mathbf{A}) = k - \sharp(E)$.

4.5.2 The Saturation Defect Series

Definition and notation 354. Let $L = [u_1, \ldots, u_s]$ ($s \geq 1$) be a list of s polynomial vectors in $\mathbf{V}[X_1, \ldots, X_k]^m$, where $\mathbf{V}$ is a residually discrete valuation domain of quotient field $\mathbf{K}$ and residue field $\mathbf{k}$.

1. For $i \in \mathbb{N}$, we denote by L_i the $\mathbf{K}$-vector space generated by the Mu_j's where $1 \leq j \leq s$ and M is a monomial at $X_1, \ldots, X_k$ of total degree at most i.

2. We denote by

$$h_{L,\mathbf{K}}(t) = \sum_{i \geq 0} (\dim_{\mathbf{K}} L_i) t^i,$$

a series (it is in fact a Hilbert series, see Lemma 355 below) that we associate to L over $\mathbf{K}$. On the other hand, we associate to L a series

$$h_{L,\mathbf{k}}(t) = \sum_{i \geq 0} (\dim_{\mathbf{k}} \bar{L}_i) t^i$$

over the residue field $\mathbf{k}$, where $\dim_{\mathbf{k}} L_i = \dim_{\mathbf{k}} \bar{L}_i$, and $\bar{L}_i$ is the $\mathbf{k}$-vector space obtained from L_i by passing modulo $\mathrm{Rad}(\mathbf{V})[X_1, \ldots, X_k]^m$.

As $\dim_{\mathbf{k}} L_i \leq \dim_{\mathbf{K}} L_i$ (Lemma 324), the series

$$\delta_L(t) := h_{L,\mathbf{K}}(t) - h_{L,\mathbf{k}}(t)$$

has nonnegative coefficients. This series will be called the *(saturation) defect series* of the list L.

Lemma 355. *We have*

$$h_{U,\mathbf{K}}(t) = \mathrm{HS}_{\mathrm{Syz}_{\mathbf{K}}(u_1, \ldots, u_s)}(t).$$

Proof. This follows from the exact sequence

$$
\begin{array}{ccccccccc}
0 & \to & \mathrm{Syz}_{\mathbf{K}}(u_1, \ldots, u_s) & \to & \mathbf{K}[X_1, \ldots, X_k]^s & \to & \langle u_1, \ldots, u_s \rangle & \to & 0 \\
& & (p_1, \ldots, p_s) & \mapsto & p_1 s_1 + \cdots + p_s u_s. & & & &
\end{array}
$$

$\square$

Remark 356. Consider a finite nonempty list $S = [s_1, \ldots, s_n]$ of primitive vectors in $\mathbf{V}[X]^m$, where $\mathbf{V}$ is a residually discrete valuation domain. It is clear that if the defect of S (with Definition 332) is zero then so is its defect series. To see this, it suffices to consider the case $m = 1$. As a matter of fact, in case $m = 1$, saying that $\delta(S) = 0$ amounts to saying that S contains only one (primitive) polynomial, and so the defect series $\delta_S(t) = 0$ as, residually, the list contains also only one (nonzero) polynomial.

The converse does not hold, i.e.,

$$\delta_S(t) = 0 \quad \nRightarrow \quad \delta(S) = 0.$$

To see this, it suffices to consider the list $[1, X]$.

Example 357. Consider the list $U = [u_1 = 1 + 2X, u_2 = 1 + 2Y]$ with $u_i \in \mathbb{Z}_{2\mathbb{Z}}[X,Y]$. We have:

$$
h_{U,\mathbb{Q}}(t) = \sum_{i=0}^{\infty} \frac{i^2 + 5i + 4}{2} t^i = \sum_{i=0}^{\infty} \left(\binom{2+i}{2} + \binom{1+i}{1} \right) t^i = \frac{1}{(1-t)^3} + \frac{1}{(1-t)^2},
$$

$$
h_{U,\mathbb{Z}/2\mathbb{Z}}(t) = \sum_{i=0}^{\infty} \frac{i^2 + 3i + 2}{2} t^i = \sum_{i=0}^{\infty} \binom{2+i}{2} t^i = \frac{1}{(1-t)^3},
$$

and, thus, the defect series of U is

$$
\delta_U(t) = \frac{1}{(1-t)^2}.
$$

Now we are in position to state that a finite list of vectors $\mathbf{V}[X_1,\ldots,X_k]^m$ ($\mathbf{V}$ a valuation domain) whose defect is null generates a $\mathbf{V}$-saturated sub-$\mathbf{V}[X_1,\ldots,X_k]$-module of $\mathbf{V}[X_1,\ldots,X_k]^m$. This result will be used as a termination condition in our Algorithm 364.

Theorem 358. *Let L be a finite list of vectors in $\mathbf{V}[X_1,\ldots,X_k]^m$, where $\mathbf{V}$ is a residually discrete valuation domain of quotient field $\mathbf{K}$ and residue field $\mathbf{k}$. If $\delta_L = 0$ then $\langle L \rangle_{\mathbf{V}[X_1,\ldots,X_k]}$ is $\mathbf{V}$-saturated.*

Proof. We keep the notation of Definition and Notation 354. As $\delta_L = 0$, we have $\dim_{\mathbf{K}} L_i = \dim_{\mathbf{k}} L_i$, and thus, by virtue of Theorem 325, $\langle L_i \rangle_{\mathbf{V}}$ is $\mathbf{V}$-saturated for each $i \in \mathbb{N}$. The desired result follows since $\langle L \rangle_{\mathbf{V}[X_1,\ldots,X_k]} = \bigcup_{i \in \mathbb{N}} \uparrow \langle L_i \rangle_{\mathbf{V}}$.

$\square$

The converse of Theorem 358 does not hold. We give hereafter two counterexamples.

Example 359. (Example 357 Continued) Let $\mathbf{V} = \mathbb{Z}_{2\mathbb{Z}}$ and consider the list $U = [u_1 = 1 + 2X, u_2 = 1 + 2Y]$ with $u_i \in \mathbf{V}[X,Y]$. The defect series of U is $\delta_U(t) = \frac{1}{(1-t)^2} \neq 0$ despite that $\langle U \rangle = \langle u_1, u_2, Yu_1 - Xu_2 \rangle = \langle u_1, u_2, Y - X \rangle = \langle u_1, Y - X \rangle$ is $\mathbf{V}$-saturated (see Example 367).

Note that, fixing any monomial order on $\mathbf{V}[X,Y]$, $\{u_1, u_2\}$ is not a Gröbner basis for $\langle u_1, u_2 \rangle$ since

$$
\mathrm{LT}(Yu_1 - Xu_2) = -X \text{ or } Y \notin \langle \mathrm{LT}(u_1), \mathrm{LT}(u_2) \rangle = \langle 2X, 2Y \rangle = 2 \langle X, Y \rangle.
$$

Example 360. Let $\mathbf{V}$ be residually discrete valuation domain with quotient field $\mathbf{K}$ and residue field $\mathbf{k}$, and consider a $L = [X, aX^2]$ of polynomials in $\mathbf{V}[X]$, with $a \in \mathrm{Rad}(\mathbf{V}) \setminus \{0\}$. We have:

$$
h_{L,\mathbf{K}}(t) = 2 + 3t + 4t^2 + \cdots = \frac{2-t}{(1-t)^2},
$$

$$h_{L,\mathbf{k}(t)} = 1 + 2t + 3t^2 + \cdots = \frac{1}{(1-t)^2},$$

and, thus, the defect series of L is

$$\delta_L(t) = \frac{1}{1-t} \neq 0$$

despite that $\langle L \rangle_{\mathbf{V}[X]} = \langle X \rangle$ is $\mathbf{V}$-saturated. Note that $\{X, aX^2\}$ is a Gröbner basis for $\langle X, aX^2 \rangle$ which is not pseudo-reduced since $aX^2 \xrightarrow{X} 0$.

Examples 359 and 360 suggest the following partial converse to Theorem 358.

Proposition 361. *Let $L = [u_1, \ldots, u_s]$ be a finite list of vectors in $\mathbf{V}[X_1, \ldots, X_k]^m$, where $\mathbf{V}$ is a residually discrete valuation domain with quotient filed $\mathbf{K}$ and residue field $\mathbf{k}$. Suppose that $G = \{u_1, \ldots, u_s\}$ is a pseudo-reduced Gröbner basis for the $\mathbf{V}[X_1, \ldots, X_k]$-module $\langle u_1, \ldots, u_s \rangle$ with respect to some monomial order $>$. If*

$$\dim_{\mathbf{K}}(\mathbf{K}u_1 + \cdots + \mathbf{K}u_s) > \dim_{\mathbf{k}}(\mathbf{k}\bar{u}_1 + \cdots + \mathbf{k}\bar{u}_s),$$

where $\bar{u}_i$ denotes the class of u_i modulo $\mathrm{Rad}(\mathbf{V})[X_1, \ldots, X_k]^m$, then $\langle u_1, \ldots, u_s \rangle$ is not $\mathbf{V}$-saturated.

Proof. Let us keep the notation of Definition and Notation 354. By hypothesis, we have $\dim_{\mathbf{K}} L_0 > \dim_{\mathbf{k}} L_0$. By virtue of Theorem 325, we know that $\langle L_0 \rangle_{\mathbf{V}}$ is not $\mathbf{V}$-saturated, that is, there exists $w = w^{(1)} \in \mathbf{V}[X_1, \ldots, X_k]^m \setminus \{(0, \ldots, 0)\}$ such that $w \in \langle L_0 \rangle_{\mathbf{K}}$ and $w \notin \langle L_0 \rangle_{\mathbf{V}}$. As $w \in \langle L_0 \rangle_{\mathbf{K}}$ and the $\mathrm{LM}(u_i)$'s are pairwise different (because G is pseudo-reduced), we infer that there exists a unique $1 \leq i_0 \leq s$ such that $\mathrm{LM}(w) = \mathrm{LM}(u_{i_0})$. If $w \in \langle u_1, \ldots, u_s \rangle$, then, as G is a Gröbner basis for the module $\langle u_1, \ldots, u_s \rangle$, there exists $1 \leq i_1 \leq s$ such that $\mathrm{LT}(u_{i_1})$ divides $\mathrm{LT}(w)$. Denote by

$$\begin{aligned}
\mathrm{LM}(w) &= \mathrm{LM}(u_{i_0}) = X^\beta X^\alpha e_r, \quad \mathrm{LT}(u_{i_1}) = abX^\alpha e_r, \\
\mathrm{LT}(u_{i_0}) &= bX^\beta X^\alpha e_r, \quad \text{and} \quad \mathrm{LT}(w) = abcX^\beta X^\alpha e_r,
\end{aligned}$$

where $a, b, c \in \mathbf{V} \setminus \{0\}$, $X^\beta, X^\alpha \in \mathbb{N}^k$, and $(e_1, \ldots, e_m)$ stands for the canonical basis of $\mathbf{V}[X_1, \ldots, X_k]^m$ (note that we used the facts that G is pseudo-reduced and $\mathbf{V}$ is a valuation domain). It follows that the vector $w^{(2)} := w - acu_{i_0}$ satisfies the properties:

$$\begin{cases}
w^{(2)} \in \mathbf{V}[X_1, \ldots, X_k]^m \\
w^{(2)} \in \langle L_0 \rangle_{\mathbf{K}} \\
w^{(2)} \notin \langle L_0 \rangle_{\mathbf{V}} \\
\mathrm{LM}(w^{(2)}) < \mathrm{LM}(w^{(1)}).
\end{cases}$$

So, as the set $\mathbb{M}_k^m$ of monomials in $\mathbf{V}[X_1,\ldots,X_k]^m$ is well-ordered (see Remark 201, Dickson's lemma (Theorem 209), and Corollary 210), we eventually find a vector $w^{(j)}$ such that

$$\begin{cases} w^{(j)} \in \mathbf{V}[X_1,\ldots,X_k]^m \\ w^{(j)} \notin \langle u_1,\ldots,u_s \rangle \\ w^{(j)} \in \langle L_0 \rangle_{\mathbf{K}}, \end{cases}$$

and, thus, $\langle u_1,\ldots,u_s \rangle$ is not saturated.

$\square$

The following possible converse to Theorem 358 is left as an open question.

Question 362. *Let $L = [u_1,\ldots,u_s]$ be a finite list of vectors in $\mathbf{V}[X_1,\ldots,X_k]^m$, where $\mathbf{V}$ is residually discrete a valuation domain. Suppose that $G = \{u_1,\ldots,u_s\}$ is a pseudo-reduced Gröbner basis for the $\mathbf{V}[X_1,\ldots,X_k]$-module $\langle u_1,\ldots,u_s \rangle$ with respect to some monomial order $>$. Is the following implication true:*

$$\delta_L \neq 0 \;\Rightarrow\; \langle u_1,\ldots,u_s \rangle \text{ is not } \mathbf{V}-\text{saturated ?}$$

In the particular case where $\mathbf{R} = \mathbb{Z}$, by virtue of Theorem 358 and Proposition 303, Question 362 becomes:

Question 363. *Let $U_0 = [u_1,\ldots,u_s]$ be a finite list of vectors in $\mathbb{Z}[X_1,\ldots,X_k]^m$. Suppose that $G_0 = \{u_1,\ldots,u_s\}$ is a pseudo-reduced Gröbner basis for the $\mathbb{Z}[X_1,\ldots,X_k]$-module $\langle u_1,\ldots,u_s \rangle$ with respect to some monomial order $>$. Denote by $\delta = \text{lcm}(\text{LC}(u_1),\ldots,\text{LC}(u_s))$ and suppose that we can compute the set $\{p_1,\ldots,p_r\}$ of the prime numbers dividing δ (the essential primes of $\langle u_1,\ldots,u_s \rangle$). For $1 \leq i \leq r$, as G_0 is also Gröbner basis for the $\mathbb{Z}_{p_i\mathbb{Z}}[X_1,\ldots,X_k]$-module generated by $u_1,\ldots,u_s$, one can transform it into a pseudo-reduced Gröbner basis $G_i = \{v_{i,1},\ldots,v_{i,t_i}\}$ for it. Denoting by $U_i = [v_{i,1},\ldots,v_{i,t_i}]$ and, for $1 \leq i \leq r$, $\mathbb{F}_{p_i} = \mathbb{Z}/p_i\mathbb{Z}$ (the "essential" finite residue fields of $\langle u_1,\ldots,u_s \rangle$), is it true that*

$$\langle u_1,\ldots,u_s \rangle \text{ is } \mathbb{Z}-\text{saturated} \;\Leftrightarrow\; h_{U_0,\mathbb{Q}}(t) = h_{U_1,\mathbb{F}_{p_1}}(t) = \cdots = h_{U_r,\mathbb{F}_{\mathfrak{q}}}(t) \text{ ?}$$

In other words: *is the fact that $\langle u_1,\ldots,u_s \rangle$ is $\mathbb{Z}$-saturated equivalent to the coincidence of $(r+1)$ Hilbert series on $\mathbb{Q}$ and the essential finite residue fields of $\langle u_1,\ldots,u_s \rangle$ (the Hilbert series of the $\text{Syz}(U_i)$'s)?*

Of course, the condition is sufficient by virtue of Proposition 303 and Theorem 358.

4.5.3   A Saturation Algorithm in the Multivariate Case

Considering a finite nonempty list $S = [s_1,\ldots,s_n]$ of vectors in $\mathbf{V}[X_1,\ldots,X_k]^m$ where $\mathbf{V}$ is a residually discrete valuation domain of quotient field $\mathbf{K}$ and residue field $\mathbf{k}$, the following algorithm computes a finite list of vectors in $\mathbf{V}[X_1,\ldots,X_k]^m$ generating $\text{Sat}(\langle s_1,\ldots,s_n \rangle)$ as a $\mathbf{V}[X_1,\ldots,X_k]$-module. During the execution of the algorithm, the $\mathbf{V}[X_1,\ldots,X_k]$-module (generated by the current list) grows every time a non-primitive vector is created by the "triangulation" and then replaced by its primitive version (for the saturation). While the generated $\mathbf{K}[X_1,\ldots,X_k]$-module (by the current list) does not change, the generated $\mathbf{k}[X_1,\ldots,X_k]$-module grows, and one is gradually approximating the saturation.

Algorithm 364. (Saturation algorithm in the multivariate case)

Input: A finite list $S = [s_1,\ldots,s_n]$ of vectors in $\mathbf{V}[X_1,\ldots,X_k]^m$, where $\mathbf{V}$ is a residually discrete valuation domain and $m \geq 1$.

Output: A finite list G of vectors in $\mathbf{V}[X_1,\ldots,X_k]^m$ generating $\mathrm{Sat}(\langle s_1,\ldots,s_n\rangle)$ as a $\mathbf{V}[X_1,\ldots,X_k]$-module.

Start by fixing a monomial order on $\mathbf{V}[X_1,\ldots,X_k]$.
Initialization: $G := S$. All along the algorithm described below, if a nonprimitive vector u is encountered during the computations, then $\mathrm{Prim}(u)$ must be added to G. Let us fix some notation. We denote by S_0 the list S put in an echelon form, and by induction $T_j = [S_0,\ldots,S_j]$ where S_{j+1} denotes $[X_1 S_j,\ldots,X_k S_j]$ put in an echelon form with respect to T_j and then put in an echelon form, with the initialization $T_0 = S_0$.

We begin by putting S in an echelon form (it becomes S_0) and then compute its defect series $\delta_{S_0}(t)$. If $\delta_{S_0}(t) = 0$ then stop; else compute S_1. If $\delta_{S_1}(t) = 0$ then stop; else compute S_2, and so on.

Theorem 365. *Algorithm 364 terminates and is correct.*

Proof. We denote by $\mathbf{K}$ the quotient field of $\mathbf{V}$ and by $\mathbf{k}$ its residue field.
First note that the primitive monomial of a primitive vector $v \in \mathbf{V}[X_1,\ldots,X_k]^m$ is nothing but the leading monomial accordingly to a POT "position over term" monomial order on $\mathbf{V}[X_1,\ldots,X_k]^m$ (see Remark 201) obtained from the monomial order on $\mathbf{V}[X_1,\ldots,X_k]$. So, the computed primitive monomials are those of a nondecreasing sequence of submodules of $\mathbf{k}[X_1,\ldots,X_k]^m$. As $\mathbf{k}[X_1,\ldots,X_k]^m$ is Noetherian, this sequence must stabilize and then we obtain the classical behavior of Hilbert series over a field saying that, after the regularity, the leading monomials obtained at total degree $k+1$ are obtained by simple translation of those obtained at degree k. Thus, the process described above will not add any new entry to G and a fortiori we ultimately obtain a defect which is zero. In a nutshell, the algorithm terminates because $\mathbf{k}[X_1,\ldots,X_k]^m$ is Noetherian and a Hilbert series is ultimately polynomial.

Why G is a generating set for $\mathrm{Sat}(\langle s_1,\ldots,s_n\rangle)$ as a $\mathbf{V}[X_1,\ldots,X_k]$-module?

This is a direct consequence of Theorem 358. □

Theorem 366. *Let $\mathbf{V}$ be a residually discrete valuation domain. Then the $\mathbf{V}$-saturation of any finitely generated submodule of $\mathbf{V}[X_1,\ldots,X_k]^m$ ($k \in \mathbb{N}$, $m \in \mathbb{N}^*$) is finitely-generated.*

Proof. This is a direct consequence of Algorithm 364.

$\square$

Example 367. (Example 357 Continued) As $\delta_U(t) = \frac{1}{(1-t)^2} \neq 0$, one has to put U in an echelon form. This can be done as follows:

$$U = [u_1 = 1 + 2X, u_2 = 1 + 2Y] \to U_0 := [u_1, \frac{1}{2}(u_2 - u_1)] = [1 + 2X, Y - X].$$

As $h_{U_0, \mathbb{Q}}(t) = h_{U_0, \mathbb{Z}/2\mathbb{Z}}(t) = \frac{1}{(1-t)^3} + \frac{1}{(1-t)^2}$, we have $\delta_{U_0}(t) = 0$. We conclude that

$$\mathrm{Sat}(\langle u_1, u_2 \rangle) = \langle 1 + 2X, Y - X \rangle.$$

As in the univariate case, we also obtain the following saturation membership test.

Algorithm 368. (Saturation membership test, the multivariate case)

Input: A finite list $[u, s_1, \ldots, s_n]$ of vectors in $\mathbf{V}[X_1, \ldots, X_k]^m$, where $\mathbf{V}$ is a residually discrete valuation domain of quotient field $\mathbf{K}$ and $m \geq 1$.
Output: An answer to the question $u \in \mathrm{Sat}(\langle s_1, \ldots, s_n \rangle)$? and, in case of positive answer, a finite list $[g_1, \ldots, g_r]$ of vectors in $\mathbf{V}[X_1, \ldots, X_k]^m$ generating $\mathrm{Sat}(\langle s_1, \ldots, s_n \rangle)$ as a $\mathbf{V}[X_1, \ldots, X_k]$-module and a list $[u_1, \ldots, u_r]$ of elements in $\mathbf{V}[X_1, \ldots, X_k]$ such that $u = u_1 g_1 + \cdots + u_r g_r$.

1. Test if $u \in \mathbf{K}[X_1, \ldots, X_k]s_1 + \ldots + \mathbf{K}[X_1, \ldots, X_k]s_n$ (with Gröbner bases techniques for example, see Proposition 228). If the answer is NO then return NO. Else, continue.

2. Write u as a $\mathbf{K}$-linear combination of the $X^\alpha s_i$, $\alpha \in \mathbb{N}^k$, $1 \leq i \leq n$.

3. Use Algorithm 364 to compute a finite list $[g_1, \ldots, g_r]$ of vectors in $\mathbf{V}[X_1, \ldots, X_k]^m$ generating $\mathrm{Sat}(\langle s_1, \ldots, s_n \rangle)$ as a $\mathbf{V}[X_1, \ldots, X_k]$-module.

4. Write each $X^\alpha s_i$ in the expression of u as a $\mathbf{K}$-linear combination of the $X^\beta g_t$, $\beta \in \mathbb{N}^k$, $1 \leq t \leq r$ (by tracing the computations done with Algorithm 364).

5. Write u as a $\mathbf{K}$-linear combination of the $X^\beta g_t$'s, $\beta \in \mathbb{N}^k$, $1 \leq t \leq r$ (using 2 and 4). Note that, by virtue of Lemma 311, the obtained $\mathbf{K}$-linear combination is a $\mathbf{V}$-linear combination if and only if $u \in \mathrm{Sat}(\langle s_1, \ldots, s_n \rangle)$.

6. Collect the $\mathbf{V}$-linear combination of the $X^\beta g_t$'s found in 5 into a $\mathbf{V}[X_1, \ldots, X_k]$-linear combination $u = u_1 g_1 + \cdots + u_r g_r$ of the g_t's.

Theorem 369. *If* $\mathbf{V}$ *is a residually discrete valuation domain then* $\mathbf{V}[X_1, \ldots, X_k]$ *is coherent.*

Proof. Let $p_1, \ldots, p_m \in \mathbf{V}[X_1, \ldots, X_k]$, and consider n vectors $s_1, \ldots, s_n \in \mathbf{V}[X_1, \ldots, X_k]^m$ generating the syzygy module of $p_1, \ldots, p_m$ over the quotient field $\mathbf{K}$ of $\mathbf{V}$ as a $\mathbf{K}[X_1, \ldots, X_k]$-module ($s_1, \ldots, s_n$ can be computed using Gröbner bases techniques, see Proposition 345). Then, the syzygy module of $p_1, \ldots, p_m$ over $\mathbf{V}$ is nothing but the $\mathbf{V}$-saturation of $\langle s_1, \ldots, s_n \rangle$ which is finitely-generated by Theorem 366.

□

As in the univariate case, we obtain the following results for Prüfer domains.

Theorem 370. *Let $\mathbf{R}$ be a Prüfer domain and $m \geq 1$. Then the $\mathbf{R}$-saturation of any finitely-generated submodule of $\mathbf{R}[X_1, \ldots, X_k]^m$ is finitely-generated.*

Proof. Do as in the proof of Theorem 350. □

Corollary 371. *If $\mathbf{R}$ is a Prüfer domain then $\mathbf{R}[X_1, \ldots, X_k]$ is coherent.*

At the end of this chapter, it is worth pointing an important work yet to be done: to prove the termination of Algorithm 364 in such an effective way that a bound could be deduced for the number of steps in the algorithm (like in the univariate case). In fact, we were obligated to utilize a Noetherianity and regularity arguments to prove the termination of Algorithm 364, which defeats all hope of computing a complexity bound.

Chapter 5

Exercises

Exercise 372. Prove constructively that a valuation domain has Krull dimension ≤ 1 if and only if its valuation group is archimedean.

Exercise 373. (H. Lombardi) For a ring $\mathbf{R}$, we denote by U the subset of $\mathbf{R}[X]$ formed by primitive polynomials (i.e., polynomials whose coefficients generate the whole ring $\mathbf{R}$). It is in fact a monoid (this is a direct consequence of the Dedekind-Mertens theorem, see Exercise 376). The Nagata ring $\mathbf{R}(X)$ is

$$\mathbf{R}(X) := U^{-1}\mathbf{R}[X].$$

Prove constructively that for any ring $\mathbf{R}$ and $d \geq 0$, we have the implication:

$$\mathrm{Kdim}(\mathbf{R}[X]) \leq d+1 \;\Rightarrow\; \mathrm{Kdim}(\mathbf{R}(X)) \leq d.$$

Exercise 374. (An Algorithm for the Divisors of Monic Polynomials and Doubly Monic Laurent Polynomials [10, 179])

(1) Prove constructively that for any ring $\mathbf{R}$, if $r^{n+1}y = r^n$ for some $r, y \in \mathbf{R}$ and $n \in \mathbb{N}$, then $r^2y - r$ is nilpotent and ry^n is idempotent. If, in addition, $\mathbf{R}$ is reduced then ry is idempotent and $r\mathbf{R} = (ry)\mathbf{R}$.

(2) Let $\mathbf{R}$ be a reduced ring, and $f = a_0 + a_1 X + \cdots + a_n X^n$, $g = b_0 + b_1 X + \cdots + b_d X^d \in \mathbf{R}[X]$ such that $fg = c_0 + c_1 X + \cdots + c_m X^m$ with $c_m = 1$.

   (a) Prove that $a_n^{n+d-m+1} b_{m-n} = a_n^{n+d-m}$.

   (b) By induction on $n + d - m$, prove that there exists a direct sum decomposition $\mathbf{R} = \mathbf{R}_0 \oplus \cdots \oplus \mathbf{R}_m$ ($m \leq n$) of $\mathbf{R}$ such that if $f = f_0 + \cdots + f_m$ is the decomposition of f with respect to the induced decomposition $\mathbf{R}[X] = \mathbf{R}_0[X] \oplus \cdots \oplus \mathbf{R}_m[X]$, then the degree coefficient of f_i is a unit of $\mathbf{R}_i$ for each i.

© Springer International Publishing Switzerland 2015
I. Yengui, *Constructive Commutative Algebra*, Lecture Notes in Mathematics 2138,
DOI 10.1007/978-3-319-19494-3_5

(3) Prove that if $\mathbf{R}$ is a nonnecessarily reduced ring, then $f = a_0 + a_1 X + \cdots + a_n X^n \in \mathbf{R}[X]$ divides a monic polynomial if and only if there exist a nilpotent polynomial N and a direct sum decomposition $\mathbf{R} = \mathbf{R}_0 \oplus \cdots \oplus \mathbf{R}_m$ $(m \leq n)$ of $\mathbf{R}$ such that if $f - N = f_0 + \cdots + f_m$ is the decomposition of $f - N$ with respect to the induced decomposition $\mathbf{R}[X] = \mathbf{R}_0[X] \oplus \cdots \oplus \mathbf{R}_m[X]$, then the degree coefficient of f_i is a unit of $\mathbf{R}_i$ for each i.

(4) Deduce that if $\mathbf{R}$ is a nonnecessarily reduced ring, then $f = a_0 + a_1 X + \cdots + a_n X^n \in \mathbf{R}[X]$ divides a monic polynomial if and only if $\langle a_0, \ldots, a_n \rangle = \mathbf{R}$ and, for each $j \in \{0, \ldots, n\}$, we can find $\beta_j \in \mathbf{R}$ and $k_j \in \mathbb{N}$ such that $(a_j(a_j\beta_j - 1))^{k_j} \equiv 0 \bmod \langle a_{j+1}, \ldots, a_n \rangle$.

(5) Let U and V be two indeterminates over a field $\mathbf{K}$, and consider the reduced ring $\mathbf{R} = \mathbf{K}[U,V]/\langle U^2 - U, UV \rangle = \mathbf{K}[u,v] = \mathbf{K}[v] \oplus \mathbf{K}[v]u$, where $u^2 = u$ and $uv = 0$. Take $f = u - (1+u)X^2 + uX^3$ and $g = v + uX^2 + (u-1)X^3$. Verify that $fg = (u - v)X^2 - 2uX^4 + X^5$. Using the algorithm coming out of your constructive proof of Question (2), find the corresponding decomposition of f.

(6) Let U and V be two indeterminates over a field $\mathbf{K}$ such that $\mathrm{Char}\mathbf{K} \neq 2$, and consider the nonreduced ring $\mathbf{R} = \mathbf{K}[U,V]/\langle U^2 - U, UV^2 \rangle = \mathbf{K}[u,v] = \mathbf{K}[v] \oplus \mathbf{K}u \oplus \mathbf{K}uv$, where $u^2 = u$ and $uv^2 = 0$. Check that the nilradical of $\mathbf{R}$ is $\mathcal{N} = \langle uv \rangle$ and that $\mathbf{R}/\mathcal{N} = \mathbf{K}[U,V]/\langle U^2 - U, UV \rangle = \mathbf{K}[u',v']$ with $u'^2 = u'$ and $u'v' = 0$.

Setting $f = u - (1+u)X^2 + uX^3 + uvX^4$ and $g = -v^4 + uX^2 + 2v^2X^3 - 2uX^4 + uX^5 + (u - 1 - uv)X^6$, verify that $fg = (u + v^4)X^2 - 4uX^4 + (u - v^2)X^5 + 4uX^6 - 4uX^7 + X^8$. Using the above proofs (reduced and nonreduced cases), find a decomposition of f in $\mathbf{R}/\mathcal{N}$ as in (2.b) and a decomposition of f in $\mathbf{R}$ as in (3).

(7) Recall that a Laurent polynomial $f \in \mathbf{R}[X, X^{-1}]$ is said to be *doubly monic* if the coefficients of the highest and lowest monomials are equal to 1.

    (a) Prove that for any ring $\mathbf{R}$, $f \in \mathbf{R}[X, X^{-1}]$ divides a doubly monic Laurent polynomial if and only if there exist $n, m \in \mathbb{N} \setminus \{0\}$ such that both $X^n f(X)$ and $X^m f(X^{-1})$ divide a monic polynomial.

    (b) Deduce that if $\mathbf{R}$ is a reduced ring, then $f \in \mathbf{R}[X, X^{-1}]$ divides a doubly monic Laurent polynomial if and only if there exists a direct sum decomposition $\mathbf{R} = \mathbf{R}_0 \oplus \cdots \oplus \mathbf{R}_m$ of $\mathbf{R}$ such that if $f = f_0 + \cdots + f_m$ is the decomposition of f with respect to the induced decomposition $\mathbf{R}[X, X^{-1}] = \mathbf{R}_0[X, X^{-1}] \oplus \cdots \oplus \mathbf{R}_m[X, X^{-1}]$, then the coefficients of the highest and lowest terms of f_i are units in $\mathbf{R}_i$ for each i.

    (c) Deduce that if $\mathbf{R}$ is a nonnecessarily reduced ring, then $f = a_k X^k + a_{k+1}X^{k+1} + \cdots + a_l X^l \in \mathbf{R}[X, X^{-1}]$, $k, l \in \mathbb{Z}$, divides a doubly monic Laurent polynomial if and only if $\langle a_k, \ldots, a_l \rangle = \mathbf{R}$ and, for each $j \in \{k, \ldots, l\}$, we can find $\beta_j, \delta_j \in \mathbf{R}$ and $m_j, n_j \in \mathbb{N}$ such that

$(a_j(a_j\beta_j - 1))^{m_j} \equiv 0 \bmod \langle a_{j+1}, \ldots, a_n \rangle$ and $(a_j(a_j\delta_j - 1))^{n_j} \equiv 0 \bmod \langle a_k, \ldots, a_{j-1} \rangle$.

(8) For any ring $\mathbf{R}$, $\mathbf{R}\langle X, X^{-1} \rangle$ will denote the localization of $\mathbf{R}[X, X^{-1}]$ at doubly monic polynomials.

(a) Prove that $\mathbf{R}\langle X, X^{-1} \rangle = \mathbf{R}\langle X \rangle \cap \mathbf{R}\langle X^{-1} \rangle$ (the rings $\mathbf{R}\langle X \rangle$ and $\mathbf{R}\langle X^{-1} \rangle$ being considered as subrings of $\mathbf{R}\langle X, X^{-1} \rangle$).

(b) Prove that $\mathbf{R}\langle X^{-1} + X \rangle \subsetneq \mathbf{R}\langle X, X^{-1} \rangle$.

(c) A Laurent polynomial $f(X) \in \mathbf{R}[X, X^{-1}]$ is said to be *symmetric at X and X^{-1}* (or, simply, symmetric) if $f(X^{-1}) = f(X)$. Prove that

$$\mathbf{R}[X^{-1} + X] = \{ f \in \mathbf{R}[X, X^{-1}] \mid f \text{ is symmetric at } X \text{ and } X^{-1} \}.$$

Deduce that any doubly monic symmetric Laurent polynomial is a monic polynomial at $X^{-1} + X$ (i.e., it can be expressed as $g(X^{-1} + X)$ with a monic polynomial g).

(d) A Laurent polynomial $f \in \mathbf{R}[X, X^{-1}]$ is said to be *doubly unitary* if the coefficients of the highest and lowest monomials are units.

Prove that for any doubly unitary Laurent polynomial $g \in \mathbf{R}[X, X^{-1}]$, there exists $h \in \mathbf{R}[X, X^{-1}]$ such that gh is a monic polynomial at $X^{-1} + X$.

Or, equivalently, prove that for any $g(X) = a_0 X^m + a_1 X^{m+1} + \cdots + a_n X^{m+n} \in \mathbb{Z}[a_0, a_1, \ldots, a_{n-1}, a_n][X, X^{-1}]$, there exists $h \in \mathbb{Z}[a_0^{\pm}, a_1, \ldots, a_{n-1}, a_n^{\pm}][X, X^{-1}]$ such gh is a monic polynomial at $X^{-1} + X$ with coefficients in $\mathbb{Z}[a_0^{\pm}, a_1, \ldots, a_{n-1}, a_n^{\pm}]$.

(e) Prove that for any ring $\mathbf{R}$, $\mathbf{R}[X, X^{-1}]$ is a finitely-generated free $\mathbf{R}[X^{-1} + X]$-module of rank 2 [67].

(f) Deduce that for any ring $\mathbf{R}$, $\mathbf{R}\langle X, X^{-1} \rangle$ is a finitely-generated free $R\langle X^{-1} + X \rangle$-module of rank 2.

Exercise 375. (Stably Free Modules Over Laurent Polynomial Rings [4])

(1) (An analogue of Proposition 56 for Laurent polynomials)

Prove constructively that for any ring $\mathbf{R}$, and $u, v \in \mathbf{R}[X]$ with u doubly monic, we have the equivalence:

$$\langle u, v \rangle = \langle 1 \rangle \text{ in } \mathbf{R}[X, X^{-1}] \quad \Longleftrightarrow \quad \mathrm{Res}_X(u, v) \in \mathbf{R}^{\times}.$$

(2) (An analogue of Theorem 57 for Laurent polynomials)

If $f \in \mathbf{R}[X, X^{-1}]$, a minimal shifted version of f is $\widetilde{f} = X^n f \in \mathbf{R}[X]$ where $n \in \mathbb{Z}$ is the minimal possible. For example a minimal shifted version of $X^{-2} + X + X^3$ is $1 + X^3 + X^5$, a minimal shifted version of $X^2 + X^3$ is $1 + X$.

Similarly, if $f \in \mathbf{R}[X_1^{\pm} \dots, X_k^{\pm}]$, a minimal shifted version of f is $\tilde{f} = X_1^{n_1} \cdots X_k^{n_k} f \in \mathbf{R}[X_1, \dots, X_k]$ where $n_1, \dots, n_k \in \mathbb{Z}$ are the minimal possible. For example a minimal shifted version of $X_1^{-2} X_2^{-1} + X_1 X_2^{-1} + X_1^2$ is $1 + X_1^3 + X_1^4 X_2$.

Prove constructively that for any ring $\mathbf{R}$, if $\langle v_1(X), \dots, v_n(X) \rangle = \mathbf{R}[X, X^{-1}]$ where v_1 is doubly monic and $n \geq 3$, then there exist $\gamma_1, \dots, \gamma_s \in \mathrm{E}_{n-1}(\mathbf{R}[X])$ such that:

$$\langle \mathrm{Res}(\tilde{v}_1, e_1 . \gamma_1{}^{\mathsf{t}}(\tilde{v}_2, \dots, \tilde{v}_n)), \dots, \mathrm{Res}(\tilde{v}_1, e_1 . \gamma_s{}^{\mathsf{t}}(\tilde{v}_2, \dots, \tilde{v}_n)) \rangle = \mathbf{R}.$$

In particular $1 \in \langle \tilde{v}_1, \dots, \tilde{v}_n \rangle$ in $\mathbf{R}[X]$. Here $e_1 . x$, where x is a column vector, stands for the first coordinate of x, and $\tilde{v}_i$ is a shifted version of v_i.

(3) (An analogue of Theorem 61 for Laurent polynomials)

Let $\mathbf{R}$ be a ring, $v_1, \dots, v_n, u_1, \dots, v_n \in \mathbf{R}[X, X^{-1}]$ such that $\sum_{i=1}^{n} u_i v_i = 1$, v_1 doubly monic, and $n \geq 3$. Denote by $\ell = \deg v_1$, $s = (n-2)\ell + 1$, and suppose that $\mathbf{R}$ contains a set $E = \{y_1, \dots, y_s\}$ such that $y_i - y_j$ is invertible for each $i \neq j$. For each $1 \leq r \leq n$ and $1 \leq i \leq s$, let $\tilde{v}_r$ be a minimal shifted version of v_r and denote by $r_i = \mathrm{Res}_X(\tilde{v}_1, \tilde{v}_2 + y_i \tilde{v}_3 + \cdots + y_i^{n-2} \tilde{v}_n)$. Prove constructively that $\langle r_1, \dots, r_s \rangle = \mathbf{R}$, that is, there exist $\alpha_1, \dots, \alpha_s \in \mathbf{R}$ such that $\alpha_1 r_1 + \cdots + \alpha_s r_s = 1$. In particular $1 \in \langle \tilde{v}_1, \dots, \tilde{v}_n \rangle$ in $\mathbf{R}[X]$.

(4) (Producing doubly unitary Laurent polynomials over a discrete field)

Let $\mathbf{R}$ a discrete ring. If $f \in \mathbf{R}[X, X^{-1}]$ is a nonzero Laurent polynomial in a single variable X, we denote $\deg(f) = \mathrm{hdeg}(f) - \mathrm{ldeg}(f)$, where $\mathrm{hdeg}(f)$ and $\mathrm{ldeg}(f)$ denote respectively the highest and lowest degrees of f. For example, $\deg(X^{-2} + X + X^3) = 3 - (-2) = 5$. Note that the degree of f can be defined as the (classical) degree of a minimal shifted version of f. We can also define the total degree of a multivariate Laurent polynomials f as the (classical) total degree of a minimal shifted version of f.

If $\mathbf{K}$ is a discrete field and $f \in \mathbf{K}[X_1^{\pm 1}, X_2^{\pm 1} \dots, X_k^{\pm 1}]$, then after a bijective change of variables $X_1 = Y_1$, $X_2 = Y_2 Y_1^m, \dots, X_k = Y_k Y_1^{m^{k-1}}$ (à la Nagata), for sufficiently large m, f becomes doubly unitary in Y_1. The problem with such a change of variables is that it explodes the degree of f at Y_1 as it is exponential. Our purpose is to change f into a doubly monic polynomial without considerably increasing its degree.

We will begin by discussing the case of two variables.

(a) Let $\mathbf{K}$ be a discrete field and consider $f = \sum_{i=1}^{t} a_i X^{n_i} Y^{m_i}$, $a_i \in \mathbf{K}$, where t is the number of monomials appearing in f. Set

$$E = \{ \frac{m_j - m_i}{n_i - n_j} \mid 1 \leq i, j \leq t \text{ and } n_i \neq n_j \}.$$

Prove that for each $\alpha \in \mathbb{Z} \setminus E$, denoting φ_α the change of variables $(X, Y) \mapsto (XY^\alpha, Y)$, the correspondence $X^{n_i} Y^{m_i} \mapsto \deg_Y(\varphi_\alpha(X^{n_i} Y^{m_i}))$ is a one-to-one. In particular, $\varphi_\alpha(f)$ is doubly unitary at Y. We denote by α_0 an element of $\mathbb{Z}\setminus E$ such that $|\alpha_0| = \min\{|\ell|, \ell \in \mathbb{Z}\setminus E\}$. Prove that if $f \in \mathbf{K}[X,Y]$ of total degree $\leq d$ and taking $\alpha_0 \geq 0$, then $\mathrm{tdeg}(\varphi_{\alpha_0}(f)) \leq d+1$.

(b) Take $f = Y + Y^2 + Y^3 + X + XY + X^2Y + X^2Y^2$. Compute E, α_0, $\varphi_{\alpha_0}(X,Y)$, and $\varphi_{\alpha_0}(f)$.

(c) What can you say about the general case (more than two variables)?

(d) From Questions (3), (4) and Algorithm 65, deduce an algorithm for unimodular completion over a Laurent polynomial ring $\mathbf{K}[X_1^{\pm 1}, X_2^{\pm 1} \ldots, X_k^{\pm 1}]$, where $\mathbf{K}$ is an infinite field.

(5) (An analogue of Lemma 182 for Laurent polynomials)

Let $\mathbf{R}$ be a ring and I an ideal of $\mathbf{R}[X,X^{-1}]$ containing a doubly monic polynomial. Prove constructively that if J is an ideal of $\mathbf{R}$ such that $I + J[X,X^{-1}] = \mathbf{R}[X,X^{-1}]$, then $(I \cap \mathbf{R}) + J = \mathbf{R}$.

(6) (An analogue of Lemma 185 for Laurent polynomials)

Let $^t(v_0(X), v_1(X), \ldots, v_n(X)) \in \mathrm{Um}_{n+1}(\mathbf{R}[X,X^{-1}])$, where $\mathbf{R}$ is an integral local ring of Krull dimension ≤ 1 and $n \geq 2$. Prove constructively that

$$^t(\widetilde{v}_0(X), \widetilde{v}_1(X), \ldots, \widetilde{v}_n(X)) \sim_{\mathrm{E}_{n+1}(\mathbf{R}[X])} {}^t(w_0(X), w_1(X), \ldots, c_2, \ldots, c_n),$$

where the c_i's are constant for $i \geq 2$, $w_i \in \mathbf{R}[X]$ with $\deg w_1(X) \leq 1$.

(7) Prove constructively that for any ring $\mathbf{R}$, if $\mathrm{Kdim}\,\mathbf{R} \leq 0$, then $\mathbf{R}(X) = \mathbf{R}\langle X\rangle = \mathbf{R}\langle X, X^{-1}\rangle$. Moreover, $\mathrm{Kdim}\,\mathbf{R}(X) = \mathrm{Kdim}\,\mathbf{R}\langle X\rangle = \mathrm{Kdim}\,\mathbf{R}\langle X, X^{-1}\rangle \leq 0$.

(8) (An analogue of Corollary 187 for Laurent polynomials)

Deduce from the previous questions that for any integral local ring $\mathbf{R}$ of Krull dimension ≤ 1 and $n \geq 2$, $\mathrm{GL}_{n+1}(\mathbf{R}[X,X^{-1}])$ acts transitively on $\mathrm{Um}_{n+1}(\mathbf{R}[X,X^{-1}])$ and, thus, that all finitely-generated stably free modules over $\mathbf{R}[X,X^{-1}]$ are free.

Exercise 376. (Dedekind-Mertens, by Th. Coquand)

If U, V are two sub-$\mathbb{Z}$-modules of a ring $\mathbf{A}$, we denote by UV the submodules generated by the uv's, with $u \in U$ and $v \in V$. For $f \in \mathbf{A}[T]$, we denote by $[f]$ the sub-$\mathbb{Z}$-module of $\mathbf{A}$ generated by the coefficients of f.

Let $f = \sum_{i\geq 0} a_i T^i$, $g = \sum_{j=0}^m b_j T^j$, $h = fg \in \mathbf{A}[T]$ and denote by $F = [f]$, $G = [g]$ and $H = [h]$.
Denote also $\widetilde{g} = \sum_{j=0}^{m-1} b_j T^j = g - b_m T_m$, and $\widetilde{G} = [\widetilde{g}]$. By induction on m (using $\widetilde{G}$), prove that

$$F^{m+1}G = F^m H.$$

Exercise 377. (Suslin's Lemma, Particular Case, by C. Quitté)

Let $\mathbf{A}$ be a ring, $d \geq 1$, $v = (X - x_1) \cdots (X - x_d) \in \mathbf{A}[X]$, $u, w \in \mathbf{A}[X]$, and take $d + 1$ elements $u_0, \ldots, y_d$ in $\mathbf{A}$.

For $0 \leq i \leq d$, set

$$r_i = \mathrm{Res}_X(v, u + y_i w) = \prod_{j=1}^{d}(u_j + y_i w_j),$$

with $u_j = u(x_j)$ and $w_j = w(x_j)$. Moreover, set

$$\pi = \prod_{i<j}(y_i - y_j)$$

and

$$c_0 + c_1 Y + \cdots + c_d Y^d = (u_1 + w_1 Y) \cdots (u_d + w_d Y).$$

(1) Prove that $\langle u_1, w_1 \rangle \langle u_2, w_2 \rangle^2 \cdots \langle u_d, w_d \rangle^d \subseteq \langle c_0, \ldots, c_d \rangle$.

(2) Prove that $\pi \langle c_0, \ldots, c_d \rangle \subseteq \langle r_0, \ldots, r_d \rangle$.

(3) Deduce that $\pi \langle u_1, w_1 \rangle \langle u_2, w_2 \rangle^2 \cdots \langle u_d, w_d \rangle^d \subseteq \langle r_0, \ldots, r_d \rangle$.

(4) Deduce that if $1 \in \langle u, v, w \rangle$ then $1 \in \langle r_0, \ldots, r_d \rangle$.

Exercise 378. (Suslin's $n!$-Theorem, with C. Quitté)

$\mathbf{R}$ denotes a ring.

(1) Let $b \in \mathbf{R}$ and $(a_1, \ldots, a_n) \in \mathbf{R}^n$ such that $(\bar{a}_1, \ldots, \bar{a}_n)$ is completable over $\mathbf{R}/b\mathbf{R}$, i.e., there exist two matrices $A, D \in \mathrm{M}_n(\mathbf{R})$ such that $AD \equiv \mathrm{I}_n \bmod \langle b \rangle$ and the fist row of A is $(a_1, \ldots, a_n)$. Let us denote by $a := \det(A)$.

    (a) Take $C, U \in \mathrm{M}_n(\mathbf{R})$ such that $AD = \mathrm{I}_n + bU$ and $DA = \mathrm{I}_n + bC$. Check that

$$\begin{pmatrix} A & b\mathrm{I}_n \\ C & D \end{pmatrix} \begin{pmatrix} D & -b\mathrm{I}_n \\ -U & A \end{pmatrix} = \begin{pmatrix} \mathrm{I}_n & 0 \\ * & \mathrm{I}_n \end{pmatrix} \in \mathrm{GL}_{2n}(\mathbf{R}).$$

    (b) Consider the diagonal matrix $B' := \mathrm{diag}(b^n, 1, \ldots, 1) \in \mathrm{M}_n(\mathbf{R})$. Prove that one can write $B' = bE + aF$ with $E \in \mathrm{E}_n(\mathbf{R})$ and $F \in \mathrm{M}_n(\mathbf{R})$.

    (c) Check that

$$\begin{pmatrix} A & b\mathrm{I}_n \\ C & D \end{pmatrix} \begin{pmatrix} \mathrm{I}_n & \tilde{A}F \\ 0 & E \end{pmatrix} = \begin{pmatrix} A & B' \\ C & D' \end{pmatrix} \in \mathrm{GL}_{2n}(\mathbf{R})$$

with $D' \in \mathrm{M}_n(\mathbf{R})$. Recall that $\tilde{A}$ is called the adjugate matrix of A (the transpose of the cofactor matrix of A) and it satisfies $A\tilde{A} = \tilde{A}A = a\mathrm{I}_n$.

    (d) Prove that $\begin{pmatrix} A & B' \\ C & D' \end{pmatrix}$ is equivalent to a matrix $\begin{pmatrix} A & B' \\ C' & D'' \end{pmatrix}$ where the last $n-1$ columns of D'' are zero. Deduce that there exists a matrix $T \in \mathrm{GL}_{n+1}(\mathbf{R})$ whose first row is $(a_1, \ldots, a_n, b^n)$.

(2) Let $(x,y,z) \in \mathrm{Um}_3(\mathbf{R})$. Prove that the row (x,y,z^2) is completable. Give an explicit completion of (x,y,z).

(3) More generally, prove the following result [166]:

$$(a_0,a_1,\ldots,a_n) \in \mathrm{Um}_{n+1}(\mathbf{R}) \Rightarrow (a_0,a_1,a_2^2,\ldots,a_n^n) \text{ is completable.}$$

Exercise 379. (An Algorithm for Unimodular Completion Over Noetherian Rings [121])

(1) (a) Let u,v,w,y_1,y_2 be elements in a strongly discrete coherent ring $\mathbf{A}$ such that $uv+w=1$ and $y_2-y_1 \in \mathbf{A}^\times$. Suppose that both $v+y_1w$ and $v+y_2w$ are zero-divisors in $\mathbf{A}$, i.e., $d_1(v+y_1w)=0$ and $d_2(v+y_2w)=0$ for some $d_1,d_2 \in \mathbf{A} \setminus \{0\}$.
Prove that $\langle d_1 \rangle \subsetneq \langle d_1,d_2 \rangle$.

(b) Generalization: Let $u,v,w,y_1,y_2,\ldots,y_n$ $(n \geq 2)$ be elements in a strongly discrete coherent ring $\mathbf{A}$ such that $uv+w=1$ and $y_j-y_i \in \mathbf{A}^\times$ for $i \neq j$. Suppose that all $v+y_1w, v+y_2w,\ldots,v+y_nw$ are zero-divisors in $\mathbf{A}$ and write $d_i(v+y_iw)=0$ with $d_i \in \mathbf{A} \setminus \{0\}$ for $1 \leq i \leq n$.
Prove that one obtains an increasing chain

$$\langle d_1 \rangle \subsetneq \langle d_1,d_2 \rangle \subsetneq \cdots \subsetneq \langle d_1,\ldots,d_n \rangle$$

of ideals in $\mathbf{A}$.

(c) Deduce that if $uv+w=1$ in a strongly discrete coherent Noetherian ring $\mathbf{A}$ containing an infinite set $E = \{y_1,y_2,\ldots\}$ such that $y_i - y_j \in \mathbf{A}^\times$ for $i \neq j$, then there exists i such that $v+y_iw$ is not a zero-divisor in $\mathbf{A}$.

(2) Suppose that $u_1v_1 + u_2v_2 + \cdots + u_nv_n = 1$ in a strongly discrete coherent Noetherian ring $\mathbf{A}$ containing an infinite set E such that $y - x \in \mathbf{A}^\times$ for all $x,y \in E$ such that $x \neq y$. Prove that there exist $y_1,\ldots,y_{n-1} \in E$ such that the sequence $(v_1 + y_1\xi_0(u_2v_2 + \cdots + u_nv_n), v_2 + y_2\xi_1(u_3v_3 + \cdots + u_nv_n),\ldots,v_{n-1} + y_{n-1}\xi_{n-2}u_nv_n)$ is regular, where $\xi_0 = 1$ and $\xi_{k+1} = \xi_k(1 - y_{k+1}u_{k+1}\xi_k)$.

(3) Let $\mathbf{A}$ be a strongly discrete coherent Noetherian ring with Krull dimension $d < \infty$ and containing an infinite set E such that $y - x \in \mathbf{A}^\times$ for all $x,y \in E$ such that $x \neq y$.
Suppose that we have $u_1v_1 + u_2v_2 + \cdots + u_nv_n = 1$ in $\mathbf{A}$ with $n \geq d+2$ and denote by $w_k = u_kv_k + \cdots + u_nv_n$.
Prove that there exist $y_1,\ldots,y_{d+1} \in E$ such that

$$1 \in \langle v_1 + y_1\xi_0w_2, v_2 + y_2\xi_1w_3,\ldots,v_{d+1} + y_{d+1}\xi_dw_{d+2} \rangle,$$

where $\xi_0 = 1$ and $\xi_{k+1} = \xi_k(1 - y_{k+1}u_{k+1}\xi_k)$. In particular, prove that there exists $M \in \mathrm{E}_n(\mathbf{A})$ such that

$$M^{\mathrm{t}}(v_1,v_2,\ldots,v_n) = {}^{\mathrm{t}}(1,0,\ldots,0).$$

(4) Recall that if $\mathbf{A}$ is a ring, the ring $\mathbf{A}\langle X \rangle$ denotes the localization of $\mathbf{A}[X]$ at the multiplicative subset of monic polynomials. By induction, we define $\mathbf{A}\langle X_1, \ldots, X_k \rangle := \mathbf{A}\langle X_1, \ldots, X_{k-1} \rangle \langle X_k \rangle$. It is in fact the localization of the polynomial ring $\mathbf{A}[X_1, \ldots, X_k]$ at the multiplicative subset

$$U_k = \{ f \in \mathbf{A}[X_1, \ldots, X_k] \text{ such that } LC(f) = 1 \},$$

where $LC(f)$ denotes the leading coefficient of f accordingly to the lexicographic monomial order with $X_k > X_{k-1} > \cdots > X_1$.

Let $\mathbf{A}$ be a strongly discrete coherent Noetherian with $\mathrm{Kdim}\,\mathbf{A} = d < \infty$ and containing an infinite set E such that $y - x \in \mathbf{A}^\times$ for all $x, y \in E$ such that $x \neq y$. Suppose that we know constructively that $\mathrm{Kdim}\,\mathbf{A}\langle X_1, \ldots, X_k \rangle = \mathrm{Kdim}\,\mathbf{A}$ (classically, this is always true for a Noetherian ring).

Let $u_1 v_1, \ldots, u_n, v_n \in \mathbf{A}[X_1, \ldots, X_k]$ such that $u_1 v_1 + u_2 v_2 + \cdots + u_n v_n = 1$ and $n \geq d + 2$. Denote by $w_k = u_k v_k + \cdots + u_n v_n$.

(a) Prove that there exist $y_1, \ldots, y_{d+1} \in E$ such that

$$\langle v_1 + y_1 \xi_0 w_2, v_2 + y_2 \xi_1 w_3, \ldots, v_{d+1} + y_{d+1} \xi_d w_{d+2} \rangle \cap U_k \neq \emptyset,$$

where $\xi_0 = 1$ and $\xi_{k+1} = \xi_k (1 - y_{k+1} u_{k+1} \xi_k)$.

(b) Deduce that, via a change of variables and by elementary operations, one can transform ${}^{t}(v_1, v_2, \ldots, v_n)$ into a unimodular row whose first coordinate is monic at X_k.

(c) Give an algorithm computing a matrix $M \in SL_n(\mathbf{A}[X_1, \ldots, X_k])$ such that

$$M\,{}^{t}(v_1, v_2, \ldots, v_n) = {}^{t}(1, 0, \ldots, 0).$$

(d) Prove that for $n \geq \max(3, d+2)$, the group $E_n(\mathbf{A}[X_1, \ldots, X_k])$ acts transitively on $\mathrm{Um}_n(\mathbf{A}[X_1, \ldots, X_k])$.

(e) Prove that $SL_n(\mathbf{R}) \equiv SL_n(\mathbf{A}) \mod E_n(\mathbf{R})$, $\forall n \geq \max(3, d+2)$.

Exercise 380. (When the Image of an Idempotent Matrix is Free)

Let $\mathbf{R}$ be a ring and $M \in \mathbf{R}^{n \times n}$. Prove that the matrix M is idempotent with rank r free image if and only if there exist $X \in \mathbf{R}^{n \times r}$ and $Y \in \mathbf{R}^{r \times n}$ such that $YX = I_r$ and $M = XY$. Moreover,

(a) $\mathrm{Im}\,M = \mathrm{Im}\,X \simeq \mathrm{Im}\,Y$, $\mathrm{Ker}\,M = \mathrm{Ker}\,Y$.

(b) For all matrices X', Y' with the same sizes as X and Y and such that $M = X'Y'$, there exists a unique matrix $U \in GL_r(\mathbf{R})$ such that $X' = UX$ and $Y = UY'$.

(c) Reformulate this result in case $r = 1$.

Exercise 381. (Seminormality [32, 107])

(1) Let $\mathbf{R}$ be a ring and $M \in \mathbf{R}^{n \times n}$ with $M^2 = M$. Prove that $\operatorname{Im} M$ has rank one if and only if M has trace one and all 2×2 minors of M are zero.

(2) Prove that for any ring $\mathbf{R}$, the natural homomorphism $\operatorname{Pic} \mathbf{R} \to \operatorname{Pic} \mathbf{R}[X]$ is an isomorphism (in short, $\operatorname{Pic} \mathbf{R} = \operatorname{Pic} \mathbf{R}[X]$) if and only if for every rank one idempotent matrix $M = (m_{i,j})_{1 \le i,j \le n}$ over $\mathbf{R}[X]$ such that

$$M(0) = \begin{pmatrix} 1 & 0 \\ 0 & 0_{n-1} \end{pmatrix} =: I_{1,n},$$

there exist $f_1, \ldots, f_n, g_1, \ldots, g_n \in \mathbf{R}[X]$ such that $m_{i,j} = f_i g_j$ for all i, j.

(3) Let $\mathbf{R} \subseteq \mathbf{T}$ be two reduced rings, and $f_1, \ldots, f_n, g_1, \ldots, g_n \in \mathbf{T}[X]$ satisfying the following hypotheses:

$$(*) \quad \begin{cases} f_1(0) = g_1(0) = 1, \ f_i(0) = g_i(0) = 0 \ \forall \, 2 \le i \le n, \\ m_{i,j} := f_i g_j \in \mathbf{R}[X] \ \forall \, 1 \le i, j \le n, \\ \sum_{i=1}^{n} f_i g_i = 1. \end{cases}$$

Denote by $M := (m_{i,j})_{1 \le i,j \le n}$. Note that the first hypothesis amounts to saying that $M(0) = I_{1,n}$. Prove that the matrix M is idempotent of rank one and that the following assertions are equivalent:

  1. The module M is free over $\mathbf{R}[X]$.

  2. $f_i, g_i \in \mathbf{R}[X] \ \forall \, 1 \le i \le n$.

  3. $f_1 \in \mathbf{R}[X]$.

(4) Let $\mathbf{R} \subseteq \mathbf{T}$ be two reduced rings with $\operatorname{Pic} \mathbf{T} = \operatorname{Pic} \mathbf{T}[X]$. Prove that the following properties are equivalent:

  1. $\operatorname{Pic} \mathbf{R} = \operatorname{Pic} \mathbf{R}[X]$.

  2. If some polynomials $f_1, \ldots, f_n, g_1, \ldots, g_n \in \mathbf{T}[X]$ satisfy the conditions $(*)$ in (4) then the f_i's and the g_i's belong to $\mathbf{R}[X]$.

  3. If some polynomials $f_1, \ldots, f_n, g_1, \ldots, g_n \in \mathbf{T}[X]$ satisfy the conditions $(*)$ then $f_1 \in \mathbf{R}[X]$.

(5) A ring $\mathbf{R}$ is called *seminormal* if for every $b, c \in \mathbf{R}$ satisfying $b^2 = c^3$ there exists $a \in \mathbf{R}$ such that $a^3 = b$ and $a^2 = c$.

  (a) Prove that seminormal $\Rightarrow$ reduced.

(b) (Schanuel's example) Let $\mathbf{R}$ be a reduced ring such that $\operatorname{Pic}\mathbf{R} = \operatorname{Pic}\mathbf{R}[X]$. Let $b, c \in \mathbf{R}$ satisfying $b^2 = c^3$. Consider $\mathbf{T} = \mathbf{R}[a] = \mathbf{R} + a\mathbf{R}$ a reduced ring containing $\mathbf{R}$ with $a^3 = b$ and $a^2 = c$ (one can take for example $\mathbf{T} = (\mathbf{R}[T]/\langle T^2 - c, T^3 - b\rangle)_{\text{red}}$). Consider the matrix $M(X) = (f_i g_j)_{1 \le i,j \le 2}$ with $f_1 = 1 + aX$, $f_2 = cX^2 = g_2$ and $g_1 = (1 - aX)(1 + cX^2)$, that is,

$$M(X) = \begin{pmatrix} (1 - a^2X^2)(1 + cX^2) & (1 + aX)cX^2 \\ (1 - aX)(1 + cX^2)cX^2 & c^2X^4 \end{pmatrix}.$$

Verify that $M(X)$ is rank one idempotent. Deduce that $a \in \mathbf{R}$.

(c) Prove that a gcd domain $\mathbf{R}$ is seminormal with $\operatorname{Pic}\mathbf{R} = \operatorname{Pic}\mathbf{R}[X] = \{1\}$.

(6) Prove that if $\mathbf{R}$ is seminormal and $\mathbf{T}$ is a reduced extension of $\mathbf{R}$ then the conductor of $\mathbf{T}$ in $\mathbf{R}$ (i.e., $\mathfrak{a} := \{r \in \mathbf{T} \mid r\mathbf{T} \subseteq \mathbf{R}\}$; this is an ideal shared by $\mathbf{R}$ and $\mathbf{T}$) is a radical ideal of $\mathbf{T}$.

(7) Let $\mathbf{R} \subseteq \mathbf{T}$ with $\mathbf{T} = \mathbf{R}[c_1, \ldots, c_q]$ reduced and finite over $\mathbf{R}$ (i.e., $\mathbf{T}$ is a finitely-generated $\mathbf{R}$-module). Let I be the conductor of $\mathbf{T}$ in $\mathbf{R}$ and suppose that it is a radical ideal. Prove that I is equal to $\{r \in \mathbf{R} \mid rc_1, \ldots, rc_q \in \mathbf{R}\}$.

(8) Let $\mathbf{R}$ be a seminormal domain. Our purpose is to prove that $\operatorname{Pic}\mathbf{R} = \operatorname{Pic}\mathbf{R}[X]$ (the Traverso-Swan-Coquand theorem [32, 167, 168]). Let $M(X) = (m_{i,j}(X))_{1 \le i,j \le n}$ be a rank one idempotent matrix over $\mathbf{R}[X]$ such that $M(0) = I_{n,1}$. Denote by $\mathbf{F}$ the field of fractions of $\mathbf{R}$. By (4.a) we know that there exist $f_1, \ldots, f_n, g_1, \ldots, g_n \in \mathbf{F}[X]$ such that $m_{i,j} = f_i g_j$ for all i, j (note that $f_1(0) = g_1(0) = 1$). Let us denote by $\mathbf{T}$ the subring of $\mathbf{F}$ generated by $\mathbf{R}$ and the coefficients of the f_i's and the g_j's and by I the conductor of $\mathbf{T}$ in $\mathbf{R}$. Our goal is to prove that $\mathbf{T} = \mathbf{R}$, or equivalently, $1 \in I$.

Let us first recall **Kronecker's theorem** [52, 81, 89] (see also [37] for a survey): *Let $\mathbf{A}$ be a ring, $f, g \in \mathbf{A}[X]$ and $h = fg$. Let a be a coefficient of f and b a coefficient of g. Then ab is integral over the subring of $\mathbf{A}$ generated by the coefficients of h.*

(a) Prove that $\mathbf{T}$ is a finitely-generated $\mathbf{R}$-module.

(b) By way of contradiction, we will suppose that $1 \notin I$. Consider a minimal prime ideal $\mathfrak{p}$ of $\mathbf{R}$ over I (that is, $\mathfrak{p}/I$ is a minimal prime ideal of $\mathbf{R}/I$). Denote by $S = \mathbf{R} \setminus \mathfrak{p}$ and S' the image of S in $\mathbf{R}/I$. We have that $\mathbf{R}/I$ is a reduced ring, $(\mathbf{R}/I)_{S'} =: \mathbf{L}$ is a field contained in the reduced ring $(\mathbf{T}/I)_{S'}$.

Using Question (6), find a contradiction (there exists $s \in S$ such that $s \in I$).

(c) Being inspired by the method explained in Sect. 2.2.4, find a method for eliminating the use of minimal prime ideals in the proof above (it will be a dual method for eliminating maximal ideals: maximal ideal $\mathfrak{m} \leftrightarrow$

minimal prime ideal $\mathfrak{p}$, $\mathbf{R}/\mathfrak{m} \leftrightarrow (\mathbf{R}/\mathfrak{p})_{\overline{\mathfrak{p}}})$. Infer a general method "by backtracking" for making the use of minimal prime ideals constructive (it will be a dual method to Elimination of Maximal Ideals by Backtracking 60).

Exercise 382. (Seminormality, an Algorithm [9])

Let $\mathbf{B}$ be a reduced ring and $f_1,\ldots,f_n,g_1,\ldots,g_n \in \mathbf{B}[X]$ such that $\sum_{i=1}^n f_i g_i = 1$, $f_1(0) = g_1(0) = 1$ and $f_i(0) = g_i(0) = 0$ for $i \geq 2$.
Let $m_{i,j}(X) = f_i(X)g_j(X)$ and $\mathbf{A}$ the ring generated by the coefficients of the $m_{i,j}(X)$'s. Let $\mathbf{B}$ be the ring generated by the coefficients of the f_i's and g_j's.
We denote by $\mathbf{A}_1$ the seminormal closure of $\mathbf{A}$ in $\mathbf{B}$, that is, the smallest subring $\mathbf{A}_1$ of $\mathbf{B}$ containing $\mathbf{A}$ such that

$$(x \in \mathbf{B},\ x^2 \in \mathbf{A}_1,\ x^3 \in \mathbf{A}_1) \implies x \in \mathbf{A}_1.$$

(1) Let $c \in \mathbf{B}$ and $m \in \mathbb{N}$ such that $c^n \in \mathbf{A}_1$ for any $n \geq m$. Prove that $c \in \mathbf{A}_1$.

(2) Denote by $f_1 = 1 + b_1 X + \cdots + b_r X^r$. Prove that $\mathbf{B} = \mathbf{A}[b_1,\ldots,b_r]$.

(3) Prove that if $a \in \mathbf{A}$ and $af_1 \in \mathbf{A}[X]$ then there exists $k \in \mathbb{N}$ such that $a^k \mathbf{B} \subseteq \mathbf{A}$.

(4) Prove that if $a \in \mathbf{A}$ and $a^m \mathbf{B} \subseteq \mathbf{A}$ for some $m \in \mathbb{N}$, then $a\mathbf{B} \subseteq \mathbf{A}_1$.

(5) Let $a \in \mathbf{B}$ and $\ell \in \mathbb{N}$ such that $a^\ell f_1 \in \mathbf{A}[X]$. Prove that $\sqrt{a\mathbf{B}} \subseteq \mathbf{A}_1$.

(6) Prove that if $\mathbf{R} \subseteq \mathbf{T}$ is a ring extension and J is an ideal of $\mathbf{T}$ then $\mathbf{R} + J$ is a ring, J is an ideal of $\mathbf{R} + J$, $\mathbf{R} \cap J$ is an ideal of $\mathbf{R}$, and the isomorphism of $\mathbf{R}$-modules $(\mathbf{R}+J)/J \simeq \mathbf{R}/(\mathbf{R}\cap J)$ is an isomorphism of rings.

(7) Let $a \in \mathbf{B}$ and $\ell \in \mathbb{N}$ such that $a^\ell f_1 \in \mathbf{A}[X]$. Denote by $J = \sqrt{a\mathbf{B}}$, $\tilde{\mathbf{A}} = (\mathbf{A} + J)/J \subseteq \mathbf{A}_1/J$, and $\tilde{\mathbf{B}} = \mathbf{B}/J$. Prove that $\mathbf{A}_1/J$ is the seminormal closure of $\tilde{\mathbf{A}}$ in $\tilde{\mathbf{B}}$.

(8) What consequence (7) has for the computation of $\mathbf{A}_1$?

Exercise 383. Let $\mathbf{T}$ be a distributive lattice, and $a, b, x \in \mathbf{T}$. prove that:

$$(x \wedge a \leq b)\ \&\ (a \leq x \vee b) \implies a \leq b.$$

Exercise 384. Prove that a Bezout domain is two-stable.

Exercise 385. Let $\mathbf{R}$ be a coherent ring, consider $f_1,\ldots,f_k \in \mathbf{R}[X_1,\ldots,X_n]$, and denote by

$$S := \mathrm{Syz}_{\mathbf{R}[X_1,\ldots,X_n]}(f_1,\ldots,f_k).$$

Prove that $S \cap \mathbf{R}^k$ is a finitely-generated $\mathbf{R}$-module.

Exercise 386. *([144])*

For a ring $\mathbf{R}$ in which one can test whether an element is a zero-divisor, we denote by $P_{(\mathbf{R})}$ the probability that an element in $\mathbf{R}$ is a zero-divisor (including zero).

(1) Prove that if $\mathbf{R}$ is a finite local ring with n elements then

    (a) $P_{(\mathbf{R})} \leq \frac{1}{2}$.

    (b) $P_{(\mathbf{R}\times\mathbf{R})} = 2P_{(\mathbf{R})} - P_{(\mathbf{R})}^2$.

    (c) If $P_{(\mathbf{R})} = \frac{1}{2}$ (that is, is maximal), then $P_{(\mathbf{R}\times\mathbf{R})} = \frac{3}{4}$.

(2) Prove that if p, q are prime numbers and $\alpha, \beta \in \mathbb{N}^*$, then

$$P_{(\mathbb{Z}/p^\alpha\mathbb{Z})} = \frac{1}{p} \ \& \ P_{((\mathbb{Z}/p^\alpha\mathbb{Z})\times(\mathbb{Z}/q^\beta\mathbb{Z}))} = \frac{1}{p} + \frac{1}{q} - \frac{1}{pq}.$$

(3) Deduce $\sup_{\{p,q\text{ prime numbers};\ \alpha,\beta\geq1\}} P_{((\mathbb{Z}/p^\alpha\mathbb{Z})\times(\mathbb{Z}/q^\beta\mathbb{Z}))}$.

Exercise 387. (With E. Pola)

Recall that a ring $\mathbf{R}$ has Krull dimension ≤ 1 if and only if

$$\forall a,b \in \mathbf{R} \ \exists n \in \mathbb{N} \ \exists \alpha, \beta \in \mathbf{R} \ | \ b^n(a^n(1+\alpha a)+\beta b) = 0. \qquad (5.1)$$

If I is an ideal of $\mathbf{R}$ and $b \in \mathbf{R}$, we denote by

$$[I : a^\infty] := \{x \in \mathbf{R} \mid \exists n \in \mathbb{N} \mid xa^n \in I\}.$$

For $a, b \in \mathbf{R}$, we denote by

$$[b : a^\infty] := [\langle b \rangle : a^\infty].$$

(1) Let $\mathbf{R}$ be a strongly discrete ring. Prove that for any $a, b \in \mathbf{R}$, we have:

$$\langle 1+aX, b \rangle \cap \mathbf{R} = [b : a^\infty] \ \& \ \mathrm{LT}(\langle 1+aX, b \rangle) = [b : a^\infty][X] + \langle aX \rangle.$$

In particular, $\mathrm{LT}(\langle 1+aX, b \rangle)$ is finitely-generated if and only if so is $[b : a^\infty]$.

(2) Prove that if $\mathbf{R}$ is a strongly discrete domain with Krull dimension ≤ 1, taking $a, b \in \mathbf{R}$ with $b \neq 0$, then, with notation of "collapse" (5.1), we have

$$[b : a^\infty] = \langle b, 1+\alpha a \rangle.$$

(3) Deduce that if $\mathbf{R}$ is a strongly discrete domain with Krull dimension ≤ 1, taking $a, b \in \mathbf{R}$ with $b \neq 0$, then with notation of collapse (5.1), we have

$$\mathrm{LT}(\langle 1+aX, b \rangle) = \langle aX \rangle + [b : a^\infty][X] = \langle aX, b, 1+\alpha a \rangle.$$

(4) Generalization:

(a) Let $\mathbf{R}$ be a strongly discrete ring, pick $a, b_1, \ldots, b_n \in \mathbf{R}$, $\ell_1, \ldots, \ell_n \in \mathbb{N}$, and consider the ideal $I = \langle 1 + aX, b_1 X^{\ell_1}, \ldots, b_n X^{\ell_n} \rangle$. Prove that

$$I \cap \mathbf{R} = [\langle b_1, \ldots, b_n \rangle : a^\infty] \quad \& \quad \mathrm{LT}(I) = [\langle b_1, \ldots, b_n \rangle : a^\infty][X] + \langle aX \rangle.$$

(b) Prove that if $\mathbf{R}$ is a strongly discrete domain with Krull dimension ≤ 1, then for any $a, b_1, \ldots, b_n \in \mathbf{R}$, $[\langle b_1, \ldots, b_n \rangle : a^\infty]$ is finitely-generated.

(c) Deduce that if $\mathbf{R}$ is a strongly discrete domain with Krull dimension ≤ 1, then the leading terms ideal of any finitely-generated ideal of $\mathbf{R}[X]$ containing an element of the form $1 + aX$ with $a \in \mathbf{R}$, is finitely-generated.

Exercise 388. (A One-Dimensional Domain Which is Not 1-Gröbner [142])

Let t, u be two independent indeterminates over the field $\mathbb{Q}$ of rationals, denote by $K = \mathbb{Q}(\sqrt{2})(u)$ and consider the following domain

$$\mathbf{A} := \mathbb{Q} + t\mathbf{K}[t] = \{f(t) \in \mathbf{K}[t] \mid f(0) \in \mathbb{Q}\}.$$

(1) Prove that the ring $\mathbf{A}$ is one-dimensional and shares the ideal $\mathfrak{m} := t\mathbf{K}[t]$ with $\mathbf{K}[t]$.

(2) Prove that $\mathfrak{m}$ is not finitely-generated as an ideal of $\mathbf{A}$.

(3) Prove that, in $\mathbf{A}$, we have $\langle t \rangle \cap \langle \sqrt{2} t \rangle = t^2 \mathbf{K}[t] = t \mathfrak{m}$.

(4) Deduce that, in $\mathbf{A}$, the ideal $\langle t \rangle \cap \langle \sqrt{2} t \rangle$ is not finitely-generated.

(5) Prove that $I \cap \mathbf{A} = \langle t \rangle \cap \langle \sqrt{2} t \rangle$.

(6) Deduce that $\mathrm{LT}(I)$ is not finitely-generated, and, thus, that $\mathbf{A}$ is not 1-Gröbner.

Exercise 389. (Continuation of Proposition 299; [173])

Let $\mathbf{R}$ be a domain with quotient field $\mathbf{K}$, take $X_1, \ldots, X_n, Z$ independent variables over $\mathbf{K}$, and consider an ideal $I = \langle f_1, \ldots, f_s \rangle$ of $\mathbf{R}[X_1, \ldots, X_n]$. Let us fix a monomial order $>$ on $\mathbf{R}[X_1, \ldots, X_n]$ and consider a normalized Gröbner basis $G = \{g_1, \ldots, g_m\}$ for $J := I \otimes_{\mathbf{R}} \mathbf{K}$. Denote by $g_i = \frac{h_i}{d_i}$ where $h_i \in \mathbf{R}[X_1, \ldots, X_n]$, $d_i = \mathrm{LC}(h_i) \in \mathbf{R}$, $H = \langle h_1, \ldots, h_s \rangle \subseteq \mathbf{R}[X_1, \ldots, X_n]$, and taking $d = \prod_{i=1}^m d_i$ (or $d = \mathrm{lcm}(d_1, \ldots, d_m)$ if such a notion exists).
For an ideal $\mathfrak{a}$ of $\mathbf{R}[X_1, \ldots, X_n]$ we denote by

$$\mathrm{Sat}(\mathfrak{a}) = \{s \in \mathbf{R}[X_1, \ldots, X_n] \mid \alpha s \in \mathfrak{a} \text{ for some } \alpha \in \mathbf{R} \setminus \{0\}\}.$$

(1) Prove that $H \neq \mathrm{Sat}(H)) \Rightarrow H \neq (H : d)$ & $I \neq \mathrm{Sat}(I)) \Rightarrow H \neq (H : d)$.

(2) Deduce that if the torsion subgroup of the additive group $\mathbf{R}[X_1, \ldots, X_n]/I$ is not trivial, then it contains an element (other than the identity element) whose order divides d.

Exercise 390. (When a Matrix Over the Integers is $\mathbb{Z}$-Saturated)

Let $A \in \mathbb{Z}^{m \times n}$ be a matrix with entries in $\mathbb{Z}$. We denote by $\mathrm{rk}_0 A$ the rank of the matrix A when considered in $\mathbb{Q}^{m \times n}$. If p is a prime number, we denote by $\mathrm{rk}_p A$ the rank of the matrix in $\mathbb{F}_p^{m \times n}$ obtained from A by passing modulo p. The set of prime numbers will be denoted by $\mathbf{P}^*$, and we denote by $\mathbf{P} := \mathbf{P}^* \cup \{0\}$ (the prime spectrum of $\mathbb{Z}$).

We denote by $v_1, \ldots, v_n$ the columns of A, $S := [v_1, \ldots, v_n]$, and $U := \mathrm{Echel}(S) = [u_1, \ldots, u_r]$ $(r = \mathrm{rk}_0 A)$ the list obtained after transforming S into a primitive triangular list over $\mathbb{Q}$ with Algorithm 319. Denote by $u_i = \frac{w_i}{\delta_i}$, where $w_i \in \mathbb{Z}^m$, $\delta_i \in \mathbb{Z} \setminus \{0\}$, take $\delta = \mathrm{lcm}(\delta_1, \ldots, \delta_r)$, and suppose that we can compute the set $\{p_1, \ldots, p_t\}$ of the prime numbers dividing δ. Prove that the following assertions are equivalent:

(i) $\mathrm{Im}(A)$ is $\mathbb{Z}$-saturated.

(ii) $\mathrm{rk}_0 A = \mathrm{rk}_{p_1} A = \cdots = \mathrm{rk}_{p_t} A$.

(iii) The map $\mathrm{rk}(A) : \mathbf{P} \to \mathbb{N}$ defined by $\mathrm{rk}(A)(q) := \mathrm{rk}_q A$, is constant.

(iv) The map $\mathbf{P}^* \to \mathbb{N}$; $p \mapsto \mathrm{rk}_p A$, is constant.

Chapter 6

Detailed Solutions to the Exercises

Exercise 372:

Recall that a ring $\mathbf{R}$ has Krull dimension ≤ 1 if and only if

$$\forall a, b \in \mathbf{R}, \ \exists n \in \mathbb{N}, \ \exists x, y \in \mathbf{R} \ | \ a^n(b^n(1+xb)+ya) = 0. \tag{6.1}$$

Suppose that $\mathbf{R}$ is a valuation domain (in particular, it is local) with valuation v and valuation group G. In the case where either a or b is zero or invertible, identity (6.1) is satisfied in a trivial way. Now, consider $a, b \in \mathrm{Rad}(\mathbf{R}) \setminus \{0\}$. As in identity (6.1), $1 + xb \in \mathbf{R}^\times$ and a^n is regular, (6.1) becomes

$$\forall a, b \in \mathrm{Rad}(\mathbf{R}) \setminus \{0\}, \ \exists n \in \mathbb{N} \ \text{such that} \ a \mid b^n,$$

or also, in other terms,

$$\forall a, b \in \mathbf{R} \text{ with } v(a), v(b) > 0, \ \exists n \in \mathbb{N} \text{ such that } nv(b) \geq v(a).$$

Exercise 373:

In the ring $\mathbf{R}[X]$, the Krull boundary monoid of X is

$$S_{\{X\}} := X^{\mathbb{N}}(1 + X\mathbf{R}[X]) = \{X^k(1 + Xf), k \in \mathbb{N}, f \in \mathbf{R}[X]\}.$$

As $\mathrm{Kdim}(\mathbf{R}[X]) \leq d + 1$, we have $\mathrm{Kdim}(S_{\{X\}}^{-1}\mathbf{R}[X]) \leq d$ (see Theorem 80). The desired result clearly follows as $\mathbf{R}(X)$ is a localization of $S_{\{X\}}^{-1}\mathbf{R}[X]$ (this is because $S_{\{X\}} \subseteq U$).

© Springer International Publishing Switzerland 2015
I. Yengui, *Constructive Commutative Algebra*, Lecture Notes in Mathematics 2138,
DOI 10.1007/978-3-319-19494-3_6

Exercise 374:

(1) Let $u = ry$. It is clear that $r^n(u-1) = 0$, $r^n(u-1)^n = 0$, and $(ry^n)^2 = ry^n$. In the reduced case, we get $r(u-1) = 0$, $u(u-1) = 0$, and $r\mathbf{R} = ru\mathbf{R} \subseteq u\mathbf{R} \subseteq r\mathbf{R}$.

(2)

(a) • If $m = n+d$ then $a_n b_d = 1$.

   • If $m < n+d$. We write all the relations between the a_i's, b_j's and c_k's in which a_n appears:

$$(S): \begin{cases} a_n b_d = \varepsilon_0 \; (= 0) \\ a_n b_{d-1} + a_{n-1} b_d = \varepsilon_1 \\ a_n b_{d-2} + a_{n-1} b_{d-1} + a_{n-2} b_d = \varepsilon_2 \\ \vdots \\ a_n b_0 + a_{n-1} b_1 + \ldots + a_{n-v} b_v = \varepsilon_d. \end{cases}$$

Where $v = \min\{d, n\}$, $\varepsilon_i = 0$ if $i < n+d-m$.

If $m < n$, then multiplying each kth equality in (S) by a_n^{k+1}, we obtain the system

$$(S'): \begin{cases} a_n b_d = 0 \\ a_n^2 b_{d-1} = 0 \\ \vdots \\ a_n^{d+1} b_0 = 0 \end{cases}$$

Thus, $a_n^{d+1} g = 0$ and $a_n^{d+1} fg = 0$. Hence, $a_n^{d+1} = 0$, and $a_n = 0$ since $\mathbf{R}$ is reduced. It follows by induction that all a_i's and b_i's with $i > m$ are zero and we can assume $n, d \leq m$.

By identification, $\varepsilon_{n+d-m} = c_m = 1$.

Considering the $(n+d-m+1)$th equality in (S) and multiplying each kth equality $(1 \leq k \leq n+d-m)$ by a_n^{k-1}, we obtain

$$a_n^{n+d-m+1} b_{m-n} = a_n^{n+d-m}.$$

(b) Using (1) and (2.a), we obtain that $e_0 = a_n b_{m-n}$ is idempotent.

   Set $e_0' = 1 - e_0$, $\mathbf{R}_0 = \mathbf{R}e_0$, $\mathbf{R}_0' = \mathbf{R}e_0'$, $f_0 = fe_0$, $g_0 = ge_0$, $f_0' = fe_0'$, $g_0' = ge_0'$. In $\mathbf{R}_0$, $e_0 a_n$ is a unit, so $\deg(f_0) = n$ and $\deg(g_0) = m - n$.

   We have $\mathbf{R} = \mathbf{R}_0 \oplus \mathbf{R}_0'$. In $\mathbf{R}_0'$, $f_0' g_0'$ is a monic polynomial with degree m, $\deg(f_0') < n$ and $\deg(g_0') \leq d$. We are done by induction on $n + d - m$.

   Concretely, if we continue the process, we find an idempotent e_1 in $\mathbf{R}_0'$ (e_1 is also an idempotent in $\mathbf{R}$) and a decomposition $\mathbf{R} = \mathbf{R}_0 \oplus \mathbf{R}_1 \oplus \mathbf{R}_1'$, and so on. So we find a priori $n + d - m + 1$ terms in the final decomposition, where $n, d \leq m$ since we first killed all a_i's and b_i's with $i > m$. In the most general case, this means $m + 1$ terms in the final decomposition. Remark that, without zero test inside $\mathbf{R}$, it is possible that we do not know which terms in the decomposition are useless, i.e., zero.

(3) Let $\mathcal{N}$ be the nilradical of $\mathbf{R}$, i.e., the ideal of nilpotent elements. The proof for the reduced case works with $\mathbf{R}/\mathcal{N}$. In the first case, we proved that $a_i = b_i = 0$ for $i > m$ and we computed idempotents $e_0, \ldots, e_m$ verifying $\sum e_i = 1$, $e_i e_j = 0$ if $j \neq i$, $e_i a_j = 0$ if $j > m - i$ (i.e., $\deg(e_i f) \leq m - i$), $e_i b_k = 0$ if $k > i$ (i.e., $\deg(e_i g) \leq i$) and $e_i(a_{m-i}b_i - 1) = 0$.

In the general case, we explicitly get with the same proof all these equalities modulo $\mathcal{N}$, i.e., we know for each previous equality $t = 0$ (in the reduced case) an exponent k for which, in the general case $t^k = 0$. This gives the desired result.

It is of interest to recall a folklore result stating that each idempotent in $\mathbf{R}/\mathcal{N}$ lifts in $\mathbf{R}$. In more details, let $r \in \mathbf{R}$ be an approximate root of the polynomial $f(X) = X^2 - X$, that is, such that $f(r) = r^2 - r \in \mathcal{N}$. Say $f(r) = r^2 - r = \eta = c_0 \eta$, where $\eta \in \mathcal{N}$ and $c_0 = 1$.

We have $f'(X) = 2X - 1$ and $f'(X)^2 = 4f(X) + 1$. Thus, $f'(r) = 1 + 4\eta$ is invertible. We replace "à la Newton" the approximate root r by $r + h$ as follows

$$f(r+h) = f(r) + hf'(r) + h^2 f_2(r,h), f_2(r,h) \in \mathbf{R}.$$

Taking $h = -\eta f'(r)^{-1}$ and setting $r_1 = r_0 - \eta f'(r)^{-1}$, we obtain $f(r_1) = c_1 \eta^2$ for some $c_1 \in \mathbf{R}$. Repeating this process, we find $r_2, c_2, \ldots, r_k, c_k \in \mathbf{R}$ such that $f(r_2) = c_2 \eta^4, \ldots, f(r_k) = c_k \eta^{2^k}$. For sufficiently large k, we get $f(r_k) = 0$ and $r - r_k \in \langle \eta \rangle \subseteq \mathcal{N}$.

Let us take an example. Let $n = 4$, $d = 5$, $m = 3$. We have $a_n^{d+1} = a_4^6 = 0$ and $b_d^{n+1} = b_5^5 = 0$. Thus, in the ring $\mathbf{R}/\langle a_4, b_5 \rangle$, the degrees are cut down at 3 and 4, and consequently $b_4^4 = 0$. Here, one may wonder if it is possible to explicitly bound the nilpotency order of b_4 in $\mathbf{R}$. Since $b_4^4 = 0$ in $\mathbf{R}/\langle a_4, b_5 \rangle$, we obtain an equality $b_4^4 = a_4 A + b_5 B$ in $\mathbf{R}$ (A and B can be computed but it is not necessary). Hence, in $\mathbf{R}$, $b_4^{4 \times 10} = a_4^6 A' + b_5^5 B' = 0$. This suggests that a function bounding the nilpotency order will be exponential at n and d.

In the ring $\mathbf{R}/\langle a_4, b_5, b_4 \rangle$, the degrees are cut down at 3 and 3, that is $n = d = m = 3$, and we are in the second case. The equality $a_n^{n+d-m+1} b_{m-n} = a_n^{n+d-m}$ signifies that $a_3^4 b_0 = a_3^3$. Item (1) says that $(r(ry - 1))^3 = 0$ with $r = a_3$ and $y = b_0$ in $\mathbf{R}/\langle a_4, b_5, b_4 \rangle$. One can precisely get $(r(ry - 1))^{3 \times (40 + 5 + 6 - 2)} = (r(ry - 1))^{147} = 0$ in $\mathbf{R}$. In $\mathbf{S} = \mathbf{R}/\langle a_4, b_5, b_4, a_3(a_3 b_0 - 1) \rangle$, $a_3 b_0 = ry$ is idempotent and corresponds to an idempotent of R. Indeed, ry is an approximate solution of the equation $X^2 - X = 0$ which lifts "à la Hensel" since $2X - 1$ is, at $X = ry$, a unit: indeed $(2X - 1)^2 = 1 + 4(X^2 - X)$ and $4(X^2 - X)$ is, at $X = ry$, nilpotent with order less than 147. Denote by $e = a_3 b_0 +$ a nilpotent element, the idempotent lifting $a_3 b_0$ in

R. This decomposes **R** and **S** into two parts. In $e\mathbf{S} \simeq \mathbf{S}/\langle e-1 \rangle$, f is quasimonic with degree 3 and $b_1 = b_2 = b_3 = 0$. This means that in $e\mathbf{R} \simeq \mathbf{R}/\langle e-1 \rangle$, f is quasimonic with degree 3 and b_1, b_2, b_3 are nilpotent. And so on ...

Another wording: with the same notations as in the reduced case, we prove the result by induction on n.

If $a_n^{n+d-m} = 0$ or $a_n^{d+1} = 0$.

Letting $k = \max\{n+d-m, d+1\}$, we have $a_n^k = 0$. Since $((f - a_n X^n)g - fg)^k = 0$, we can explicitly find a polynomial h in $\mathbf{R}[X]$ such that $(f - a_n X^n)gh = (fg)^k$ is a monic polynomial with degree mk, and we are done by the induction hypothesis.

If $a_n^{n+d-m} \neq 0$ and $a_n^{d+1} \neq 0$. By the calculations done in the reduced case, we have

$$a_n^{n+d-m+1} b_{m-n} = a_n^{n+d-m}.$$

By item (1), $e_0 = (a_n b_{m-n})^{n+d-m}$ is idempotent and $\alpha = a_n(a_n b_{m-n} - 1)$ is nilpotent. We have $a_n = a_n^2 b_{m-n} - \alpha$, where $\alpha^{n+d-m} = 0$. Hence $a_n^2 = a_n^3 b_{m-n} - \alpha a_n$, and $a_n = a_n^3 b_{m-n}^2 - \alpha a_n b_{m-n} - \alpha$. And so on, we can see that $a_n = b_{m-n}^{n+d-m} a_n^{n+d-m+1} + \beta = a_n(a_n b_{m-n})^{n+d-m} + \beta$, where $\beta^{n+d-m} = 0$. Thus, with $a_n' = a_n e_0 = b_{m-n}^{n+d-m} a_n^{n+d-m+1}$, we have

$$b_{m-n}^2 (a_n')^2 - b_{m-n} a_n' = b_{m-n}^2 a_n^2 (a_n b_{m-n})^{2(n+d-m)} - b_{m-n} a_n (a_n b_{m-n})^{n+d-m} = 0$$

as $b_{m-n} a_n^{n+d-m+1} = a_n^{n+d-m}$. Since $b_{m-n} a_n'$ is idempotent, $b_{m-n} a_n' = e_0$, and

$$e_0 \mathbf{R} = b_{m-n} a_n' \mathbf{R} \subseteq a_n' \mathbf{R} = a_n e_0 \mathbf{R} \subseteq e_0 \mathbf{R},$$

that is, $a_n' \mathbf{R}$ is generated by the idempotent e_0.

Denote by $f_1 = f - a_n X^n + a_n' X^n$, $f = f_1 - N$, where N is nilpotent. We have $f_1 g = fg + Ng$, and, thus, $(f_1 g - fg)^{n+d-m} = 0$ and we can explicitly find a polynomial D in $\mathbf{R}[X]$ such that $f_1 g D = (fg)^{n+d-m}$ is a monic polynomial of degree $m(n+d-m)$. Of course, the degree coefficient of f_1 is a_n'. It remains only to do as in the reduced case, just replace f by f_1, a_n by a_n', $a_n b_{m-n}$ by $(a_n b_{m-n})^{n+d-m}$, g by gD, and m by $(n+d-m)m$.

(4) Suppose that $f \in U(X)^*$. It is clear that one easily obtains an equality asserting that $\langle a_0, \ldots, a_n \rangle = \mathbf{R}$. For each $j \in \{0, \ldots, n\}$, considering the ring $\mathbf{R}/\langle a_{j+1}, \ldots, a_n \rangle$ and reviewing the proof of (3), we see that the first step of the algorithm produces an equality of the form $\bar{a}_j^{k_j} = \bar{0}$ or $\bar{a}_j^{k_j+1} \bar{\beta}_j = \bar{a}_j^{k_j}$ for some $\beta_j \in \mathbf{R}$. Hence, $(a_j(a_j \beta_j - 1))^{k_j} \equiv 0 \bmod \langle a_{j+1}, \ldots, a_n \rangle$.

Conversely, suppose that $\langle a_0, \ldots, a_n \rangle = \mathbf{R}$ and, that for each $j \in \{0, \ldots, n\}$, we can find $\beta_j \in \mathbf{R}$ and $k_j \in \mathbb{N}$ such that $(a_j(a_j \beta_j - 1))^{k_j} \equiv 0 \bmod \langle a_{j+1}, \ldots, a_n \rangle$. Since $(a_n(a_n \beta_n - 1))^{k_n} = 0$, we have $a_n^{k_n+1} \gamma_n = a_n^{k_n}$, where $\gamma_n = \sum_{i=1}^{k_n} C_{k_n}^i (-1)^{k_n-i} a_n^{i-1} \beta_n^i$. Now, as in the proof of (3), we can write $f = f_1 - N$, where $f_1 = f - a_n X^n +$

$\gamma_n^{kn} a_n^{kn+1} X^n$, and $N^{kn} = 0$. To prove that f divides some monic polynomial, it suffices to do the same for f_1.

Denoting by $e_0 = (a_n \gamma_n)^{kn}$, e_0 is idempotent by (1), $\mathbf{R} = \mathbf{R}e_0 \oplus \mathbf{R}(1 - e_0)$, $f_1 = f_1 e_0 + f_1(1 - e_0)$, and the degree coefficient of $f_1 e_0$ is a unit of $\mathbf{R}e_0[X]$. Our task is then reduced to prove that $f_1(1 - e_0)$ divides some monic polynomial in $\mathbf{R}(1 - e_0)[X]$. Since $\deg(f_1(1 - e_0)) < n$ and all the hypotheses on f are inherited by $f_1(1 - e_0)$, the desired result can be obtained by induction on n. Note that the condition $\langle a_0, \dots, a_n \rangle = R$ is needed to get the induction started.

(5) Using the algorithm described above, we find:
$e_0 = a_3 b_2 = u^2 = u$, $\mathbf{R}_0 = \mathbf{R}u = u\mathbf{K}[u, v]$, $f_0 = uf = u - 2uX^2 + uX^3$, $g_0 = uX^2$, $\mathbf{R}_0' = \mathbf{R}_1 = \mathbf{R}(1-u) = (1-u)\mathbf{K}[u, v]$, $f_0' = f_1 = (u-1)X^2$, $g_0' = g_1 = v + (u-1)X^3$.
Thus, in $\mathbf{K}[u, v] = u\mathbf{K}[u, v] \oplus (1 - u)\mathbf{K}[u, v]$, the decomposition of f is

$$f = (u - 2uX^2 + uX^3) + ((u-1)X^2).$$

Of course, $\mathbf{R}_0 = u\mathbf{K}[u, v] \simeq \mathbf{R}$, by this isomorphism $f_0 \leftrightarrow 1 - 2X^2 + X^3$, $g_0 \leftrightarrow X^2$; $\mathbf{R}_1 = (1-u)\mathbf{K}[u, v] \simeq \mathbf{R}$, by this isomorphism $f_1 \leftrightarrow -X^2$, $g_1 \leftrightarrow v - X^3$.

(6) Let us decompose f in $\mathbf{R}/\mathcal{N}$. Consider the images modulo $\mathcal{N}$, $f' = u' - (1 + u')X^2 + u'X^3$, $g' = -v'^4 + u'X^2 + 2v'^2 X^3 - 2u'X^4 + u'X^5 + (u'-1)X^6$, $f'g' = (u' + v'^4)X^2 - 4u'X^4 + (u' - v'^2)X^5 + 4u'X^6 - 4u'X^7 + X^8$, respectively of f, g, and fg.

As in (5), the algorithm yields to the direct sum decompositions:
$\mathbf{R}/\mathcal{N} = u'\mathbf{K}[u', v'] \oplus (1 - u')\mathbf{K}[u', v']$,
$f' = (u' - 2u'X^2 + u'X^3) + ((u'-1)X^2)$.
$f - uvX^4 = (u - 2uX^2 + uX^3) + ((u-1)X^2)$, where $(uvX^4)^2 = 0$.

If we want to decompose f in $\mathbf{R}$, using the algorithm described in (3) for the nonreduced case, we have:
$((f - uvX^4)g - fg)^2 = 0$ and, thus, $(f - uvX^4)(g^2(f + uvX^4)) = (fg)^2$.
Note that g^2 has degree 12 and highest coefficient $1 - u$, $f + uvX^4$ has degree 4 and highest coefficient $2uv$, whereas $g^2(f + uvX^4)$ has degree 14 and highest coefficient $u - 1 - uv$.

The first idempotent element found is $e_0 = (a_3 b_{13})^{17-16} = a_3 b_{13} = a_3((g^2)_{12}(f + uvX^4)_1 + (g^2)_{11}(f + uvX^4)_2 + (g^2)_{10}(f + uvX^4)_3 + (g^2)_9(f + uvX^4)_4) = uu = u$.
Thus, $f_0 = (f - uvX^4)u = u - 2uX^2 + uX^3$, $f_0' = f_1 = (f - uvX^4)(1 - u) = (u-1)X^2$, $\mathbf{R} = \mathbf{K}[u, v] = u\mathbf{K}[u, v] \oplus (1 - u)\mathbf{K}[u, v]$, and $f - uvX^4 = (u - 2uX^2 + uX^3) + ((u-1)X^2)$, where $(uvX^4)^2 = 0$.

(7) Denote by $U(X) = \{f \in \mathbf{R}[X], f \text{ is monic}\}$, U^* the saturation of U,

$$\mathcal{V} = \{f \in \mathbf{R}[X, X^{-1}], \text{ the coefficient of the highest}$$
$$\text{and lowest monomials are equal to } 1\},$$

and $\mathcal{V}^*$ the saturation of $\mathcal{V}$.

(a) It is clear that the condition is necessary. For the sufficiency, suppose that we can find two polynomials $g, h \in \mathbf{R}[X]$ such that $X^n f(X)h(X) \in U(X)$ and $X^m f(X^{-1})g(X) \in U(X)$. Then, $(X^n h(X) + X^{-m} g(X^{-1}))f \in \mathscr{V}$ and $f \in \mathscr{V}^*$.

(b) Using (7.a) and (2.b), if $x_0, \ldots, x_p$ and $y_0, \ldots, y_q$ are two systems of nonzero orthogonal idempotents associated respectively to $X^n f(X)$ and $X^m f(X^{-1})$, then denoting $\{x_i y_j, 0 \le i \le p, 0 \le j \le q\} = \{\varepsilon_0, \ldots, \varepsilon_m\}$, we take $\mathbf{R}_i = \mathbf{R}\varepsilon_i$. For the sufficiency, for each i, denote by α_i and β_i respectively the inverses of the lowest and highest coefficients of f_i in $\mathbf{R}_i$, and by k_i and l_i respectively the lowest and highest degrees of f_i ($k \le k_i$, $l_i \le l$). Then

$$\left(\sum_{i=0}^{m} (\alpha_i X^{k-k_i} + \beta_i X^{l-l_i}) \right) f$$

has 1 as lowest and highest coefficient and $f \in \mathscr{V}^*$.

(c) Do as in the proof of (4).

(8)

(a) It is clear that $\mathbf{R}\langle X, X^{-1} \rangle \subseteq \mathbf{R}\langle X \rangle \cap \mathbf{R}\langle X^{-1} \rangle$. Conversely, let $f_1, f_2 \in \mathbf{R}[X, X^{-1}]$, g_1 a monic polynomial in $\mathbf{R}[X]$, and g_2 a monic polynomial in $\mathbf{R}[X^{-1}]$ such that $\frac{f_1}{g_1} = \frac{f_2}{g_2}$ in the total quotient ring of $\mathbf{R}[X, X^{-1}]$. Then, $f_1 g_2 = g_1 f_2$ and, thus, $f_1(Xg_1 + X^{-1}g_2) = g_1(Xf_1 + X^{-1}f_2)$ and $\frac{f_1}{g_1} \in \mathbf{R}\langle X, X^{-1} \rangle$ since $Xg_1 + X^{-1}g_2 \in \mathscr{V}$.

(b) The inclusion is clear. To see that $\mathbf{R}\langle X^{-1} + X \rangle \ne \mathbf{R}\langle X, X^{-1} \rangle$, it suffices to consider the polynomial $X^{-1} + X^2$. It is invertible in $\mathbf{R}\langle X, X^{-1} \rangle$, but not in $\mathbf{R}\langle X^{-1} + X \rangle$ as this would imply the existence of $f \in \mathbf{R}[X^{-1} + X]$ and $g \in U(X^{-1} + X)$ such that $(X^{-1} + X^2) f = g$, and then $m = n - 1 = n + 2$.

(c) Notation: Let $f = a_m X^m + a_{m+1} X^{m+1} + \cdots + a_{m+n} X^{m+n}$, with $a_m, a_{m+n} \in \mathbf{R} \setminus \{0\}$, $n \in \mathbb{N}$, and $m \in \mathbb{Z}$. The nonnegative integer n will be called the degree of f, and denoted by $\deg(f)$.

Of course, we have $\mathbf{R}[X^{-1} + X] \subseteq \{f \in \mathbf{R}[X, X^{-1}] \mid f \text{ is symmetric at } X \text{ and } X^{-1}\}$.

Conversely, let $f \in \mathbf{R}[X, X^{-1}] \setminus \{0\}$ be a symmetric Laurent polynomial at X and X^{-1} of degree $2n$ (the degree of a symmetric Laurent polynomial is necessarily even). We proceed by induction on n. If $n = 0$ then $f = aX^m$ for some $a \in \mathbf{R} \setminus \{0\}$. As it is symmetric, necessarily $m = 0$, and, thus, $f \in \mathbf{R} \subseteq \mathbf{R}[X^{-1} + X]$.

Now, suppose that $n \ge 1$. The polynomial $g = f - a(X^{-1} + X)^n$, where a is the leading coefficient of f, is also symmetric with $\deg(g) < \deg(f)$. The induction hypothesis applies and gives the desired result.

The final deduction is immediate.

(d) If $g = X^n$ for some $n \in \mathbb{Z}$, then $X^n(X^{-n-1}+X^{-n+1}) = X^{-1}+X \in U(X+X^{-1})$. So, we can suppose that $g \in U(X)$ and $g(0) \in \mathbf{R}^\times$.

Recall that an element b of a ring $\mathbf{B}$ is said to be integral over $\mathbf{A}$, a subring of $\mathbf{B}$, if there are $n \geq 1$ and $a_j \in \mathbf{A}$ such that $b^n + a_{n-1}b^{n-1} + \cdots + a_1 b + a_0 = 0$.

We have the inclusions

$$\begin{aligned}
\mathbf{R} &\subseteq \mathbf{R}[X^{-1}+X]/(g\mathbf{R}[X,X^{-1}]\cap\mathbf{R}[X^{-1}+X]) \subseteq \mathbf{R}[X,X^{-1}]/g\mathbf{R}[X,X^{-1}] \\
&= S^{-1}\mathbf{R}[X]/S^{-1}g\mathbf{R}[X] \cong \overline{S}^{-1}(\mathbf{R}[X]/g\mathbf{R}[X]) \cong \mathbf{R}[\theta,\theta^{-1}],
\end{aligned}$$

where $\overline{S}$ is the multiplicative set generated by the class $\theta = \overline{X}$ of X modulo $g\mathbf{R}[X]$. Since g is a doubly unitary polynomial, both of θ and θ^{-1} are integral over $\mathbf{R}$ and, thus, $\mathbf{R}[\theta,\theta^{-1}]$ is integral over $\mathbf{R}$. It follows that $\mathbf{R}[X^{-1}+X]/(g\mathbf{R}[X,X^{-1}]\cap\mathbf{R}[X^{-1}+X])$ is integral over $\mathbf{R}$, that is $g\mathbf{R}[X,X^{-1}]\cap\mathbf{R}[X^{-1}+X]$ contains a monic polynomial ($\in U(X^{-1}+X)$), the desired conclusion.

That is good, but not enough. Imagine than we choose a polynomial g in $\mathbf{R}[X,X^{-1}]$, say $g = X^{-2}+2X^{-2}+3-X$, and want to explicitly find $h \in \mathbf{R}[X,X^{-1}]$ such that gh is a monic polynomial at $X^{-1}+X$. How can we find h?

The solution is (as often) to find the algorithm behind the above proof. In fact, in our situation, it is just a "universal" polynomial identity ensuing from an equality to zero modulo g in the ring $\mathbf{R}[X,X^{-1}]$. In more details, denoting by $g(X) = a_0 X^m + a_1 X^{m+1} + \cdots + a_n X^{m+n} = X^m(a_0 + a_1 X + \cdots + a_{n-1}X^{n-1} + a_n X^n) = X^m \tilde{g}$ with $m \in \mathbb{Z}$, and

$$\begin{aligned}
\mathscr{B} &= ((X^{-1})^{n-1}, (X^{-1})^{n-2}, \ldots, (X^{-1})^2, X^{-1}, 1, X, X^2, \ldots, X^{n-2}, X^{n-1}) \\
&= (u_1, \ldots, u_{2n-1}),
\end{aligned}$$

we have:

$$\begin{aligned}
L_1 &:= (X^{-1}+X)\cdot(X^{-1})^{n-1} - a_0^{-1}\tilde{g}(X)X^{-n} \\
&= (-a_0^{-1}a_1, 1 - a_0^{-1}a_2, -a_0^{-1}a_3, \ldots, -a_0^{-1}a_{n-1}, -a_0^{-1}a_n, 0, \ldots, 0)_\mathscr{B}, \\
L_2 &:= (X^{-1}+X)\cdot(X^{-1})^{n-2} = (1, 0, 1, \ldots, 0, \ldots, 0)_\mathscr{B},
\end{aligned}$$

$$\vdots$$

$$L_{n-1} := (X^{-1}+X)\cdot(X^{-1}) = (\overbrace{0,\ldots,0}^{n-3}, 1, 0, 1, \overbrace{0\ldots,0}^{n-1})_\mathscr{B},$$

$$L_n := (X^{-1}+X)\cdot 1 = (\overbrace{0,\ldots,0}^{n-2}, 1, 0, 1, \overbrace{0\ldots,0}^{n-2})_\mathscr{B},$$

$$L_{n+1} := (X^{-1}+X)\cdot X = (\overbrace{0,\ldots,0}^{n-1},1,0,1,\overbrace{0\ldots,0}^{n-3})_{\mathscr{B}},$$

$$\vdots$$

$$L_{2n-2} := (X^{-1}+X)\cdot X^{n-2} = (0,\ldots,0,1,0,1)_{\mathscr{B}},$$

$$L_{2n-1} := (X^{-1}+X)\cdot X^{n-1} - a_n^{-1}\tilde{g}(X)$$
$$= (0,\ldots,0,-a_n^{-1}a_0,-a_n^{-1}a_1,\ldots,-a_n^{-1}a_{n-3},1-a_n^{-1}a_{n-2},-a_n^{-1}a_{n-1})_{\mathscr{B}}.$$

Thus, for $1 \le i \le 2n-1$, denoting by $L_i = (b_{i,1},\ldots,b_{i,2n-1})_{\mathscr{B}}$, and setting

$$B := (b_{i,j})_{1\le i,j\le 2n-1}$$

$$=\begin{pmatrix}
-a_0^{-1}a_1 & 1-a_0^{-1}a_2 & a_0^{-1}a_3 & \cdots & -a_0^{-1}a_{n-1} & -a_0^{-1}a_n & 0 & & \cdots & & 0 \\
1 & 0 & 1 & & & & & & & & \\
& \ddots & & \ddots & & & & & & & \\
& & & & 1 & 0 & 1 & & & & \\
& & & & & & \ddots & & \ddots & & \\
& & & & & & & & 1 & 0 & 1 \\
0 & \cdots & & 0 & -a_n^{-1}a_0 & -a_n^{-1}a_1 & \cdots & -a_n^{-1}a_{n-3} & 1-a_n^{-1}a_{n-2} & -a_n^{-1}a_{n-1}
\end{pmatrix},$$

and $A := (X^{-1}+X)\mathbf{I}_{2n-1} - B$, we have :

$$B\,{}^t(u_1,\ldots,u_{n-1},1,u_{n+1},\ldots,u_{2n-1}) = {}^t(a_0^{-1}\tilde{g}(X)X^{-n},0,\ldots,0,a_n^{-1}\tilde{g}(X)).$$

It follows from Cramer's rule that $\det A$ (which is a monic polynomial at $(X^{-1}+X)$) is equal to the determinant of the matrix obtained from A by replacing its nth column by ${}^t(a_0^{-1}\tilde{g}(X)X^{-n},0,\ldots,0,a_n^{-1}\tilde{g}(X))$. Thus, denoting by $\tilde{h}$ the determinant of the matrix obtained from A by replacing its nth column by ${}^t(a_0^{-1}X^{-n},0,\ldots,0,a_n^{-1})$, we obtain $\det A = \tilde{g}\tilde{h}$, where $\det A$ is a monic polynomial at $(X^{-1}+X)$ with coefficients in $\mathbb{Z}[a_0^{\pm},a_1,\ldots,a_{n-1},a_n^{\pm}]$ and of degree $2n-1$. As $X^m(X^{-m-1}+X^{-m+1}) = (X^{-1}+X)$, we conclude that

$$(X^{-1}+X)\det A = g(X^{-m-1}+X^{-m+1})\tilde{h},$$

a monic polynomial at $(X^{-1} + X)$ with coefficients in $\mathbb{Z}[a_0^{\pm}, a_1, \ldots, a_{n-1}, a_n^{\pm}]$ and of degree $2n$.

Now, let's go back to our example $g = X^{-2} + 2X^{-1} + 3 - X = X^{-2}(1 + 2X + 3X^2 - X^3) = X^{-2}\tilde{g}$ with $\tilde{g} = 1 + 2X + 3X^2 - X^3$. Keeping the above notation, we obtain:

$$\det A =$$

$$\begin{vmatrix} 2+(X^{-1}+X) & 2 & -1 & 0 & 0 \\ -1 & (X^{-1}+X) & -1 & 0 & 0 \\ 0 & -1 & (X^{-1}+X) & -1 & 0 \\ 0 & 0 & -1 & (X^{-1}+X) & -1 \\ 0 & 0 & -1 & -3 & -3+(X^{-1}+X) \end{vmatrix}$$

$$= 1 - X - 4X^2 - 16X^3 - 9X^4 - 17X^5 - 9X^6 - 16X^7 - 4X^8 - X^9 + X^{10}$$

$$= \tilde{g}(X) \begin{vmatrix} 2+(X^{-1}+X) & 2 & X^{-3} & 0 & 0 \\ -1 & (X^{-1}+X) & 0 & 0 & 0 \\ 0 & -1 & 0 & -1 & 0 \\ 0 & 0 & 0 & (X^{-1}+X) & -1 \\ 0 & 0 & -1 & -3 & -3+(X^{-1}+X) \end{vmatrix}$$

$$= \tilde{g}(X)(1 - 3X - X^2 - 4X^3 - X^4 - 4X^5 - 2X^6 - X^7),$$

and finally,

$$(1 - 3X - X^2 - 4X^3 - X^4 - 4X^5 - 2X^6 - X^7)(X + X^3)g$$
$$= (X^{-1} + X)\det A = p(X^{-1} + X)$$
with $p(t) = t^6 - t^5 - 9t^4 - 12t^3 + 8t^2 + 13t.$

(d) For $i \in \{-1, 1\}$, we have

$$X^{2i} - (X^{-1} + X)X^i + 1,$$

and, thus, by induction on $n \in \mathbb{N}$, X^{-n} and X^n belong to $\mathbf{R}[X^{-1} + X] \cdot 1 + \mathbf{R}[X^{-1} + X] \cdot X$. We deduce that $\mathbf{R}[X, X^{-1}]$ is generated by 1 and X as an $\mathbf{R}[X^{-1} + X]$-module.

Let $f = a_0 + a_1(X^{-1} + X) + \cdots + a_k(X^{-1} + X)^k$, $g = b_0 + b_1(X^{-1} + X) + \cdots + b_k(X^{-1} + X)^k \in \mathbf{R}[X^{-1} + X]$ such that $f + g \cdot X = 0$. Multiplying by X^k, we obtain the following equality:

$$a_0 X^k + a_1 X^{k-1}(1 + X^2) + \cdots + a_k(1 + X^2)^k = -b_0 X^{k+1} - b_1 X^k(1 + X^2) - \cdots - b_k X(1 + X^2)^k.$$

Comparing coefficients of powers of X on both sides we obtain:

$$b_k = 0, \ a_k = 0, \ b_{k-1} = 0, \ a_{k-1} = 0, \ldots, b_0 = 0, \ a_0 = 0.$$

(e) This is an immediate consequence of the previous two questions.

Exercise 375:

(1) "$\Leftarrow$" This is an immediate consequence of Proposition 52.

"$\Rightarrow$" $\langle u,v \rangle = \langle 1 \rangle$ in $\mathbf{R}[X,X^{-1}]$ $\iff$ $\exists n \in \mathbb{N}, f,g \in \mathbf{R}[X] / fu + gv = X^n$. Since u is a monic polynomial, we have $\mathrm{Res}(u,gv) = \mathrm{Res}(u,g)\,\mathrm{Res}(u,v)$ and $\mathrm{Res}(u,gv) = \mathrm{Res}(u,gv+fu) = \mathrm{Res}(u,X^n) = \mathrm{Res}(u,X)^n = ((-1)^{\deg u}u(0))^n \in \mathbf{R}^\times$. The desired conclusion follows.

(2) Let $u_1(X),\dots,u_n(X) \in \mathbf{R}[X]$ such that $\tilde{v}_1 u_1 + \cdots + \tilde{v}_n u_n = X^q$ for some $q \in \mathbb{N}$, where the $\tilde{v}_i$'s are shifted versions of the v_i's. Set $w = \tilde{v}_3 u_3 + \cdots + \tilde{v}_n u_n$ and $V = {}^t(\tilde{v}_2,\dots,\tilde{v}_n)$. We suppose that $\tilde{v}_1$ has degree d and for $2 \le i \le n$, the formal degree of $\tilde{v}_i$ is $d_i < d$.

We proceed by induction on $\min_{2 \le i \le n}\{d_i\}$. To simplify, we always suppose that $d_2 = \min_{2 \le i \le n}\{d_i\}$.

For $d_2 = -1$, $\tilde{v}_2 = 0$ and by one elementary operation, we put w in the second coordinate. We have $\mathrm{Res}(\tilde{v}_1,w) = \mathrm{Res}(\tilde{v}_1,\tilde{v}_1 u_1 + w) = \mathrm{Res}(\tilde{v}_1,X^q) = ((-1)^{\deg \tilde{v}_1}\tilde{v}_1(0))^n \in \mathbf{R}^\times$ and we are done.

Now, suppose that we can find the desired elementary matrices for $d_2 = m-1$ and let show that we can do the job for $d_2 = m$.

Let a be the coefficient of degree m of $\tilde{v}_2$ and consider the ring $\mathbf{T} = \mathbf{R}/\langle a \rangle$. In $\mathbf{T}$, all the induction hypotheses are satisfied without changing the $\tilde{v}_i$ nor the u_i. Thus, we can obtain $\Gamma_1,\dots,\Gamma_k \in E_{n-1}(\mathbf{T}[X])$ such that

$$\langle \mathrm{Res}(\tilde{v}_1,e_1.\Gamma_1 V),\dots,\mathrm{Res}(\tilde{v}_1,e_1.\Gamma_k V) \rangle = \mathbf{T}.$$

It follows that, denoting by $\Upsilon_1,\dots,\Upsilon_k$ the matrices in $E_{n-1}(\mathbf{R}[X])$ lifting respectively $\Gamma_1,\dots,\Gamma_k$, we have

$$\langle \mathrm{Res}(\tilde{v}_1,e_1.\Upsilon_1 V),\dots,\mathrm{Res}(\tilde{v}_1,e_1.\Upsilon_k V),a \rangle = \mathbf{R}.$$

Let $b \in \mathbf{R}$ such that

$$ab \equiv 1 \bmod \langle \mathrm{Res}(\tilde{v}_1,e_1.\Upsilon_1 V),\dots,\mathrm{Res}(\tilde{v}_1,e_1.\Upsilon_k V) \rangle = J$$

and consider the ring $\mathbf{C} = \mathbf{R}/J$. Note that in $\mathbf{C}$, we have $ab = 1$.

By an elementary operation, we replace $\tilde{v}_3$ by its remainder modulo $\tilde{v}_2$, say $\tilde{v}_3'$, and then we exchange $\tilde{v}_2$ and $-\tilde{v}_3'$. The new column V' obtained has as first coordinate a polynomial with formal degree $m-1$. The induction hypothesis applies and we obtain $\Delta_1,\dots,\Delta_r \in E_{n-1}(\mathbf{C}[X])$ such that

$$\langle \mathrm{Res}(\tilde{v}_1,e_1.\Delta_1 V'),\dots,\mathrm{Res}(\tilde{v}_1,e_1.\Delta_r V') \rangle = \mathbf{C}.$$

Since V' is the image of V by a matrix in $E_{n-1}(\mathbf{C}[X])$, we obtain matrices $\Lambda_1,\dots,\Lambda_r \in E_{n-1}(\mathbf{C}[X])$ such that

$$\langle \mathrm{Res}(\tilde{v}_1,e_1.\Lambda_1 V),\dots,\mathrm{Res}(\tilde{v}_1,e_1.\Lambda_r V) \rangle = \mathbf{C}.$$

The matrices Λ_j lift in $E_{n-1}(\mathbf{R}[X])$ as, say $\Psi_1,\dots,\Psi_r$.

Finally, we obtain

$$\langle \mathrm{Res}(\tilde{v}_1,e_1.\Psi_1 V),\dots,\mathrm{Res}(\tilde{v}_1,e_1.\Psi_r V) \rangle + J = \mathbf{R},$$

the desired conclusion.

(3) *A nonconstructive proof:* To prove that $\langle r_1,\dots,r_s \rangle = \mathbf{R}$ it suffices to prove that, for each maximal ideal $\mathfrak{M}$ of $\mathbf{R}$, there exists $1 \le i \le s$ such that $r_i \notin \mathfrak{M}$. For this, let $\mathfrak{M}$ be a maximal ideal of $\mathbf{R}$ and by way of contradiction suppose that $\overline{r_1},\dots,\overline{r_s} = 0$

in the residue field $\mathbf{K} := \mathbf{R}/\mathfrak{M}$. It is worth pointing out that, denoting by $w_i = \tilde{v}_2 + y_i\tilde{v}_3 + \cdots + y_i^{n-2}\tilde{v}_n$, $\overline{\mathrm{Res}_X(\tilde{v}_1, w_i)} = \mathrm{Res}_X(\overline{\tilde{v}_1}, \overline{w}_i)$ since $\tilde{v}_1$ is monic.

This means that for each i there exists $\xi_i \in \mathbf{K}$ such that $\overline{\tilde{v}_1}(\xi_i) = \overline{w}_i(\xi_i) = \overline{0}$. But since $\deg_X \tilde{v}_1 = l$, $\overline{\tilde{v}_1}$ has at most l distinct roots and hence there exists at least one root among the ξ_i repeated $n-1$ times. We can suppose that $\xi_1 = \xi_2 = \cdots = \xi_{n-1} := \xi$. Thus, we have:

$$\begin{pmatrix} 1 & y_1 & \cdots & y_1^{n-2} \\ 1 & y_2 & \cdots & y_2^{n-2} \\ \vdots & \vdots & \vdots & \vdots \\ 1 & y_{n-1} & \cdots & y_{n-1}^{n-2} \end{pmatrix} \begin{pmatrix} \tilde{v}_2(\xi) \\ \tilde{v}_3(\xi) \\ \vdots \\ \tilde{v}_n(\xi) \end{pmatrix} = \begin{pmatrix} 0 \\ 0 \\ \vdots \\ 0 \end{pmatrix}.$$

Since the matrix above is a Vandermonde matrix, its determinant is equal to

$$\prod_{1 \le i < j \le n-1} (y_j - y_i),$$

which is invertible in $\mathbf{R}$. Thus, $\overline{\tilde{v}_1}(\xi) = \overline{\tilde{v}_2}(\xi) = \cdots = \overline{\tilde{v}_n}(\xi) = 0$. Now, using the fact that $1 \in \langle v_1, \ldots, v_n \rangle$ in $\mathbf{R}[X, X^{-1}]$, we infer that, in $\mathbf{R}[X]$, $\langle \tilde{v}_1, \ldots, \tilde{v}_1 \rangle$ contains a monomial X^q for some $q \in \mathbb{N}$. This forces ξ into being zero, in contradiction with the fact that $\overline{\tilde{v}_1}(0) \neq 0$. The last claim that $1 \in \langle \tilde{v}_1, \ldots, \tilde{v}_n \rangle$ in $\mathbf{R}[X]$ follows easily from the fact that $r_i \in \langle \tilde{v}_1, \ldots, \tilde{v}_n \rangle$, for all $1 \le i \le s$.

A Constructive Proof: do as in the constructive proof of Theorem 61.

(4) (Producing doubly monic Laurent polynomials over a field)

(a) $\deg_Y(\varphi_\alpha(X^{n_i}Y^{m_i})) = \deg_Y(\varphi_\alpha(X^{n_j}Y^{m_j}))$

$\Rightarrow \deg_Y((XY^\alpha)^{n_i}Y^{m_i})) = \deg_Y(((XY^\alpha)^{n_j}Y^{m_j}))$

$\Rightarrow \alpha n_i + m_i = \alpha n_j + m_j$

$\Rightarrow (n_i = n_j$ and $m_i = m_j)$ because the case $(n_i \neq n_j$ and $\alpha = \frac{m_j - m_i}{n_i - n_j})$ is impossible.

Now, suppose that $f \in \mathbf{K}[X, Y]$ with total degree $\le d$. For $n_i \neq n_j$, we have $\frac{|m_j - m_i|}{|n_i - n_j|}) \le |m_j - m_i| \le d$, and, thus, $\mathbb{Z} \setminus E \supseteq (]-\infty, -d-1] \cup [d+1, +\infty[) \cap \mathbb{Z}$, $\alpha_0 \le d+1$, and $\mathrm{tdeg}(\varphi_{\alpha_0}(f)) \le d+1$.

(b)

$$E = \{\frac{0-1}{0-1}, \frac{0-2}{0-1}, \frac{0-3}{0-1}, \frac{1-1}{0-1}, \frac{1-2}{0-1}, \frac{1-3}{0-1}, \frac{1-1}{0-2}, \frac{1-2}{0-2}, \frac{1-3}{0-2},$$
$$\frac{2-1}{0-2}, \frac{2-2}{0-2}, \frac{2-3}{0-2}, \frac{1-0}{1-2}, \frac{2-0}{1-2}, \frac{1-1}{1-2}, \frac{2-1}{1-2}\}$$
$$= \{1, 2, 3, 0, \frac{1}{2}, -\frac{1}{2}, -1, -2\}.$$

Thus, $\alpha_0 = \pm 3$. Taking $\alpha_0 = -3$, $\varphi_{\alpha_0}(X,Y) = (\frac{X}{Y^3}, Y)$, and

$$\varphi_{-3}(f) = \frac{X^2}{Y^5} + \frac{X^2}{Y^4} + \frac{X}{Y^3} + \frac{X}{Y^2} + Y + Y^2 + Y^3,$$

which is not only doubly unitary but also has the same number of monomials (7) as f and each monomial of $\varphi_{\alpha_0}(f)$ has a different degree at Y. If we just want to transform f into a doubly unitary polynomial at Y then one can take $\alpha = 2$ and then obtain

$$\varphi_2(f) = Y + Y^2 + XY^2 + Y^3 + XY^3 + X^2Y^5 + X^2Y^6.$$

(c) The general case ($k \geq 2$) can easily be deduced from the case of two variables. Let $f = \sum_j X_1^{n_{1,j}} X_2^{n_{2,j}} \cdots X_k^{n_{k,j}} \in \mathbf{K}[X_1^{\pm 1}, X_2^{\pm 1} \ldots, X_k^{\pm 1}]$, $\mathbf{K}$ a field. For each $1 \leq i \leq k$, we set $L(i) := \max_{j,j'}\{|n_{i,j} - n_{i,j'}|\}$. We will call it the length of the variable X_i in f. Suppose X_1 has the greatest length and X_2 has the lowest one. Then fixing the variables $X_3, \ldots, X_k$ and doing as in case of two variables, we can transform f into a doubly unitary Laurent polynomial at X_2.

(d) **An algorithm for unimodular completion over Laurent polynomial ring: general case.**

Input: A column $\mathscr{V} = \mathscr{V}(X) = {}^t(v_1(X), \ldots, v_n(X)) \in \mathbf{R}[X^{\pm 1}]^n$ such that v_1 is doubly unitary and $1 \in \langle v_1, \ldots, v_n \rangle$. We assume the "size" of an element $a \in \mathbf{R}$ is measured by $\deg(a) \in \mathbb{N}$, the function deg sharing the usual properties of a total degree function in a polynomial ring: $\deg(a+b) \leq \max(\deg(a), \deg(b))$, $\deg(ab) \leq \deg(a) + \deg(b)$, $\max_{1 \leq i \leq n}\{\deg v_i\} \leq d$ (where $d \geq 2$). We assume that the ring $\mathbf{R}$ contains infinitely many y_i of degree 0 such that $y_i - y_j$ is invertible for $i \neq j$.

Output: A matrix $G = \mathscr{B}\mathscr{D} \in M_n(\mathbf{R}[X^{\pm 1}])$ such that $\mathscr{B}\widetilde{\mathscr{V}} = \widetilde{\mathscr{V}}(0)$, and $\mathscr{D}$ is a diagonal matrix with suitable powers of X on the diagonal, such that $\mathscr{D}\mathscr{V} = \widetilde{\mathscr{V}} = {}^t(\widetilde{v}_1, \ldots, \widetilde{v}_n)$.

Step 1: Shift $\mathscr{V}$ into $\widetilde{\mathscr{V}}$ so that $\widetilde{\mathscr{V}} \in \mathbf{R}[X]^n$. This operation can be performed via multiplying $\mathscr{V}$ by a diagonal matrix $\mathscr{D}$ with suitable powers of X on the diagonal.

Step 2: For $1 \leq i \leq s = (n-2)d+1$, where $d = \deg_X v_1$, set $w_i = \widetilde{v}_2 + y_i\widetilde{v}_3 + \cdots + y_i^{n-2}\widetilde{v}_n$, compute $r_i := \text{Res}_X(\widetilde{v}_1, w_i)$ and find $\alpha_1, \ldots, \alpha_s \in \mathbf{R}$ such that $\alpha_1 r_1 + \cdots + \alpha_s r_s = 1$ (here we use the constructive proof of Question (3) or Gröbner bases techniques (Chapter 3)).

For $1 \leq i \leq s$, compute $f_i, g_i \in \mathbf{R}[X]$ such that $f_i\widetilde{v}_1 + g_iw_i = r_i$ (use Proposition 52).

Step 3: Set
$b_s := 0,$

$b_{s-1} := \alpha_s \, r_s X,$

$b_{s-2} := b_{s-1} + \alpha_{s-1} r_{s-1} X,$

$\vdots$

$b_0 := b_1 + \alpha_1 r_1 X = X$ (this follows from the fact that $X = \sum_{i=1}^{s} \alpha_i r_i X$).

Step 4: For $1 \leq i \leq s$, find $\mathscr{B}_i \in \mathrm{SL}_n(\mathbf{R}[X])$ such that $\mathscr{B}_i \widetilde{\mathscr{V}}(b_{i-1}) = \widetilde{\mathscr{V}}(b_i)$.

In more details, let γ_i be the matrix corresponding to the elementary operation $L_2 \to L_2 + \sum_{j=3}^{n} y_i^{j-2} L_j$, that is, $\gamma_i := E_{2,n}(y_i^{n-2}) \cdots E_{2,3}(y_i)$.

For $3 \leq j \leq n$, set $F_{i,j} := \frac{\widetilde{v}_j(b_{i-1}) - \widetilde{v}_j(b_i)}{b_{i-1} - b_i} = \frac{\widetilde{v}_j(b_{i-1}) - \widetilde{v}_j(b_i)}{\alpha_i r_i X} \in \mathbf{R}[X]$, so that one obtains

$$
\begin{aligned}
\widetilde{v}_j(b_{i-1}) - \widetilde{v}_j(b_i) &= \alpha_i r_i X F_{i,j} = \alpha_i X F_{i,j} f_i(b_{i-1}) \widetilde{v}_1(b_{i-1}) \\
&\quad + \alpha_i X F_{i,j} g_i(b_{i-1}) w_i(b_{i-1}) \\
&= \sigma_{i,j} \widetilde{v}_1(b_{i-1}) + \tau_{i,j} w_i(b_{i-1}),
\end{aligned}
$$

with $\sigma_{i,j} := \alpha_i X F_{i,j} f_i(b_{i-1}), \tau_{i,j} := \alpha_i X F_{i,j} g_i(b_{i-1}) \in \mathbf{R}[X]$.

Let $\Gamma_i \in E_n(\mathbf{R}[X])$ be the matrix corresponding to the elementary operations: $L_j \to L_j - \sigma_{i,j} L_1 - \tau_{i,j} L_2, 3 \leq j \leq n$, that is

$$
\Gamma_i := \prod_{j=3}^{n} E_{j,1}(-\sigma_{i,j}) E_{j,2}(-\tau_{i,j}).
$$

Set

$$
B_{i,2} := \Gamma_i \gamma_i \in E_n(\mathbf{R}[X]),
$$

so that we have

$$
B_{i,2} \widetilde{\mathscr{V}}(b_{i-1}) = \begin{pmatrix} \widetilde{v}_1(b_{i-1}) \\ w_i(b_{i-1}) \\ \widetilde{v}_3(b_i) \\ \vdots \\ \widetilde{v}_n(b_i) \end{pmatrix}.
$$

Following Lemma 64, set

$s_{i,1}(X,Y,Z) := \frac{\widetilde{v}_1(X+YZ) - \widetilde{v}_1(X)}{Y} \in \mathbf{R}[X,Y,Z],$

$s_{i,2}(X,Y,Z) := \frac{w_i(X+YZ) - w_i(X)}{Y} \in \mathbf{R}[X,Y,Z],$

$t_{i,1}(X,Y,Z) := \frac{f_i(X+YZ) - f_i(X)}{Y} \in \mathbf{R}[X,Y,Z],$

$t_{i,2}(X,Y,Z) := \frac{g_i(X+YZ) - g_i(X)}{Y} \in \mathbf{R}[X,Y,Z],$

$C_{i,1,1} := 1 + s_{i,1}(b_{i-1}, r_i, -\alpha_i X) f_i(b_{i-1}) + t_{i,2}(b_{i-1}, r_i, -\alpha_i X) w_i(b_{i-1}) \in \mathbf{R}[X],$

$$C_{i,1,2} = s_{i,1}(b_{i-1}, r_i, -\alpha_i X)\, g_i(b_{i-1}) - t_{i,2}(b_{i-1}, r_i, -\alpha_i X)\, \widetilde{v}_1(b_{i-1}) \in \mathbf{R}[X],$$

$$C_{i,2,1} = s_{i,2}(b_{i-1}, r_i, -\alpha_i X)\, f_i(b_{i-1}) - t_{i,1}(b_{i-1}, r_i, -\alpha_i X)\, w_i(b_{i-1}) \in \mathbf{R}[X],$$

$$C_{i,2,2} = 1 + s_{i,2}(b_{i-1}, r_i, -\alpha_i X)\, g_i(b_{i-1}) + t_{i,1}(b_{i-1}, r_i, -\alpha_i X)\, \widetilde{v}_1(b_{i-1}) \in \mathbf{R}[X],$$

$$C_i := \begin{pmatrix} C_{i,1,1} & C_{i,1,2} \\ C_{i,2,1} & C_{i,2,2} \end{pmatrix} \in \mathrm{SL}_2(\mathbf{R}[X]).$$

Note that

$$C_i \begin{pmatrix} \widetilde{v}_1(b_{i-1}) \\ w_i(b_{i-1}) \end{pmatrix} = \begin{pmatrix} \widetilde{v}_1(b_i) \\ w_i(b_i) \end{pmatrix}.$$

Set

$$B_{i,1} := \gamma_i^{-1} \begin{pmatrix} C_i & 0 \\ 0 & I_{n-2} \end{pmatrix},$$

with

$$\gamma_i^{-1} = E_{2,3}(-y_i) \cdots E_{2,n}(-y_i^{n-2}).$$

Set

$$\mathscr{B}_i := B_{i,1} B_{i,2} \in \mathrm{SL}_n(\mathbf{R}[X]),$$

so that $\mathscr{B}_i \widetilde{\mathscr{V}}(b_{i-1}) = \widetilde{\mathscr{V}}(b_i)$.

Step 5: $\mathscr{B} := \mathscr{B}_s \cdots \mathscr{B}_1$ and $G := \mathscr{B}\mathscr{D}$.

Complexity bounds: The matrix $\mathscr{B}$ is the product of at most $(n-2)d + 1$ matrices in $\mathrm{SL}_2(\mathbf{R}[X])$ and $4\,[(n-2)d + 1]\,(n-2) = \mathrm{O}(n^2 d)$ elementary matrices in $\mathrm{M}_n(\mathbf{R}[X])$. Moreover, $\deg \mathscr{B}$ is bounded by $nd^{\mathrm{O}(k)}$ and the sequential complexity of this algorithm amounts to $\mathrm{O}(n^4 d)$ arithmetic operations in $\mathbf{R}$ on elements of degree bounded by $nd^{\mathrm{O}(k)}$.

As a matter of fact:

In Step 1: $\deg \widetilde{v}_i \leq d$

In Step 2: $\deg w_i \leq d$, $\deg r_i \leq d^2$, $\deg(\alpha_i r_i) \leq d^{O(k)}$, $\deg f_i \leq d^{O(k)}$, and $\deg g_i \leq d$.

In Step 3: $\deg b_i \leq d^{O(k)}$.

In Step 4: $\deg \mathscr{B}_i \leq d^{O(k)}$.

In Step 5: $\deg G \leq nd^{O(k)}$.

It is immediate that $B_{i,2} \in E_n(\mathbf{R}[X])$ is the product of $3(n-2)$ elementary matrices in $M_n(\mathbf{R}[X])$, while $B_{i,1}$ is the product of one matrix in $SL_2(\mathbf{R}[X])$ by $(n-2)$ elementary matrices. Thus, $\mathscr{B}$ is the product of $[(n-2)d+1](4(n-2)+1)$ matrices, among them, $4[(n-2)d+1](n-2)$ are elementary and $(n-2)d+1$ in $SL_2(\mathbf{R}[X])$.

An algorithm for unimodular completion: case of $K[X_1^{\pm}, \dots, X_k^{\pm}]$ where K is an infinite field.

We will use the notation $\underline{X} = (X_1^{\pm 1}, \dots, X_k^{\pm 1})$ and $\widetilde{v}_i(0) = \widetilde{v}_i(X_k = 0)$. Note that, contrary to the paper [133], the following algorithm for unimodular completion does not convert the given Laurent polynomial vector to a "regular" polynomial vector, eliminates all the variables at one time, and does not use the fact that the base ring is Noetherian.

Input: One column $\mathscr{V} = \mathscr{V}(\underline{X}) = {}^t(v_1(\underline{X}), \dots, v_n(\underline{X})) \in K[\underline{X}^n]$ such that $1 \in \langle v_1, \dots, v_n \rangle$, with $\max_{1 \leq i \leq n}\{\deg v_i\} = d$ (where $d \geq 2$).

Output: A matrix M in $M_n(K[\underline{X}])$, whose determinant is a monomial, such that $M\mathscr{V} = {}^t(1, 0, \dots, 0)$.

Step 1: Make a change of variables so that v_1 becomes doubly unitary at X_k.

Step 2: Perform the general algorithm with $\mathbf{A} = K[X_1^{\pm 1}, \dots, X_{k-1}^{\pm 1}]$ and $X = X_k$. Output the matrix $\mathscr{B}$ such that $\mathscr{B}\widetilde{\mathscr{V}} = \widetilde{\mathscr{V}}(0)$.

Step 3: Output the final matrix

$$M := E_{2,1}(-1)E_{1,2}(1-\widetilde{v}_1(0))E_{2,1}((1-\widetilde{v}_2(0))(\widetilde{v}_1(0))^{-1})$$
$$E_{3,1}(-\widetilde{v}_3(0)(\widetilde{v}_1(0))^{-1})\dots E_{n,1}(-\widetilde{v}_n(0)(\widetilde{v}_1(0))^{-1})\mathscr{B}\mathscr{D}.$$

Here $\mathscr{D}$ is a diagonal matrix corresponding to the shift step of the general algorithm.

Complexity bounds: The final matrix M is the product of at most $(n-2)d+1$ matrices in $SL_2(\mathbf{A}[X])$, $4[(n-2)d+1](n-2)+n+1 = O(n^2 d)$ elementary matrices in $M_n(\mathbf{A}[X])$, and one diagonal matrix. The sequential complexity of this algorithm amounts to $n^4 d^{O(k^2)}$ field operations in $\mathbf{K}$.

Example 1: Let $\mathscr{V} = \begin{pmatrix} v_1 \\ v_2 \\ v_3 \end{pmatrix} = \begin{pmatrix} yx^{-2}+1+yx^{-1} \\ 1+yx^{-1} \\ -yx+x \end{pmatrix} \in \mathrm{Um}_3$

$(\mathbb{Q}[x^{\pm 1}, y^{\pm 1}])$.

The first step consists in eliminating x. With the above notation, we get:

$$\widetilde{\mathscr{V}} = \begin{pmatrix} \widetilde{v}_1 \\ \widetilde{v}_2 \\ \widetilde{v}_3 \end{pmatrix} = \begin{pmatrix} x^2+yx+y \\ x+y \\ -y+1 \end{pmatrix}, d=2, n=3, s=3, y_1=0, y_2=1, y_3=2,$$

$r_1=y$, $r_2=1$, $r_3=2y^2-5y+4$, $\alpha_1=0$, $\alpha_2=1$, $\alpha_3=0$, $f_1=1$, $g_1=-x$, $f_2=1$, $g_2=1-x-y$, $f_3=1$, $g_3=2-x-2y$, $w_1=x+y$, $w_2=x+1$, $w_3=x-y+2$, $b_3=0$, $b_2=0$, $b_1=x$, $b_0=x$, $F_{2,3}=0$, $\sigma_{2,3}=0$, $\tau_{2,3}=0$, $\Gamma_2=I_3$,

$$\gamma_2 = \begin{pmatrix} 1 & 0 & 0 \\ 0 & 1 & 1 \\ 0 & 0 & 1 \end{pmatrix}, \mathscr{B}_{2,2}=\gamma_2,$$

$$\mathscr{B}_{2,1} = \begin{pmatrix} 1-yx+x & -x^2+x^2y-2yx+xy^2 & 0 \\ -x & 1-x+x^2+yx & -1 \\ 0 & 0 & 1 \end{pmatrix},$$

$$\mathscr{B} = \mathscr{B}_2 = \mathscr{B}_{2,1}\mathscr{B}_{2,2}$$
$$= \begin{pmatrix} 1-yx+x & -x^2+x^2y-2yx+xy^2 & -x^2+x^2y-2yx+xy^2 \\ -x & 1-x+x^2+yx & -x+x^2+yx \\ 0 & 0 & 1 \end{pmatrix}.$$

Note that $\mathscr{B}\widetilde{\mathscr{V}}(x,y) = \widetilde{\mathscr{V}}(0,y) = \begin{pmatrix} y \\ y \\ -y+1 \end{pmatrix}$.

Letting $\mathscr{D} = \begin{pmatrix} x^2 & 0 & 0 \\ 0 & x & 0 \\ 0 & 0 & x^{-1} \end{pmatrix}$ be the diagonal matrix corresponding to the shift step, the final matrix M such that $M\mathscr{V} = {}^t(1,0,0)$, is:

$$M = E_{2,1}(-1)E_{1,2}(1-\widetilde{v}_1(0))E_{2,1}((1-\widetilde{v}_2(0))(\widetilde{v}_1(0))^{-1})E_{3,1}(-\widetilde{v}_3(0)(\widetilde{v}_1(0))^{-1})$$
$$\ldots E_{n,1}(-\widetilde{v}_n(0)(\widetilde{v}_1(0))^{-1})\mathscr{B}\mathscr{D},$$

that is,

$$M = E_{2,1}(-1)E_{1,2}(1-y)E_{2,1}(\frac{1-y}{y})E_{3,1}(\frac{y-1}{y})\mathscr{B}\mathscr{D}$$

$$= \begin{pmatrix} -\frac{x^2(y-3y^2x+3yx-y^2+y^3x-1-x)}{y} \\ yx^2(-1+yx-2x) \\ (\frac{1}{y}-1)(-x^2+x^3y-x^3) \\ \frac{x(5y^2x-3y^2x^2-4y^3x+3yx^2+y^3x^2+xy^4-3yx-x^2-y^2+y)}{y} \\ -yx(-3yx+yx^2+y^2x-2x^2-1+x) \\ (1-\frac{1}{y})(-x^3+x^3y-2x^2y+x^2y^2) \\ \frac{5y^2-3y^2x-4y^3+3yx+y^3x+y^4-3y-x}{y} \\ -y(-3y+yx+y^2-2x+1) \\ -2x+xy-3y+y^2+\frac{x}{y}+2+\frac{1}{x} \end{pmatrix}$$

with determinant x^2. Thus,

$$M^{-1} = \begin{pmatrix} \frac{y+xy+x^2}{x^2} & \frac{-1+x+y}{x^2} & -\frac{-2y+xy+y^2-x}{x} \\ \frac{x+y}{x} & \frac{-y+y^2+1}{xy} & 1-y \\ -(y-1)x & -\frac{x(y-1)^2}{y} & x \end{pmatrix}$$

is a completion of our vector $\mathscr{V}$ to an invertible matrix.

Example 2: Now, let $\mathscr{V} = \begin{pmatrix} yx^2+x \\ 1+y \\ -yx+x \\ xy+1 \end{pmatrix} \in \mathrm{Um}_4(\mathbb{Q}[x^{\pm 1}, y^{\pm 1}]).$

All the computations have been done with the Computer Algebra System
MAPLE. The code of the algorithm unimodlaurent gives the matrix B
corresponding to the first step. These results allow us to find the matrix M
such that $M\mathscr{V} = {}^t(1,0,0,0)$:

```
> v:=[y*x^2+x,1+y,x-x*y,y*x+1];B:=unimodlaurent(v,x,y);
```

$$v = [y\ x^2 + x,\ 1 + y,\ x - x\ y,\ x\ y + 1];$$

$$B = \begin{pmatrix} 1-\frac{1}{2}y^2x & -xy+\frac{1}{2}y^2x & \frac{1}{2}y^2x & \frac{1}{2}y^2x \\ 0 & 1 & 0 & 0 \\ 0 & 0 & 1 & 0 \\ -\frac{1}{2}y^2x & -xy+\frac{1}{2}y^2x & \frac{1}{2}y^2x & 1+\frac{1}{2}y^2x \end{pmatrix};$$

$$M = \begin{pmatrix} -\frac{1}{2}\frac{-2+y^2x}{x} & -xy+\frac{1}{2}y^2x & \frac{1}{2}y^2 & \frac{1}{2}y^2x \\ \frac{1}{2}\frac{(1+y)(-2+y^2x)}{x} & \frac{1}{2}y^2x-\frac{1}{2}y^3x+xy+1 & -\frac{1}{2}(1+y)^2 & -\frac{1}{2}(1+y)y^2x \\ -\frac{1}{2}\frac{(y-1)(-2+y^2x)}{x} & \frac{1}{2}(y-1)xy(-2+y) & \frac{1}{2}\frac{y^3x-y^2x+2}{x} & \frac{1}{2}(y-1)y^2x \\ -x^{-1} & 0 & 0 & 1 \end{pmatrix};$$

$$\det(M) = x^{-2}.$$

Thus,

$$M^{-1} = \begin{pmatrix} x(xy+1) & -\frac{1}{2}y(-2+y)x^2 & -\frac{1}{2}x^2y^2 & -\frac{1}{2}x^2y^2 \\ 1+y & 1 & 0 & 0 \\ -(y-1)x & 0 & x & 0 \\ xy+1 & -\frac{1}{2}xy(-2+y) & -\frac{1}{2}y^2x & 1-\frac{1}{2}y^2x \end{pmatrix}$$

is a completion of our vector $\mathcal{V}$ to an invertible matrix.

(5) Let us denote by f a doubly monic polynomial in $I \cap \mathbf{R}[X]$. Since $I+J[X,X^{-1}] = \mathbf{R}[X,X^{-1}]$, there exist $g \in I$, $h \in J[X]$, and $q \in \mathbb{N}$ such that $g+h=X^q$. It follows that $X^q \in \langle \bar{f}, \bar{g} \rangle$ where the classes are taken modulo $J[X]$. By virtue of Question (1), we obtain that $\mathrm{Res}(\bar{f}, \bar{g}) \in (\mathbf{R}/J)^\times$. As f is a monic polynomial, $\mathrm{Res}(\bar{f}, \bar{g}) = \overline{\mathrm{Res}(f,g)}$, and thus $\langle \mathrm{Res}(f,g) \rangle + J = \mathbf{R}$. The desired conclusion follows from the fact that $\mathrm{Res}(f,g) \in I \cap \mathbf{R}$.

(6) This can be obtained by a close adaptation of the proof of Lemma 185.

(7) By Remark 138, we know that if $\mathrm{Kdim}\,\mathbf{R} \le 0$ then $\mathbf{R}\langle X \rangle = \mathbf{R}(X)$. Thus, since $\mathbf{R}(X) = \mathbf{R}(X^{-1})$, we have $\mathbf{R}\langle X^{-1} \rangle \subseteq \mathbf{R}\langle X \rangle$. By Exercise 374, we know that $\mathbf{R}\langle X, X^{-1} \rangle = \mathbf{R}\langle X \rangle \cap \mathbf{R}\langle X^{-1} \rangle$. It follows that $\mathbf{R}\langle X, X^{-1} \rangle = \mathbf{R}\langle X^{-1} \rangle = \mathbf{R}(X^{-1}) = \mathbf{R}(X) = \mathbf{R}\langle X \rangle$. For the assertion concerning the Krull dimension, one can suppose that $\mathbf{R}$ is local, and, thus, every element in $\mathbf{R}$ is either nilpotent or invertible. Clearly, this property is inherited by the ring $\mathbf{R}(X)$, and, thus, $\mathrm{Kdim}\,\mathbf{R}(X) \le 0$.

(8) This can be obtained by a close adaptation of the proof of Corollary 187 (for related results, see [1, 2]).

Exercise 376:

Clearly, we have $F^m H \subseteq F^{m+1} G$. For the converse, since $f\tilde{g} = h - b_m f T^m$, we have $[f\tilde{g}] \subseteq H + b_m F$. By the induction hypothesis, we have

$$F^m \tilde{G} \subseteq F^{m-1}[f\tilde{g}] \subseteq F^{m-1} H + b_m F^m. \qquad (6.2)$$

Now, as $a_i b_m = c_{i+m} - (a_{i+1} b_{m-1} + a_{i+2} b_{m-2} + \cdots)$, we have

$$a_i b_m \in c_{i+m} + \sum_{j>i} a_j \tilde{G}. \qquad (6.3)$$

Multiplying (6.3) by F^m and using (6.2), we obtain

$$a_ib_mF^m \subseteq c_{i+m}F^m + \sum_{j>i}a_jF^m\tilde{G} \subseteq (c_{i+m}F^m + \sum_{j>i}a_jF^{m-1}H) + \sum_{j>i}a_jb_mF^m.$$

Setting $E_i = a_ib_mF^m$, we get

$$E_i \subseteq F^mH + \sum_{j>i+1} E_j.$$

As $E_j = 0$ for $j \gg 0$, the above containment gives $E_i \subseteq F^mH$, that is,

$$a_ib_mF^m \subseteq F^mH. \tag{6.4}$$

Multiplying (6.2) by a_i and using (6.4), we get

$$a_iF^m\tilde{G} \subseteq a_iF^{m-1}H + a_ib_mF^m \subseteq a_iF^{m-1}H + F^mH \subseteq F^mH. \tag{6.5}$$

Combining (6.4) and (6.5), we infer that $a_iF^mG \subseteq F^mH$, and, thus, $F^{m+1}G \subseteq F^mH$, as desired.

Exercise 377:

(1) Use Dedekind-Mertens, Exercise 376.

(2) The vectors ${}^t(r_0,\ldots,r_d)$ and ${}^t(c_0,\ldots,c_d)$ are relied by the Vandermonde matrix as follows:

$$\begin{pmatrix} 1 & y_0 & \cdots & y_0^d \\ 1 & y_1 & \cdots & y_1^d \\ \vdots & \vdots & \vdots & \vdots \\ 1 & y_d & \cdots & y_d^d \end{pmatrix}\begin{pmatrix} c_0 \\ c_1 \\ \vdots \\ c_d \end{pmatrix} = \begin{pmatrix} r_0 \\ r_1 \\ \vdots \\ r_d \end{pmatrix}.$$

It follows that

$$\pi\langle c_0,\ldots,c_d\rangle \subseteq \langle r_0,\ldots,r_d\rangle.$$

(4) There exist $\tilde{u},\tilde{v},\tilde{w} \in \mathbf{A}[X]$ such that $u\tilde{u}+v\tilde{v}+w\tilde{w} = 1$. Replacing X by x_i, we obtain

$$1 \in \langle u_1,w_1\rangle\langle u_2,w_2\rangle^2\cdots\langle u_d,w_d\rangle^d \subseteq \langle r_0,\ldots,r_d\rangle.$$

Exercise 378:

(1)

    (b) First note that b is invertible modulo a. Consider, over $\mathbf{R}/a\mathbf{R}$, the matrix $b^{-1}B'$: it is a diagonal matrix with determinant 1, and, thus, it is in $E_n(\mathbf{R}/a\mathbf{R})$ by virtue of Proposition 164. It lifts as a matrix $E \in E_n(\mathbf{R})$.

    (d) The matrix T is obtained from $\begin{pmatrix} A & B' \\ C & D'' \end{pmatrix}$ by supressing rows from 2 to n, and the last $n-1$ columns.

(2) Write $xu + yv + zw = 1$ for some $u, v, w \in \mathbf{R}$. Modulo z, the row (x, y) is completable to $A = \begin{pmatrix} x & y \\ -v & u \end{pmatrix}$ whose determinant is $a := ux + vy \equiv 1 \mod \langle z \rangle$. Then, use (1).

Hereafter, some of the computation details.

    (i) Denoting $b = z$, $c = w$, $D = \tilde{A} = \begin{pmatrix} u & -y \\ v & x \end{pmatrix}$, and $C = U = -wI_2$, we have:

$$AD = DA = aI_2 = (1 - wz)I_2 = I_2 - wzI_2 = I_2 + bU = I_2 + bC,$$

$$\begin{pmatrix} A & bI_2 \\ C & D \end{pmatrix}\begin{pmatrix} D & -bI_2 \\ -U & A \end{pmatrix} = \begin{pmatrix} x & y & z & 0 \\ -v & u & 0 & z \\ -w & 0 & u & -y \\ 0 & -w & v & x \end{pmatrix}\begin{pmatrix} u & -y & -z & 0 \\ v & x & 0 & -z \\ w & 0 & x & y \\ 0 & w & -v & u \end{pmatrix}$$

$$= \begin{pmatrix} I_2 & 0 \\ * & I_2 \end{pmatrix} \in GL_4(\mathbf{R}).$$

    (ii) Modulo $\langle 1 - wz \rangle$, we have

$$\begin{pmatrix} z & 0 \\ 0 & w \end{pmatrix} \xrightarrow{L_1 \to L_1 + zL_2} \begin{pmatrix} z & 1 \\ 0 & w \end{pmatrix} \xrightarrow{C_1 \to C_1 + (1-z)C_2} \begin{pmatrix} 1 & 1 \\ (1-z)w & w \end{pmatrix}$$

$$\xrightarrow{L_2 \to L_2 + (z-1)wL_1} \begin{pmatrix} 1 & 1 \\ 0 & 1 \end{pmatrix} \xrightarrow{C_2 \to C_2 - C_1} \begin{pmatrix} 1 & 0 \\ 0 & 1 \end{pmatrix} = I_2.$$

Thus, denoting by

$$\begin{aligned} E &= E_{1,2}(-z)\,E_{2,1}((1-z)w)\,E_{1,2}(1)\,E_{2,1}(z-1) \\ &= \begin{pmatrix} z - (z^2 - z)(1 - wz) & -(z-1)(1-wz) \\ -(-z+1)(1-wz) & w + (1-wz) \end{pmatrix}, \end{aligned}$$

we have

$$\begin{pmatrix} z & 0 \\ 0 & w \end{pmatrix} = E + (1 - wz)T,$$

with $T = \begin{pmatrix} z^2 - z & z - 1 \\ -z + 1 & -1 \end{pmatrix}$. It follows that

$$
\begin{aligned}
\begin{pmatrix} z^2 & 0 \\ 0 & 1 \end{pmatrix} &= \begin{pmatrix} z^2 & 0 \\ 0 & wz \end{pmatrix} + \begin{pmatrix} 0 & 0 \\ 0 & 1 - wz \end{pmatrix} \\
&= zE + (1 - wz)zT + (1 - wz)\mathrm{diag}(0, 1) \\
&= zE + (1 - wz)F,
\end{aligned}
$$

with $F = zT + \mathrm{diag}(0, 1)$.

(iii) We have:

$$
\begin{aligned}
\tilde{A}F &= \begin{pmatrix} u & -y \\ v & x \end{pmatrix} \begin{pmatrix} z^3 - z^2 & z^2 - z \\ -z^2 + z & -z + 1 \end{pmatrix} \\
&= \begin{pmatrix} uz^3 - uz^2 + yz^2 - yz & uz^2 - uz + yz - y \\ vz^3 - z^2v - xz^2 + xz & z^2v - zv - xz + x \end{pmatrix}.
\end{aligned}
$$

Modulo $\langle ux + vy + wz - 1 \rangle$, we have:

$$
\begin{aligned}
&\begin{pmatrix} A & bI_2 \\ C & D \end{pmatrix} \begin{pmatrix} I_2 & \tilde{A}F \\ 0 & E \end{pmatrix} \\
&= \begin{pmatrix} x & y & z^2 & 0 \\ -v & u & 0 & 1 \\ -w & 0 & 2uz - uz^2 - yz + y & -uz + u - y \\ 0 & -w & 2zv - z^2v + xz - x & -zv + v + x \end{pmatrix} \\
&\begin{array}{l} L_3 \to L_3 - (-uz + u - y)L_2 \\ L_4 \to L_4 - (-zv + v + x)L_2 \end{array} \\
&\hookrightarrow \\
&= \begin{pmatrix} x & & y \\ -v & & u \\ -w + v(-uz + u - y) & & -u(-uz + u - y) \\ v(-zv + v + x) & & -w - u(-zv + v + x) \\ \\ & z^2 & 0 \\ & 0 & 1 \\ & 2uz - uz^2 - yz + y & 0 \\ & 2zv - z^2v + xz - x & 0 \end{pmatrix}
\end{aligned}
$$

Finally, we see that (x, y, z^2) is the first row of the matrix

$$M = \begin{pmatrix} x & y \\ -w+v(-uz+u-y) & -u(-uz+u-y) \\ v(-zv+v+x) & -w-u(-zv+v+x) \\ z^2 \\ 2uz-uz^2-yz+y \\ 2zv-z^2v+xz-x \end{pmatrix},$$

with $\det(M) \equiv 1 \mod \langle ux+vy+wz-1 \rangle$. More precisely, we obtain

$$\det(M) = \begin{vmatrix} x & y & z^2 \\ -yv-w-uv(z-1) & u(y+u(z-1)) & y(1-z)+zu(2-z) \\ v(x-v(z-1)) & -xu-w+uv(z-1) & x(z-1)-zv(z-2) \end{vmatrix}$$

$$= (ux+vy+wz)^2.$$

(3) Induct on n.

Exercise 379:

(1)

(a) Suppose that both $v+y_1w$ and $v+y_2w$ are zero-divisors, with $y_2 - y_1 \in \mathbf{A}^\times$. Write $d_1(v+y_1w) = 0$ and $d_2(v+y_2w) = 0$ with $d_1, d_2 \neq 0$. If $d_2 = d_1\delta_1$ for some $\delta_1 \in \mathbf{A}$ then $d_2(v+y_1w) = 0$. Together with $d_2(v+y_2w) = 0$, this would imply that $(y_1 - y_2)d_2w = 0$, $d_2w = 0$ and $d_2v = 0$. Since $1 = uv + w$ then we would have $d_2 = 0$ which is not true. Thus, $\langle d_1 \rangle \subsetneq \langle d_1, d_2 \rangle$.

(b) We induct on n. The case $n = 2$ is done in (a). For $n \geq 3$, suppose that all $v+y_1w, v+y_2w, \ldots, v+y_nw$ are zero-divisors, that is, there exist $d_i \in \mathbf{A} \setminus \{0\}$, $1 \leq i \leq n$, such that $d_i(v+y_iw) = 0$. We can suppose that for any $\sigma \in S_n$ (the permutation group of $\{1, \ldots, n\}$), $d_{\sigma(n)} \notin \langle d_{\sigma(1)}, \ldots, d_{\sigma(n-1)} \rangle$. Passing to the ring $\mathbf{A}/\langle d_1 \rangle$ and using the induction hypothesis, we infer that $\bar{d}_n \notin \langle \overline{d_1}, \ldots, \overline{d_{n-1}} \rangle$ and, thus, $d_n \notin \langle d_1, \ldots, d_{n-1} \rangle$. Together with the fact that we have an increasing chain $\langle d_1 \rangle \subsetneq \langle d_1, d_2 \rangle \subsetneq \cdots \subsetneq \langle d_1, \ldots, d_{n-1} \rangle$ (by induction hypothesis), we obtain an increasing chain

$$\langle d_1 \rangle \subsetneq \langle d_1, d_2 \rangle \subsetneq \cdots \subsetneq \langle d_1, \ldots, d_n \rangle$$

of ideals in $\mathbf{A}$.

(c) This is an immediate consequence of (b) as in a Noetherian ring, every non-decreasing sequence of finitely-generated ideals pauses.

(2) Denote $w_i = u_i v_i + \cdots + u_n v_n$. Applying (1.c), we can find $y_1 \in E$ such that $v_1 + y_1 w_2$ is not a zero-divisor. Note that in $\mathbf{A}$ we have

$$(v_1 + y_1 w_2) u_1 + (1 - y_1 u_1) w_2 = 1.$$

In the ring $\mathbf{A}/\langle v_1 + y_1 w_2 \rangle$, we have:

$$(1 - y_1 u_1) u_2 v_2 + (1 - y_1 u_1) u_3 v_3 + \cdots + (1 - y_1 u_1) u_n v_n = 1.$$

Applying (1.c) again, we can find $y_2 \in E$ such that $v_2 + y_2 (1 - y_1 u_1) w_3$ is not a zero-divisor in $\mathbf{A}/\langle v_1 + y_1 w_2 \rangle$, and so on, we construct the desired regular sequence. Suppose that in the ring $\mathbf{A}/\langle v_1 + y_1 \xi_0 w_2, v_2 + y_2 \xi_1 w_3, \dots, v_k + y_k \xi_{k-1} w_{k+1} \rangle$, we have:

$$\xi_k w_{k+1} = 1,$$

and then

$$(v_{k+1} + y_{k+1} \xi_k w_{k+2}) \xi_k u_{k+1} + \xi_k (1 - y_{k+1} \xi_k u_{k+1}) w_{k+2} = 1.$$

Thus, in the ring $\mathbf{A}/\langle v_1 + y_1 \xi_0 w_2, v_2 + y_2 \xi_1 w_3, \dots, v_k + y_k \xi_{k-1} w_{k+1}, v_{k+1} + y_{k+1} \xi_k w_{k+2} \rangle$, we have:

$$\xi_k (1 - y_{k+1} \xi_k u_{k+1}) w_{k+2} = 1.$$

It follows that the sequence (ξ_k) satisfies the relation:

$$\begin{cases} \xi_0 = 1 \\ \xi_{k+1} = \xi_k (1 - y_{k+1} u_{k+1} \xi_k). \end{cases}$$

(3) Using (2), there exist $y_1, \dots, y_{d+1} \in E$ such that the sequence $(v_1 + y_1 \xi_0 w_2, v_2 + y_2 \xi_1 w_3, \dots, v_{d+1} + y_{d+1} \xi_d w_{d+2})$ is regular. Since $\text{Kdim} \mathbf{A} = d$, this sequence is also singular (see Definition 85). From Proposition 84, we infer that

$$1 \in \langle v_1 + y_1 \xi_0 w_2, v_2 + y_2 \xi_1 w_3, \dots, v_{d+1} + y_{d+1} \xi_d w_{d+2} \rangle.$$

For the computation of M, considering $q_1, \dots, q_{d+1} \in \mathbf{A}$ such that

$$q_1 (v_1 + y_1 \xi_0 w_2) + \cdots + q_{d+1} (v_{d+1} + y_{d+1} \xi_d w_{d+2}) = 1,$$

and denoting by

$$
\begin{aligned}
M_1 &:= \prod_{k=d+1}^{1} E_{k,k+1}(y_k \xi_{k-1} u_{k+1}) \cdots E_{k,n}(y_k \xi_{k-1} u_n), \\
M_2 &:= E_{n,1}((v_n - 1)q_1) \cdots E_{n,d+1}((v_n - 1)q_{d+1}), \\
M_3 &:= E_{1,n}(-v_1 - y_1 \xi_0 w_2) \cdots E_{d+1,n}(-v_{d+1} - y_{d+1} \xi_d w_{d+2}) \\
&\qquad E_{d+2,n}(-v_{d+2}) \cdots E_{n-1,n}(-v_{n-1}),
\end{aligned}
$$

we have

$$
E_{n,1}(-1) E_{1,n}(1) M_3 M_2 M_1 {}^t(v_1, v_2, \ldots, v_n) = {}^t(1, 0, \ldots, 0).
$$

This is because:

$$
M_1 {}^t(v_1, v_2, \ldots, v_n) = {}^t(v_1 + y_1 \xi_0 w_2, \ldots, v_{d+1} + y_{d+1} \xi_d w_{d+2}, v_{d+2}, \ldots, v_n) := \mathcal{V}_1,
$$

$$
M_2 \mathcal{V}_1 = {}^t(v_1 + y_1 \xi_0 w_2, \ldots, v_{d+1} + y_{d+1} \xi_d w_{d+2}, v_{d+2}, \ldots, v_{n-1}, 1) := \mathcal{V}_2, \text{ and}
$$

$$
M_3 \mathcal{V}_2 = {}^t(0, \ldots, 0, 1).
$$

(4)

(a) Just use (3) and the fact that $\mathrm{Kdim}\, \mathbf{A}\langle X_1, \ldots, X_k \rangle = \mathrm{Kdim}\, \mathbf{A} = d < \infty$.

(b) By (a), one can transform ${}^t(v_1, v_2, \ldots, v_n)$ into a unimodular row whose first coordinate is $F \in \langle v_1 + y_1 \xi_0 w_2, v_2 + y_2 \xi_1 w_3, \ldots, v_{d+1} + y_{d+1} \xi_d w_{d+2} \rangle \cap U_k$. By a change of variables "à la Nagata", that is of type $(X_1, \ldots, X_{k-1}, X_k) \to (Y_1, \ldots, Y_{k-1}, X_k)$ with $X_{k-1} = Y_{k-1} + X_k^m, X_{k-2} = Y_{k-2} + X_k^{m^2}, \ldots, X_1 = Y_1 + X_k^{m^{k-1}}$ for sufficiently large m, F becomes monic at X_k. In fact, in order to avoid the explosion of the degrees of the considered polynomials, one has to make a change of variables of type $(X_1, \ldots, X_{k-1}, X_k) \to (Y_1, \ldots, Y_{k-1}, X_k)$ with $X_{k-1} = Y_{k-1} + X_k^{n_1}, X_{k-2} = Y_{k-2} + X_k^{n_2}, \ldots, X_1 = Y_1 + X_k^{n_{k-1}}$ and $(n_1, \ldots, n_{k-1}) \in \mathbb{N}^{k-1}$ as small as possible. Of course, if possible, it should be better to use a linear change of variables so that the polynomials considered keep the same total degree.

Now, suppose that F is monic at X_k, $\deg_{X_k} F = \delta$, $\deg_{X_k} v_n = \delta'$, $\delta'' = \max(\delta, \delta')$, and let $q_1, \ldots, q_{d+1} \in \mathbf{A}$ such that

$$
q_1(v_1 + y_1 \xi_0 w_2) + \cdots + q_{d+1}(v_{d+1} + y_{d+1} \xi_d w_{d+2}) = F.
$$

Set

$$
M_1 := \prod_{k=d+1}^{1} E_{k,k+1}(y_k \xi_{k-1} u_{k+1}) \cdots E_{k,n}(y_k \xi_{k-1} u_n),
$$

$$
M_2 := E_{n,1}(X_k^{\delta''-\delta+1} q_1) \cdots E_{n,d+1}(X_k^{\delta''-\delta+1} q_{d+1}),
$$

so that

$$M_1{}^t(v_1, v_2, \ldots, v_n) = {}^t(v_1 + y_1 \xi_0 w_2, \ldots, v_{d+1} + y_{d+1} \xi_d w_{d+2}, v_{d+2}, \ldots, v_n),$$

and

$$E_{1,n}(1) E_{n,1}(-1) M_2 M_1{}^t(v_1, v_2, \ldots, v_n)$$

has as first coordinate a monic polynomial at X_k with degree $\delta'' + 1$.

(c) An algorithm for unimodular completion over Noetherian rings

Input: A unimodular column vector $\mathcal{V} = \mathcal{V}(\underline{X}) = {}^t(v_1(\underline{X}), \ldots, v_n(\underline{X})) \in \mathbf{A}[\underline{X}]^n = \mathbf{A}[X_1, \ldots, X_k]^n$.

We assume that:

- $\mathbf{A}$ is a strongly discrete Noetherian ring (and, thus, $\mathbf{A}$ is a Gröbner ring).

- $\mathbf{A}$ contains infinitely many y_i such that $y_i - y_j$ is invertible for $i \neq j$

- $\mathrm{Kdim}\,\mathbf{A}\langle X_1, \ldots, X_k \rangle = \mathrm{Kdim}\,\mathbf{A} = d < \infty$, and $n \geq \max(3, d+2)$.

Output: A matrix M in $\mathrm{SL}_n(\mathbf{A}[\underline{X}])$ such that $M\mathcal{V} = {}^t(1, 0, \ldots, 0)$.

Step 1: Make a change of variables and elementary operations on $\mathcal{V}$ so that v_1 becomes monic at X_k (follow the algorithm given in (d)).

Step 2: Use Algorithm 65 to compute $\mathcal{B} \in \mathrm{SL}_n(\mathbf{A}[\underline{X}])$ such that $\mathcal{B}\mathcal{V}(X_1, \ldots, X_{k-1}, X_k) = \mathcal{V}(X_1, \ldots, X_{k-1}, 0)$.

Step 3: Repeat for $\mathcal{V}(X_1, \ldots, X_{k-1}, 0)$.

Step 4: (basic step) Follow the algorithm given in (3) to transform $\mathcal{V}(0, \ldots, 0)$ into ${}^t(1, 0, \ldots, 0)$ using elementary operations.

An example: Let $V = \begin{pmatrix} v_1 \\ v_2 \\ v_3 \end{pmatrix} = \begin{pmatrix} ax + 1 - a \\ (1-a)x + a \\ 1 - ax^2 - x + ax - a \end{pmatrix}, U = \begin{pmatrix} u_1 \\ u_2 \\ u_3 \end{pmatrix} = \begin{pmatrix} x \\ 1 \\ a \end{pmatrix} \in \mathrm{Um}_3((\mathbb{Q}[a]/\langle u \rangle)[x])$ with $u = a^2 - a$. Note that we have ${}^tVU = 1$.

The first step consists in making a change of variables and elementary operations on V so that it becomes monic at x. We obtain a matrix M such that $MV = v$

$$M = \begin{bmatrix} x^3 - x^2 + x & x^2 - x & 1 \\ 0 & 1 & 0 \\ -1 & 0 & 0 \end{bmatrix}, MV = \begin{bmatrix} x^2 + x - ax + a \\ 1 - ax^2 - x + ax - a \\ -ax - 1 + a \end{bmatrix}.$$

We will finish the computations with the Computer Algebra System MAPLE. The code, called unimodNoether, implements almost the whole of the algorithm.

```
> G:=normalfmatrix(multiply(multiply(E,unimodNoether(v,u,x,a)),M));

 G=[-1/2*x^4+3/2*x^3-3/2*x^2+x+(1/2*x^4-1/2*x^3+1/2*x^2)*a,
 -1/2*x^3+x^2-x+1/2*x^3*a+1, -1/2*x+1+1/2*a*x],
 [451/76*x^4-527/76*x^3+201/38*x^2-163/38*x-125/76*x^5+
 (757/76*x^3-4874/361*x^2-8039/1444*x^5+1845/1444*x^4
 +3696/361*x^6+43/19*x)*a,-125/76*x^4+201/38*x-163/38*x^2
 +163/38*x^3+(139/19*x^2-8039/1444*x^4-62/19*x+1055/361*x^3
 +1+3696/361*x^5)*a,163/38*x-1-125/76*x^2
 +(-163/38*x+355/76*x^2+3696/361*x^3+1)*a],
 [-125/76*x^5+413/76*x^4-413/76*x^3+72/19*x^2-1-125/38*x
 +(-25749/1444*x^5+21721/1444
 *x^4+413/76*x^3+7392/361*x^6-6936/361*x^2+125/38*x)*a,
 -125/76*x^4+72/19*x^3-125/38*x^2+1+163/38*x
 +(-25749/1444*x^4+6024/361*x^3+7392/361*x^5+125/38*x^2+29/38*x)*a,
 -125/76*x^2+72/19*x+(201/76*x^2+7392/361*x^3-53/19*x)*a]
```

```
> F:=normalfvector1(multiply(G, V));
                    F := [1, 0, 0]
```

(d) By The Stable Range Theorem 92, the result holds for $k = 0$. Assume, by induction, that the statement holds for $k - 1$. Let $\mathbf{B} = \mathbf{A}[X_1,\ldots,X_{k-1}]$, $X = X_k$,

and $\mathscr{V} = \begin{pmatrix} v_1 \\ \vdots \\ v_n \end{pmatrix} \in \mathrm{Um}_n(\mathbf{B}[X])$.

We may assume that v_1 is monic by multiplying $\mathscr{V}$ by an elementary matrix and changing variables (here we use the algorithmic proof of (4.b)). Now use Lemma 177 and proceed as in the proof of Theorem 178.

(e) Use (d) and reduce the columns of the given matrix column by column.

Exercise 380:

Let $X \in \mathbf{R}^{n \times r}$ be a matrix whose columns form a free basis $\mathcal{B}$ for $\operatorname{Im} M$. Then, there exists a unique matrix $Y \in \mathbf{R}^{r \times n}$ such that $M = XY$. In more details, denoting by c_i the ith column of M and writing $c_i = (\alpha_{1,i}, \ldots, \alpha_{r,i})_{\mathcal{B}}$, we have $Y = (\alpha_{j,i})_{1 \leq j \leq r, 1 \leq i \leq n}$ (it is unique since X has rank r).
As $MX = X$ (this is because $\operatorname{Im} X \subseteq \operatorname{Im} M$ and $M^2 = M$), we have $XYX = X$, and, thus, $X(\mathbf{I}_r - YX) = 0$, and, finally, $YX = \mathbf{I}_r$ (this is because the columns of X are linearly independent).
Conversely, supposing that $YX = \mathbf{I}_r$ and $M = XY$, then $M^2 = XYXY = X\mathbf{I}_r Y = XY = M$, $MX = XYX = X$, $\operatorname{Im} M = \operatorname{Im} X$, and the columns of X are linearly independent (this because a relation $XZ = 0$ would imply $Z = YXZ = 0$).

(a) The sequence $\mathbf{R}^n \xrightarrow{\mathbf{I}_n - M} \mathbf{R}^n \xrightarrow{Y} \mathbf{R}^r$ is exact. This is because $Y(\mathbf{I}_n - M) = 0$, and, if $YZ = 0$, then $MZ = 0$, and, thus, $Z = (\mathbf{I}_n - M)Z$. It follows that $\operatorname{Ker} Y = \operatorname{Im}(\mathbf{I}_n - M) = \operatorname{Ker} M$, and $\operatorname{Im} Y \simeq \mathbf{R}^n / \operatorname{Ker} Y = \mathbf{R}^n / \operatorname{Ker} M \simeq \operatorname{Im} M$.

(b) Setting $U = YX'$ and $V = Y'X$, we have:
$$UV = YX'Y'X = YMX = YX = \mathbf{I}_r,$$
$$X'V = X'Y'X = MX = X, \text{ and, thus, } X' = XU,$$
$$UY' = YX'Y' = YM = Y, \text{ and, thus, } Y' = VY,$$
$$Y'X' = VYXU = VU = \mathbf{I}_r.$$

(c) The image of a rank one projection matrix M is free if and only if there exist a column vector C and a row vector L such that $LC = 1$ and $CL = M$. Moreover, C and L are unique up to a multiplication by a unit of $\mathbf{R}$ under the sole condition $CL = M$.

Exercise 381:

(1) "$\Leftarrow$" As M has trace one, we have $\mathcal{D}_1(M) = \langle 1 \rangle$.

"$\Rightarrow$" As, locally, M is conjugate to a standard projection matrix $\mathbf{I}_{1,n}$ (see Theorem 10), necessarily, M has trace one.

(2) The group homomorphisms $\operatorname{Pic} \mathbf{R} \xrightarrow{i : M \mapsto M} \operatorname{Pic} \mathbf{R}[X] \xrightarrow{\rho : M \mapsto M(0)} \operatorname{Pic} \mathbf{R}$ are such that $\rho \circ i = \operatorname{id}_{\operatorname{Pic} \mathbf{R}}$. Of course, ρ is surjective and i is injective. The homomorphism ρ is an isomorphism if and only if it is injective, or equivalently, if and only if for every rank one idempotent matrix $M(X)$ if $\operatorname{Im} M(0)$ is free then so is $\operatorname{Im} M(X)$. Therefore, the desired result follows from Exercise 381.

(3) Denoting by $C = {}^{t}(f_1, \ldots, f_n)$ and $L = (g_1, \ldots, g_n)$, we have $M = CL$, $LC = 1$, $\det(M) = 0$, and M has trace one. Thus, M is rank one idempotent by Exercise 381 and (1).

"3 $\Rightarrow$ 2" As $f_1(0) = 1$, the fact that $f_1 g_j \in \mathbf{R}[X]$ forces the coefficients of g_j into being in $\mathbf{R}$ for all $1 \leq j \leq n$. As well, as $g_1(0) = 1$, the fact that $f_i g_1 \in \mathbf{R}[X]$ forces the coefficients of f_i into being in $\mathbf{R}$ for all $1 \leq i \leq n$.

"$2 \Rightarrow 3$" This is trivial.

"$1 \Leftrightarrow 2$" This is a direct consequence of Exercise 381.c. Note that the units of $A[X]$ are those of A ($A = R$ or T) as A is reduced.

(4) This follows from (2) and (3).

(5) (a) $b^2 = 0 \Rightarrow b^2 = 0^3 \Rightarrow \exists a \in R \mid a^3 = b$ and $a^2 = 0 \Rightarrow b = 0$.

   (b) Denoting by $C = {}^t(f_1, f_2)$ and $L = (g_1, g_2)$, we have $M(X) = CL$, $LC = 1$, $\det(M(X) = 0$, and M has trace one. Thus, $M(X)$ is rank one idempotent. By virtue of (3), the image of the matrix $M(X)$ is free if and only if $f_1 \in R[X]$, i.e., $a \in R$.

   (c) Since if R is a gcd domain then so is $R[X]$, it suffices to prove that $\mathrm{Pic}\, R = \{1\}$. For this aim, we will use the characterization given in Exercise 381. Consider a rank one idempotent matrix $M = (m_{i,j})_{1 \le i,j \le n}$. As $\sum_{i=1}^n m_{i,i} = 1$ (by (1)), one of the $m_{i,i}$'s is regular. We can suppose that $m_{1,1}$ is regular. Denoting by $f = \gcd(m_{1,1}, \ldots, m_{1,n})$, we can write $m_{1,j} = f g_j$ with $\gcd(g_1, \ldots, g_n) = 1$. Simplifying the equality $m_{1,1} m_{i,j} = m_{1,j} m_{i,1}$, one gets $g_1 m_{i,j} = m_{1,j} g_j$. Thus, g_1 divides all the $m_{1,j} g_j$, and a fortiori their gcd $m_{i,1}$. Write $m_{i,1} = g_1 f_i$. Since $g_1 f_1 = m_{1,1} = f g_1$, we obtain $f_1 = f$. Finally, the equality $m_{1,1} m_{i,j} = m_{1,j} m_{i,1}$ yields to $m_{i,j} = f_i g_j$.

(6) We have to prove that if $u \in T$ and $u^2 \in \mathfrak{a}$, then $u \in \mathfrak{a}$. For this, consider $c \in T$, and let us prove that $uc \in R$. As $u^2 \in \mathfrak{a}$, we have $u^2 c^2, u^3 c^3 = u^2(uc^3) \in R$. Since $(u^3 c^3)^2 = (u^2 c^2)^3$, there exists $a \in R$ such that $a^2 = (uc)^2$ and $a^3 = (uc)^3$, and thus, $(a - uc)^3 = (a^3 - (uc)^3) - 3uc(a^2 - (uc)^2) = 0$. As T is reduced, we infer that $uc = a \in R$.

(7) Let $x \in T$ such that $xc_1, \ldots, xc_q \in R$. As $x^\ell c_i^\ell \in R$ for all $\ell \in \mathbb{N}$, for a sufficiently large N (one can take $N = q(d-1)$ where d is the maximum of the degrees of the integral dependence relations of the c_i's over R), we have $x^N y \in R$ for all $y \in T$. It follows that $x \in \sqrt{I} = I$.

(8) (a) This is an immediate consequence of Kronecker's theorem above as the coefficients of the f_i's and the g_j's are integral over R (here we used the fact that $f_1(0) = g_1(0) = 1$).

   (b) For $x \in R$, we denote by $\bar{x}$ what it becomes after the change of rings: $R \to L$. Since L is a field, using Exercise 381.c, we obtain (by uniqueness) that the polynomials $\bar{f}_i$ and $\bar{g}_j$ are in $L[X]$. This means that there exists $s \in R \setminus \mathfrak{p}$ such that the polynomials $s f_i$ and $s g_j$ have their coefficients in R. By (7) we infer that $s \in I$, a contradiction.

   (c) Suppose that a classical proof (like the proof above) shows by way of contradiction that a ring R is trivial as follows. One assumes that R is not trivial, considers a minimal prime ideal $\mathfrak{p}$ of R, and then performs the computations in the localization $R_\mathfrak{p}$ of R ($R_\mathfrak{p}$ is local and

zero-dimensional, thus, it is a field in the reduced case) and finds a contradiction $1 = 0$. In order to reread constructively such a proof, one has
to proceed as follows. First, ensure that the proof becomes a constructive
proof of the equality $1 = 0$ under the further hypothesis that $\mathbf{R}$ is local
and zero-dimensional. Second, remove the hypothesis "$\mathbf{R}$ is local and
zero-dimensional" and follow step by step the previous proof by forcing
x into being invertible (by passing to the ring $\mathbf{A}[\frac{1}{x}]$, where $\mathbf{A}$ is the current ring) each time a disjunction "x nilpotent or x invertible" is required
for pursuing the computations. Each time one proves that $1 = 0$ in the
current ring, one actually proves that the last element tested is nilpotent,
allowing to go back to the last disjunction but now following the branch
"x nilpotent". If the considered proof is sufficiently uniform, then the
constructed binary tree is finite and one eventually obtains the desired
constructive proof.

Exercise 382:

(1) Let us explain the proof with an example. Suppose that $m = 2^4 = 16$.
$(c^{16}, c^{24} \in \mathbf{A}_1 \Rightarrow c^8 \in \mathbf{A}_1)$, $(c^{18}, c^{27} \in \mathbf{A}_1 \Rightarrow c^9 \in \mathbf{A}_1)$, and so on, for any
$n \geq 2^3$, $a^n \in \mathbf{A}_1$. Then we pass from 2^3 to 2^2, and from 2^3 to 2, and finally,
$(c^2, c^3 \in \mathbf{A}_1 \Rightarrow c \in \mathbf{A}_1)$.

(2) As $m_{1,j} = f_1 g_j \in \mathbf{A}[X]$ and $f_1(0) = 1$, we obtain by identifying the coefficients
of $m_{1,j}$ degree by degree, that $g_j \in \mathbf{B}[X]$. Similarly, as $g_1 \in \mathbf{B}[X]$, $g_1(0) = 1$,
and $m_{i,1} = f_i g_1 \in \mathbf{A}[X]$, we obtain that $f_i \in \mathbf{B}[X]$.

(3) Use (2) and the fact that every b_i is integral over $\mathbf{A}$ (Kronecker's Theorem,
Exercise 381). Denoting by d_i the degree of an integral dependence relation
of b_i over $\mathbf{A}$, one can take $k = \sum_{i=1}^{r}(d_i - 1)$.

(4) For $b \in \mathbf{B}$, we have $(ab)^m \mathbf{B} \subseteq \mathbf{A}$. Thus, $(ab)^n \in \mathbf{A}_1$ for any $n \geq m$. Using (1),
we infer that $ab \in \mathbf{A}_1$.

(5) This follows from (4) and (3).

(7) Let $\mathbf{C}$ be the seminormal closure of $\tilde{\mathbf{A}}$ in $\tilde{\mathbf{B}}$. Write $\mathbf{C} = \mathbf{A}_2/J$, with $J \subseteq \mathbf{A}_2$,
as a subring of $\mathbf{B}/J$. It is clear that $\mathbf{A}_1 \subseteq \mathbf{A}_2$. Let $a \in \mathbf{A}_2$ and assume first
that $\bar{a}^2, \bar{a}^3 \in \tilde{\mathbf{A}}$. Then $a^2, a^3 \in \mathbf{A}_1$, and so $a \in \mathbf{A}_1$. Reasoning inductively, we
replace $\mathbf{A}$ by $\mathbf{A}[a]$. Since any element in $\mathbf{C}$ can be reached in a finite number
of steps, we obtain that $\mathbf{A}_2 = \mathbf{A}_1$.

(8) The concrete consequence of (7) for the computation of $\mathbf{A}_1$ is that, whenever
we find an $a \in \mathbf{B}$ such that $a^\ell f_1 \in \mathbf{A}[X]$ for some integer ℓ, we are allowed to
replace $\mathbf{A}$ and $\mathbf{B}$ by $\tilde{\mathbf{A}}$ and $\tilde{\mathbf{B}}$. Indeed, it is clear that if further computations
show that the seminormal closure of $\tilde{\mathbf{A}}$ in $\tilde{\mathbf{B}}$ is equal to $\tilde{\mathbf{B}}$, (7) says that $\mathbf{A}_1 = \mathbf{B}$.
In short, "we are allowed to continue the computation modulo J".

Exercise 383: Writing $x \wedge a \wedge b = x \wedge a$ and $a = a \wedge (x \vee b)$, we get

$$a = (a \wedge x) \vee (a \wedge b) = (x \wedge a \wedge b) \vee (a \wedge b) = a \wedge b.$$

Exercise 384: Let $\mathbf{R}$ be a Bezout domain. We want to prove that for any $a, b_1, b_2 \in \mathbf{R}$ there exist $x_1, x_2 \in \mathbf{R}$ such that $a \in \langle b_1 + ax_1, b_2 + ax_2 \rangle$, or, equivalently, $\langle a, b_1, b_2 \rangle = \langle b_1 + ax_1, b_2 + ax_2 \rangle$.

Denoting by $d = \gcd(b_1, b_2)$, we have $b_1 = dc_1$, $b_2 = dc_2$, with $c_1, c_2 \in \mathbf{R}$ and $\gcd(c_1, c_2) = 1$. As $\mathbf{R}$ is a Bezout domain, we can find $y_1, y_2 \in \mathbf{R}$ such that $y_1 c_1 + y_2 c_2 = 1$. Thus,

$$a = a + c_1 b_2 - c_2 b_1 = a y_1 c_1 + a y_2 c_2 + c_1 b_2 - c_2 b_1 = c_1 (a y_1 + b_2) + c_2 (a y_2 - b_1),$$

and one can take $x_1 = -y_2$ and $x_2 = y_1$.

Exercise 385: Write $(f_1, \ldots, f_k) = \sum_{i=1}^{r} X^{\alpha_i} (a_{i,1} \ldots, a_{i,k})$, where the α_i's are pairwise different elements in $\mathbb{N}^n$, $(a_{i,1} \ldots, a_{i,k}) \in \mathbf{R}^k$, and $r \geq 1$. Clearly, we have

$$S \cap \mathbf{R}^k = \cap_{i=1}^{r} \mathrm{Syz}_{\mathbf{R}}(a_{i,1} \ldots, a_{i,k}).$$

So, if $\mathbf{R}$ is coherent, we have a finite generating set for $S \cap \mathbf{R}^k$.

Exercise 386: For a ring $\mathbf{R}$, we will denote its subset of zero-divisors by $Z(\mathbf{R})$ (including zero).

(1) (a) As $\mathbf{R}$ is local, we have:

$$x \in Z(\mathbf{R}) \Rightarrow 1 + x \in \mathbf{R}^{\times}, \text{ and thus,}$$

$$\sharp Z(\mathbf{R}) \leq \sharp U(\mathbf{R}), \text{ and}$$

$$P_{(\mathbf{R})} \leq \frac{1}{2}.$$

(b) Note that If $\mathbf{R}$ and $\mathbf{T}$ are two finite rings, then:

$$\sharp (\mathbf{R} \times \mathbf{T})^{\times} = \sharp \mathbf{R}^{\times} \cdot \sharp \mathbf{T}^{\times},$$

$$\sharp Z(\mathbf{R} \times \mathbf{T}) = \sharp Z(\mathbf{R}) \cdot \sharp(\mathbf{T}) + \sharp(\mathbf{R}) \cdot \sharp Z(\mathbf{T}) - \sharp Z(\mathbf{R}) \cdot \sharp Z(\mathbf{T}).$$

In particular, if $\mathbf{R}$ is a finite local ring with n elements, denoting $\sharp Z(\mathbf{R}) = k$ (necessarily, $2k \leq n$), we have:

$$\sharp (\mathbf{R} \times \mathbf{R})^{\times} = (n-k)^2, \ \sharp Z(\mathbf{R} \times \mathbf{R}) = k(2n-k), \text{ and thus,}$$

$$\frac{\sharp Z(\mathbf{R} \times \mathbf{R})}{\sharp(\mathbf{R} \times \mathbf{R})} = \frac{k(2n-k)}{n^2} = P_{(\mathbf{R} \times \mathbf{R})} = 2\frac{k}{n} - \frac{k^2}{n^2} = 2P_{(\mathbf{R})} - P_{(\mathbf{R})}^2.$$

(2) From the Euler indicator formula

$$\Phi(p^\alpha) = \sharp(\mathbb{Z}/p^\alpha\mathbb{Z})^\times = p^\alpha - p^{\alpha-1},$$

we infer that $\sharp Z(\mathbb{Z}/p^\alpha\mathbb{Z}) = p^{\alpha-1}$, and hence, $P_{(\mathbb{Z}/p^\alpha\mathbb{Z})} = \frac{1}{p}$. In addition,

$$P_{((\mathbb{Z}/p^\alpha\mathbb{Z})\times(\mathbb{Z}/q^\beta\mathbb{Z}))} = \frac{p^{\alpha-1}q^\beta + p^\alpha q^{\beta-1} - p^{\alpha-1}q^\beta - 1}{p^\alpha q^\beta} = \frac{1}{p} + \frac{1}{q} - \frac{1}{pq}.$$

(2) It is clear that $P_{(\mathbb{Z}/p^\alpha\mathbb{Z})}$ is maximal when $p = 2$. Taking $p = q = 2$, we have $P_{((\mathbb{Z}/2^\alpha\mathbb{Z})\times(\mathbb{Z}/2^\beta\mathbb{Z}))} = \frac{3}{4}$. For $p, q \geq 3$, we have $P_{((\mathbb{Z}/p^\alpha\mathbb{Z})\times(\mathbb{Z}/q^\beta\mathbb{Z}))} \leq \frac{1}{p} + \frac{1}{q} \leq \frac{2}{3} < \frac{3}{4}$. For $p = 2$ and $q \geq 3$, we have $P_{((\mathbb{Z}/p^\alpha\mathbb{Z})\times(\mathbb{Z}/q^\beta\mathbb{Z}))} = \frac{1}{2} + \frac{1}{q} - \frac{1}{2q} = \frac{1}{2} + \frac{1}{2q} \leq \frac{2}{3} < \frac{3}{4}$. We conclude that

$$\sup_{\{p,q \text{ prime numbers}; \; \alpha,\beta \geq 1\}} P_{((\mathbb{Z}/p^\alpha\mathbb{Z})\times(\mathbb{Z}/q^\beta\mathbb{Z}))} = \frac{3}{4}.$$

It is in fact a maximum reached only when $p = q = 2$.

Exercise 387:

(1) Just pass to the ring $\mathbf{R}/\langle b \rangle$ and use Lemma 260.

(2) First, we have $b, 1 + \alpha a \in [b : a^\infty]$. Second, letting $x \in [b : a^\infty]$, there exist $m \in \mathbb{N}$ and $\theta \in \mathbf{R}$ such that $xa^m = \theta b$. As the multiplicative subsets $a^\mathbb{N}$ and $1 + a\mathbf{R}$ are comaximal, there exist $u, v \in \mathbf{R}$ such that $ua^m + v(1 + \alpha a) = 1$, and hence

$$x = xua^m + xv(1 + \alpha a) = \theta bu + xv(1 + \alpha a) \in \langle b, 1 + \alpha a \rangle.$$

(3) This follows from (1) and (2).

(4)

(a) First note that if an ideal of $\mathbf{R}[X]$ contains $f = 1 + aX$ and $g = bX^\ell$, with $a, b \in \mathbf{R}$ and $\ell \in \mathbb{N}^*$, then it contains $bX^{\ell-1}f - ag = bX^{\ell-1}$. So, by iteration,

$$I = \langle 1 + aX, b_1 X^{\ell_1}, \dots, b_n X^{\ell_n} \rangle = \langle 1 + aX, b_1, \dots, b_n \rangle.$$

One can obtain the desired result by passing to the ring $\mathbf{R}/\langle b_1, \dots, b_n \rangle$ and using Lemma 260.

(b) We induct on n. For $n = 1$, this is (2). If one of the b_i's is zero, then the result follows by the induction hypothesis. Now, suppose that all the b_i's are nonzero. As $\mathbf{R}$ is a domain of Krull dimension ≤ 1, for $1 \leq i \leq n$, there exists $m_i \in \mathbb{N}$ and $x_i, y_i \in \mathbf{R}$ such that $a^{m_i}(1 - ax_i) + b_i y_i = 0$. We will prove that $[\langle b_1, \ldots, b_n \rangle : a^\infty] = \langle b_1, \ldots, b_n, 1 - ax_i \rangle$ for any $1 \leq i \leq n$. The inclusion $\langle b_1, \ldots, b_n, 1 - ax_i \rangle \subseteq [\langle b_1, \ldots, b_n \rangle : a^\infty]$ is clear. For the converse, letting $u \in [\langle b_1, \ldots, b_n \rangle : a^\infty]$, there exists $m \in \mathbb{N}$ such that $ua^m = \theta_1 b_1 + \cdots + \theta_n b_n$ with $\theta_j \in \mathbf{R}$. As the multiplicative subsets $a^{\mathbb{N}}$ and $1 + a\mathbf{R}$ are comaximal, there exist $c, d \in \mathbf{R}$ such that $ca^m + d(1 - ax_i) = 1$, and hence

$$u = uca^m + ud(1 - ax_i) = c\theta_1 b_1 + \cdots + c\theta_n b_n + ud(1 - ax_i) \in \langle b_1, \ldots, b_n, 1 - ax_i \rangle.$$

(c) Let $\mathbf{R}$ be a strongly discrete domain with Krull dimension ≤ 1, and consider an ideal $I = \langle 1 + aX, f_1, \ldots, f_n \rangle$ of $\mathbf{R}[X]$ with $a \in \mathbf{R}$. For $1 \leq i \leq n$, performing a division according to the ascending powers of f_i by $1 + aX$, we find $g_i \in \mathbf{R}[X]$, $b_i \in \mathbf{R}$, and $\ell_i \in \mathbb{N}$ such that $f_i = (1 + aX)g_i + b_i X^{\ell_i}$, and, thus,

$$I = \langle 1 + aX, b_1 X^{\ell_1}, \ldots, b_n X^{\ell_n} \rangle.$$

The desired conclusion follows from (a) and (b).

Exercise 388:

(1) It is clear that the ring $\mathbf{A}$ shares the ideal $\mathfrak{m} := t\mathbf{K}[t]$ with $\mathbf{K}[t]$.

The fact that $\mathbf{A}$ has Krull dimension 1 follows "classically" from the fact that the prime spectrum of $\mathbf{A}$ is $\{(f\mathbf{K}[t]) \cap \mathbf{A} \mid f \in \mathbf{K}[t] \text{ and } f \text{ irreducible}\}$. Let us now give a constructive proof.

Let $f, g \in \mathbf{A}$. We want to find $h, h' \in \mathbf{A}$ such that $f^n(1 - hf) + gh' = 0$ for some $n \in \mathbb{N}$. Two cases may arise:

Case 1: $f(0) \neq 0$. As $\mathbf{K}[t]$ has Krull dimension one (see Theorem 88), there exists $n \in \mathbb{N}$ such that $f^n(1 - \varphi f) + (tg)\psi = 0$, with $\varphi, \psi \in \mathbf{K}[t]$. Putting $t = 0$, we obtain $\varphi(0)f(0) = 1$, and, thus, $\varphi(0) \in \mathbb{Q}$ and $\varphi \in \mathbf{A}$. It suffices to take $h = \varphi$ and $h' = t\psi$.

Case 2: $f(0) = 0$. Set $f = t^p f_1$, $g = t^q g_1$, with $p > 0$, $q \geq 0$, $f_1(0) \neq 0$, and $g_1(0) \neq 0$. As $\mathbf{K}[t]$ has Krull dimension one, there exists $n \in \mathbb{N}$ such that $(tf)^n(1 - \varphi tf) + g_1 \psi = 0$, with $\varphi, \psi \in \mathbf{K}[t]$. We have $t^{n+np} f_1^n(1 - \varphi tf) = -g_1 \psi$. As t^{n+np} does not divide g_1, it divides ψ, that is, there exists $\eta \in \mathbf{K}[t]$ such that $\psi = t^{n+np}\eta$. It follows that $f_1^n(1 - \varphi tf) = -g_1 \eta$. Let N be a positive integer greater than n and q. Multiplying the equality $f_1^n(1 - \varphi tf) = -g_1 \eta$ by $t^{pN} f_1^{N-n}$, we obtain $t^{pN} f_1^N(1 - t\varphi f) = -(t^q g_1)(t^{pN-q} f_1^{N-n}\eta)$, or also, $f^N(1 - t\varphi f) + g(t^{pN-q} f_1^{N-n}\eta) = 0$. It suffices to take $h = t\varphi$ and $h' = t^{pN-q} f_1^{N-n}\eta$.

(2) Take $\alpha \in \mathbf{K}$, and suppose that $\alpha t \in t h_1 \mathbf{A} + \cdots + t h_r \mathbf{A} \subseteq \mathfrak{m}$, with $h_1, \ldots, h_r \in \mathbf{K}[t]$. Since the only terms of degree 1 at t in $t h_1 \mathbf{A} + \cdots + t h_r \mathbf{A}$ are of the form $(h_1(0)q_1 + \cdots + h_r(0)q_r)t$ with $q_1, \ldots, q_r \in \mathbb{Q}$, one gets $\alpha = h_1(0)q_1 + \cdots + h_r(0)q_r \in \mathbb{Q}$.

$h_1(0) + \cdots + \mathbb{Q} \cdot h_r(0)$. As $\mathbf{K}$ is not finitely-generated as $\mathbb{Q}$-vector space (because u is transcendental over $\mathbb{Q}$), we infer that $\mathfrak{m}$ is not finitely-generated as an ideal of $\mathbf{A}$.

(3) First, $t^2\mathbf{K}[t] = t(t\mathbf{K}[t]) = \sqrt{2}t(t\mathbf{K}[t]) \subseteq \langle t \rangle \cap \langle \sqrt{2}t \rangle$. Second, if $x \in \langle t \rangle \cap \langle \sqrt{2}t \rangle$, then

$$x = t(q_1 + tf(t)) = \sqrt{2}t(q_2 + tg(t)),$$

where $q_1, q_2 \in \mathbb{Q}$ and $f, g \in \mathbf{K}[t]$. This implies that $q_1 - \sqrt{2}q_2 = t(f(t) - g(t))$, and thus $q_1 = q_2 = 0$ (as $\sqrt{2} \notin \mathbb{Q}$), and $x = t^2 f(t) \in t^2\mathbf{K}[t]$.

(4) By (3), $\langle t \rangle \cap \langle \sqrt{2}t \rangle$ is isomorphic as $\mathbf{A}$-module to $\mathfrak{m}$, and, thus, it is not finitely-generated by virtue of (2).

(5) If $x = ta = \sqrt{2}tb \in \langle t \rangle \cap \langle \sqrt{2}t \rangle$, where $a, b \in \mathbf{A}$, then $x = xX + x(1-X) = ta(1-X) + \sqrt{2}tbX \in I \cap \mathbf{A}$. Conversely, if $y = \sqrt{2}tXU(X) + t(1-X)V(X) \in I \cap \mathbf{A}$, for some $U, V \in \mathbf{A}[X]$, then by successively taking $X = 0$ and $X = 1$, one gets $y \in \langle t \rangle \cap \langle \sqrt{2}t \rangle$.

(6) This follows from Proposition 219 by taking $n = 0$.

Exercise 389:

(1) Suppose that $H \neq \mathrm{Sat}(H)$ and consider an $f \in \mathrm{Sat}(H) \setminus H$ of minimal multidegree (this is possible by Dickson's lemma (Theorem 209)) among the elements of $\mathrm{Sat}(H) \setminus H$. Dividing f by the Gröbner basis G and multiplying by a suitable power of d, we get $d^n f = u_1 h_1 + \cdots + u_m h_m$ with $n \in \mathbb{N}^*$ and $u_i \in \mathbf{R}[X_1, \ldots, X_n]$. We can suppose that when dividing f by G, the first division step consists in dividing $\mathrm{LM}(f)$ by $\mathrm{LM}(g_1)$ so that $\mathrm{LT}(d^n f) = \mathrm{LT}(u_1 g_1)$ and $\mathrm{mdeg}(d^n f - \mathrm{LT}(u_1 g_1)) < \mathrm{mdeg}(d^n f) = \mathrm{mdeg}(f)$.

We have $\mathrm{LT}(u_1) = \frac{\mathrm{LT}(d^n f)}{\mathrm{LT}(g_1)} = \frac{d^n}{d_1} \frac{\mathrm{LT}(f)}{\mathrm{LM}(g_1)}$, and

$$d^n f - \mathrm{LT}(u_1)g_1 = d^{n-1}\left(df - \frac{d}{d_1} \frac{\mathrm{LT}(f)}{\mathrm{LM}(g_1)} g_1\right) \in H.$$

It follows that the polynomial $g := df - \frac{d}{d_1} \frac{\mathrm{LT}(f)}{\mathrm{LM}(g_1)} g_1$ is in $\mathrm{Sat}(H)$. Since $\mathrm{mdeg}(g) < \mathrm{mdeg}(f)$, by minimality of $\mathrm{mdeg}(f)$, g is necessarily in H and hence $f \in (H : d) \setminus H$.

(2) This is an immediate consequence of (1).

Exercise 390: The equivalence (i) $\Leftrightarrow$ (ii) is given by Proposition 329.
(iii) $\Rightarrow$ (ii) and (iii) $\Rightarrow$ (ii). Nothing to prove.
(ii) $\Rightarrow$ (iii). If p is a prime number which does not divide δ then we have $\mathrm{rk}_p A = \mathrm{rk}_0 A$.
(iv) $\Rightarrow$ (iii). It suffices to consider a prime number which does not divide δ.

Notation List

$\mathrm{Im}(\varphi)$	The image of φ
$\mathrm{Ker}(\varphi)$	The kernel of φ
$\mathrm{Coker}(\varphi)$	The cokernel of φ
$\mathrm{Proj}\,\mathbf{R}$	The isomorphism class of finitely-generated projective $\mathbf{R}$-modules
$\mathrm{K}_0(\mathbf{R})$	The Groethendieck group of $\mathbf{R}$
$\mathrm{Syz}(a_1,\ldots,a_n)$	The syzygy module of $(a_1,\ldots,a_n)$
$\mathrm{GL}_s(\mathbf{R})$	The group of invertible matrices of size $s \times s$ with entries in $\mathbf{R}$
$\mathrm{SL}_s(\mathbf{R})$	The subgroup of $\mathrm{GL}_s(\mathbf{R})$ formed by matrices of determinant 1
$\mathrm{M}_{n,m}(\mathbf{R})$	The set of matrices of size $n \times m$ with entries in $\mathbf{R}$
$\mathrm{M}_n(\mathbf{R})$	The set of matrices of size $n \times n$ with entries in $\mathbf{R}$
$\mathrm{M}(\mathbf{R})$	The set $\cup_{n\geq 1}\mathrm{M}_n(\mathbf{R})$
$\mathrm{Idem}(\mathbf{R})$	The set of idempotent matrices in $\mathrm{M}(\mathbf{R})$
$\mathrm{GL}(\mathbf{R})$	The group $\cup_{n\geq 1}\mathrm{GL}_n(\mathbf{R})$
$\mathrm{E}_s(\mathbf{R})$	The subgroup of $\mathrm{SL}_s(\mathbf{R})$ generated by elementary matrices
$\mathscr{D}_k(G)$	The determinantal ideal of order k of G
$\mathscr{F}_n(T)$	The nth Fitting ideal of T
$\mathbf{R}^\times$	The group of units of $\mathbf{R}$
$\mathrm{I}_{r,m}$	The standard projection matrix $\begin{pmatrix} \mathrm{I}_r & 0_{r,m-r} \\ 0_{m-r,r} & 0_{m-r,m-r} \end{pmatrix}$
$\mathrm{Um}_n(\mathbf{R})$	The set of unimodular rows (or vectors) of length n with entries in $\mathbf{R}$
$\mathrm{Rad}(\mathbf{R})$	The set of all $x \in \mathbf{R}$ such that $1 + x\mathbf{R} \subseteq \mathbf{R}^\times$
$a^{\mathbb{N}}$	The monoid $\{a^n;\ n \in \mathbb{N}\}$
$\mathscr{M}(a)$	The monoid $a^{\mathbb{N}}$
$S^{-1}\mathbf{R}$ or $\mathbf{R}_S$	The localization of $\mathbf{R}$ at S

© Springer International Publishing Switzerland 2015
I. Yengui, *Constructive Commutative Algebra*, Lecture Notes in Mathematics 2138,
DOI 10.1007/978-3-319-19494-3

$\mathrm{Spec}(\mathbf{R})$	The set of prime ideals of $\mathbf{R}$
$\mathcal{M}(U)$	The monoid generated by U
$\mathcal{I}_{\mathbf{R}}(I)$ or $\mathcal{I}(I)$	The ideal generated by I
$\mathcal{S}(I;U)$	The monoid $\mathcal{M}(U) + \mathcal{I}(I)$
$\mathcal{M}(u_1,\ldots,u_\ell)$	The monoid $\mathcal{M}(\{u_1,\ldots,u_\ell\})$
$\mathcal{I}(a_1,\ldots,a_k)$	The ideal $\mathcal{I}(\{a_1,\ldots,a_k\})$
$\mathcal{S}(a_1,\ldots,a_k;u_1,\ldots,u_\ell)$	The monoid $\mathcal{M}(u_1,\ldots,u_\ell) + \mathcal{I}(a_1,\ldots,a_k)$
$\mathbf{R}_a$	The localization of the ring $\mathbf{R}$ at the monoid $a^{\mathbb{N}}$
M_a	The localization of the module M at the monoid $a^{\mathbb{N}}$
$\mathrm{Res}(f,g)$	The resultant of f and g
$\mathrm{Res}_X(f,g)$	The resultant of f and g with respect to X
$\gcd(f,g)$	The greatest common divisor of f and g
$\mathbf{R}\langle X \rangle$	The localization of $\mathbf{R}[X]$ at the monoid of monic polynomials
$\mathbf{R}(X)$	The localization of $\mathbf{R}[X]$ at the monoid of primitive polynomials
$\mathbf{R}_{\mathfrak{p}}$	The localization of $\mathbf{R}$ at the monoid $\mathbf{R} \setminus \mathfrak{p}$, where $\mathfrak{p}$ is a prime ideal
$\mathbf{R}_{\mathrm{red}}$	The reduced ring associated to the ring $\mathbf{R}$
$\mathrm{LC}(f)$	The leading coefficient of the polynomial f
$\mathrm{LM}(f)$	The leading monomial of the polynomial f
$\mathrm{LT}(f)$	The leading term of the polynomial f
$\mathrm{LT}(I)$	The ideal $\langle \mathrm{LT}(f) : f \in I \rangle$
$\mathrm{mdeg}(f)$	The multidegree of the polynomial f
$\mathrm{tdeg}(f)$	The total degree of the polynomial f
$E_{i,j}(a)$	The matrix with 1s on the diagonal, a on position (i,j) and 0s elsewhere
$\bar{S}$	The saturation of the monoid S
$\mathrm{D}_{\mathbf{R}}(\mathfrak{a})$	The radical $\sqrt{\mathfrak{a}}$ of the ideal $\mathfrak{a}$
$\mathrm{D}_{\mathbf{R}}(x_1,\ldots,x_n)$	The radical ideal $\mathrm{D}_{\mathbf{R}}(\langle x_1,\ldots,x_n \rangle)$
$\mathrm{Zar}\,\mathbf{R}$	The Zariski lattice of the ring $\mathbf{R}$, i.e., the set $\{\mathrm{D}_{\mathbf{R}}(x_1,\ldots,x_n) \mid n \in \mathbb{N} \ \& \ x_1,\ldots,x_n \in \mathbf{R}\}$
$I : J$	The conductor of J in I, i.e., $\{x \in \mathbf{R} \mid xJ \subseteq I\}$, where I, J are ideals of the ring $\mathbf{R}$
$I : a$	The conductor $I : \langle a \rangle$ of $\langle a \rangle$ in I
$\mathrm{Ann}(x)$	The annihilator of x, i.e., $\langle 0 \rangle : \langle x \rangle$
$\mathrm{K}_{\mathbf{R}}(x)$	The Krull boundary ideal of x, i.e., $\langle x \rangle + (\mathrm{D}_{\mathbf{R}}(0) : x)$
$\mathbf{R}^{\{x\}}$	The upper Krull boundary of x in $\mathbf{R}$, i.e., $\mathbf{R}/\mathrm{K}_{\mathbf{R}}(x)$
$S_{\{x\}}$	The Krull boundary monoid of x., i.e., $x^{\mathbb{N}}(1 + x\mathbf{R})$
$\mathbf{R}_{\{x\}}$	The lower Krull boundary of x in $\mathbf{R}$, i.e., $\mathbf{R}_{S_{\{x\}}}$
$\mathrm{Kdim}\,\mathbf{R}$ or $\dim\,\mathbf{R}$	The Krull dimension of $\mathbf{R}$

$I_{\mathbf{R}}(a,b) =$
$\cup_{n \in \mathbb{N}}(a^n b^{n+1}\mathbf{R} +$
$a^{n+1}\mathbf{R} : a^n b^n \mathbf{R})$

$H \lhd G$	H is a normal subgroup of G
$H \ntriangleleft G$	H is not a normal subgroup of G
$\mathrm{diag}(u_1,\ldots,u_n)$	The matrix in $M_n(\mathbf{R})$ with u_i on position (i,i) for $1 \le i \le n$, and 0s elsewhere
$e_{i,j}$	The matrix with 1 on position (i,j) and 0s elsewhere
$u \sim_G u'$	There exists $A \in G$ such that $uA = u'$
$\mathrm{GL}_n(\mathbf{A}, J)$	The normal subgroup of $\mathrm{GL}_n(\mathbf{A})$ consisting of matrices M which are $\equiv I_n \bmod M_n(J)$
$\mathrm{E}_n(\mathbf{A}, J)$	The normal subgroup of $\mathrm{E}_n(\mathbf{A})$ generated by the elementary matrices $\{E_{i,j}(a);\ a \in J,\ 1 \le i \ne j \le n\}$
$\mathrm{SL}_n(\mathbf{A}, J)$	The group $\mathrm{SL}_n(\mathbf{A}) \cap \mathrm{GL}_n(\mathbf{A}, J)$
$\mathbb{M}_n^m$	The set of monomials in $\mathbf{R}[X_1,\ldots,X_n]^m$
$\mathbb{M}_n$	The set of monomials in $\mathbf{R}[X_1,\ldots,X_n]$
$Y^{\uparrow}$	The set $\{Z \in E \mid Z \ge Y\}$
$\mathscr{M}_E^+(Y_1,\ldots,Y_m)$	The final subset of E of finite type $\cup_{i=1}^m Y_i^{\uparrow}$
$\mathscr{F}(E)$	The set of final subsets of finite type of E, including the empty subset considered as generated by the empty family
$\mathscr{M}_d$	The set $\mathscr{F}(\mathbb{N}^d) \setminus \{\emptyset\}$
$\bar{f}^F$	A remainder of f on division by F
$\mathrm{LC}_n(I)$	The ideal generated by the leading coefficients of the elements of I of degree n
$\mathrm{LC}_\infty(I)$	The ideal $\cup_{n \in \mathbb{N}}\mathrm{LC}_n(I)$
$S(f,g)$	The S-polynomial of f and g
$S(f,f)$	The auto-S-polynomial of f
$\mathbb{Z}_{p\mathbb{Z}}$	The localization $\{\frac{a}{b} \in \mathbb{Q} \mid a \in \mathbb{Z} \text{ and } b \in \mathbb{Z} \setminus p\mathbb{Z}\}$ of $\mathbb{Z}$
$\mathbf{R}\langle X_1,\ldots,X_n\rangle$	The ring $(\mathbf{R}\langle X_1,\ldots,X_{n-1}\rangle)\langle X_n\rangle$
$\mathrm{Ann}(a^\infty)$	The ideal $\cup_{n \in \mathbb{N}}\mathrm{Ann}(a^n)$
$\mathbb{F}_2$	The field with two elements
$\mathbf{R}_{a_1.a_2.\ldots.a_n}$	The ring $\mathscr{M}(a_1,\ldots,a_n)^{-1}\mathbf{R} = \mathbf{R}[\frac{1}{a_1\cdots a_n}]$
$\mathrm{Sat}(S)$	The saturation of S
$(I : a^\infty)$	The ideal $\{x \in \mathbf{A} \mid \exists n \in \mathbb{N} \mid xa^n \in I\}$
$\mathrm{index}(u)$	The position of the last primitive component of u
$\mathrm{PrimMon}(u)$	The last monomial of u which has an invertible coefficient
$\mathrm{PrimCoeff}(u)$	The coefficient of $\mathrm{PrimMon}(u)$
$\mathrm{Prim}(u)$	The primitive version of u
$\mathscr{I}(u)$	The height of u, i.e., the couple $(\mathrm{index}(u), \mathrm{mdeg}(\mathrm{PrimMon}(u)))$

Echel(S)	The list S put in an echelon form
PrimRed($s; S$)	The reduction u of s modulo S so that $[S, u]$ becomes in an echelon form
HS$_I(t)$	The Hilbert series of I
$\delta(S)$	The (saturation) defect of the list S
$\delta_S(t)$	The (saturation) defect series of the list S
tdeg(f)	The total degree of f

Bibliography

[1] Abedelfatah, A.: On stably free modules over Laurent polynomial rings. Proc. Am. Math. Soc. **139**, 4199–4206 (2011)

[2] Abedelfatah, A.: On the action of the elementary group on the unimodular rows. J. Algebra **368**, 300–304 (2012)

[3] Adams, W.W., Loustaunau, P.: An Introduction to Gröbner Bases. Graduate Studies in Mathematics, vol. 3. American Mathematical Society, Providence (1994)

[4] Amidou, M., Yengui, I.: An algorithm for unimodular completion over Laurent polynomial rings. Linear Algebra Appl. **429**, 1687–1698 (2008)

[5] Arnold, E.A.: Modular algorithms for computing Gröbner bases. J. Symb. Comput. **35**, 403–419 (2003)

[6] Aschenbrenner, M.: Ideal membership in polynomial rings over the integers. J. Am. Math. Soc. **17**, 407–441 (2004)

[7] Ayoub, C.: On constructing bases for ideals in polynomial rings over the integers. J. Number Theory **17**(2), 204–225 (1983)

[8] Barhoumi, S.: Seminormality and polynomial ring. J. Algebra **322**, 1974–1978 (2009)

[9] Barhoumi, S., Lombardi, H.: An algorithm for the Traverso-Swan theorem over seminormal rings. J. Algebra **320**, 1531–1542 (2008)

[10] Barhoumi, S., Yengui, I.: On a localization of the Laurent polynomial ring. JP. J. Algebra Number Theory Appl. **5**(3), 591–602 (2005)

[11] Barhoumi, S., Lombardi, H., Yengui, I.: Projective modules over polynomial rings: a constructive approach. Math. Nach **282**, 792–799 (2009)

[12] Bass, H.: Algebraic K-Theory. W.A. Benjamin Inc., New York/Amsterdam (1968)

[13] Bass, H.: Libération des modules projectifs sur certains anneaux de polynômes, Sém. Bourbaki 1973/74, exp. 448. Lecture Notes in Mathematics, vol. 431, pp. 228–254. Springer, Berlin/New York (1975)

[14] Basu, S., Pollack, R., Roy, M.-F.: Algorithms in Real Algebraic Geometry. Algorithms and Computation in Mathematics, 2nd edn. Springer, Berlin (2006)

[15] Bayer, D.: The division algorithm and the Hilbert scheme. Ph.D. dissertation, Harvard University (1982)

© Springer International Publishing Switzerland 2015
I. Yengui, *Constructive Commutative Algebra*, Lecture Notes in Mathematics 2138,
DOI 10.1007/978-3-319-19494-3

[16] Bernstein, D.: Fast ideal arithmetic via lazy localization. In: Cohen, H. (ed.) Algorithmic Number Theory. Proceeding of the Second International Symposium, ANTS-II, Talence, France, 18–23 May 1996. Lecture Notes in Computer Science, vol. 1122, pp. 27–34. Springer, Berlin (1996)

[17] Bernstein, D.: Factoring into coprimes in essentially linear time. J. Algorithms **54**, 1–30 (2005)

[18] Boileau, A., Joyal, A.: La Logique des Topos. J. Symb. Log. **46**, 6–16 (1981)

[19] Bourbaki, N.: Algèbre Commutative. Chapitres 5–6. Masson, Paris (1985)

[20] Brewer, J., Costa, D.: Projective modules over some non-Noetherian polynomial rings. J. Pure Appl. Algebra **13**, 157–163 (1978)

[21] Brickenstein, M., Dreyer, A., Greuel, G.-M., Wedler, M., Wienand, O.: New developments in the theory of Groebner bases and applications to formal verification. J. Pure Appl. Algebra **213**, 1612–1635 (2009)

[22] Buchberger, B.: Ein Algorithmus zum Auffinden der Basiselemente des Restklassenringes nach einem nulldimensionalen polynomideal. Ph.D. thesis, University of Innsbruck (1965)

[23] Buchberger, B.: A critical pair/completion algorithm for finitely-generated ideals in rings. In: Logic and Machines: Decision Problems and Complexity. Springer Lectures Notes in Computer Science, vol. 171, pp. 137–161. Springer, New York (1984)

[24] Buchmann, J., Lenstra, H.: Approximating rings of integers in number fields. J. Théor. Nombres Bordeaux **6**(2), 221–260 (1994)

[25] Byrne, E., Fitzpatrick, P.: Gröbner bases over Galois rings with an application to decoding alternant codes. J. Symb. Comput. **31**, 565–584 (2001)

[26] Cahen, P.-J.: Construction B,I,D et anneaux localement ou résiduellement de Jaffard. Archiv. Math. **54**, 125–141 (1990)

[27] Cahen, P.-J., Elkhayyari, Z., Kabbaj, S.: Krull and valuative dimension of the Serre conjecture ring $R\langle n\rangle$. In: Commutative Ring Theory. Lecture Notes in Pure and Applied Mathematics, vol. 185, pp. 173–180. Marcel Dekker, New York (1997)

[28] Cai, Y., Kapur, D.: An algorithm for computing a Gröbner basis of a polynomial ideal over a ring with zero divisors. Math. Comput. Sci. **2**, 601–634 (2009)

[29] Caniglia, L., Cortiñas, G., Danón, S., Heintz, J., Krick, T., Solernó, P.: Algorithmic aspects of Suslin's proof of Serre's conjecture. Comput. Complex. **3**, 31–55 (1993)

[30] Cohn, P.M.: On the structure of GL_n of a ring. Publ. Math. I.H.E.S. **30**, 5–54 (1966)

[31] Coquand, T.: Sur un théorème de Kronecker concernant les variétés algébriques. C. R. Acad. Sci. Paris, Ser. I **338**, 291–294 (2004)

[32] Coquand, T.: On seminormality. J. Algebra **305**, 577–584 (2006)

[33] Coquand, T.: A refinement of Forster's theorem. Preprint (2007)

[34] Coquand, T., Lombardi, H.: Hidden constructions in abstract algebra (3) Krull dimension of distributive lattices and commutative rings. In: Fontana, M., Kabbaj, S.-E., Wiegand, S. (eds.) Commutative Ring Theory and Applications. Lecture Notes in Pure and Applied Mathematics, vol. 131, pp. 477–499. Marcel Dekker, New York (2002)

[35] Coquand, T., Lombardi, H.: A short proof for the Krull dimension of a polynomial ring. Am. Math. Mon. **112**, 826–829 (2005)

[36] Coquand, T., Quitté, C.: Constructive finite free resolutions. Manuscripta Math. **137**, 331–345 (2011)

[37] Coquand, T., Ducos, L., Lombardi, H., Quitté, C.: L'idéal des coefficients du produit de deux polynômes. Rev. Math. Enseign. Supér. **113**, 25–39 (2003)

[38] Coquand, T., Lombardi, H., Quitté, C.: Generating nonnoetherian modules constructively. Manuscripta Math. **115**, 513–520 (2004)

[39] Coquand, T., Lombardi, H., Roy, M.-F.: An elementary characterisation of Krull dimension. In: Corsilla, L., Schuster, P. (eds.) From Sets and Types to Analysis and Topology: Towards Practicable Foundations for Constructive Mathematics. Oxford University Press, Oxford (2005)

[40] Coquand, T., Lombardi, H., Schuster, P.: A nilregular element property. Arch. Math. **85**, 49–54 (2005)

[41] Coquand, T., Lombardi, H., Schuster, P.: The projective spectrum as a distributive lattice. Cahiers de Topologie et Géométrie différentielle catégoriques **48**, 220–228 (2007)

[42] Coste, M., Lombardi, H., Roy, M.-F.: Dynamical method in algebra: effective Nullstellensätze. Ann. Pure Appl. Logic **111**, 203–256 (2001)

[43] Cox, D., Little, J., O'Shea, D.: Ideals, Varieties and Algorithms, 2nd edn. Springer, New York (1997)

[44] Decker, W., Greuel, G.-M., Pfister, G., Schönemann, H.: SINGULAR 4-0-2 – A Computer Algebra System for Polynomial Computations. http://www.singular.uni-kl.de (2015)

[45] Decker, W., Pfister, G.: A First Course in Computational Algebraic Geometry. AIMS Library Series. Cambridge University Press, Cambridge (2013)

[46] Della Dora, J., Dicrescenzo, C., Duval, D.: About a new method for computing in algebraic number fields. In: Caviness, B.F. (ed.) EUROCAL '85. Lecture Notes in Computer Science, vol. 204, pp. 289–290. Springer, Berlin (1985)

[47] Diaz-Toca, G.M., Lombardi, H.: Dynamic Galois theory. J. Symb. Comput. **45**, 1316–1329 (2010)

[48] Ducos, L., Monceur, S., Yengui, I.: Computing the V-saturation of finitely-generated submodules of $V[X]^m$ where V is a valuation domain. J. Symb. Comput. **72**, 196–205 (2016)

[49] Ducos, L., Quitté, C., Lombardi, H., Salou, M.: Théorie algorithmique des anneaux arithmétiques, de Prüfer et de Dedekind. J. Algebra **281**, 604–650 (2004)

[50] Ducos, L., Valibouze, A., Yengui, I.: Computing syzygies over $V[X_1, \ldots, X_k]$, V a valuation domain. J. Algebra **425**, 133–145 (2015)

[51] Duval, D., Reynaud, J.-C.: Sketches and computation (part II) dynamic evaluation and applications. Math. Struct. Comput. Sci. **4**, 239–271 (1994)
(see http://www.Imc.imag.fr/Imc-cf/Dominique.Duval/evdyn.html)

[52] Edwards, H.: Divisor Theory. Birkhäuser, Boston (1989)

[53] Eisenbud, D.: Commutative Algebra with a View Toward Algebraic Geometry. Springer, New York (1995)

[54] Ebert, G.L.: Some comments on the modular approach to Gröbner-bases. ACM SIGSAM Bull. **17**, 28–32 (1983)

[55] Ellouz, A., Lombardi, H., Yengui, I.: A constructive comparison of the rings $\mathbf{R}(X)$ and $\mathbf{R}\langle X \rangle$ and application to the Lequain-Simis induction theorem. J. Algebra **320**, 521–533 (2008)

[56] Español, L.: Dimensión en álgebra constructiva. Doctoral thesis, Universidad de Zaragoza, Zaragoza (1978)

[57] Español, L.: Constructive Krull dimension of lattices. Rev. Acad. Cienc. Zaragoza **37**, 5–9 (1982)

[58] Español, L.: Le spectre d'un anneau dans l'algèbre constructive et applications à la dimension. Cahiers de topologie et géométrie différentielle catégorique **24**, 133–144 (1983)

[59] Español, L.: Dimension of Boolean valued lattices and rings. J. Pure Appl. Algebra **42**, 223–236 (1986)

[60] Español, L.: Finite chain calculus in distributive lattices and elementary Krull dimension. In: Lamban, L., Romero, A., Rubio, J. (eds.) Contribuciones científicas en honor de Mirian Andres Gomez Servicio de Publicaciones. Universidad de La Rioja, Logroño (2010)

[61] Fabiańska, A.: A Maple QuillenSuslin package. http://wwwb.math.rwth-aachen.de/QuillenSuslin/ (2007)

[62] Fabiańska, A., Quadrat, A.: Applications of the Quillen-Suslin theorem to the multidimensional systems theory. INRIA Report 6126 (2007), Published in Gröbner Bases in Control Theory and Signal Processing. In: Park, H., Regensburger, G. (eds.) Radon Series on Computation and Applied Mathematics, vol. 3, pp. 23–106. de Gruyter, Berlin (2007)

[63] Faugère, J.-C.: A new efficient algorithm for computing Gröbner bases without reduction to zero (F_5). In: Proceedings of the International Symposium on Symbolic and Algebraic Computation, ISSAC (2002)

[64] Fitchas, N., Galligo, A.: Nullstellensatz effectif et conjecture de Serre (Théorème de Quillen-Suslin) pour le calcul formel. Math. Nachr. **149**, 231–253 (1990)

[65] Gallo, S., Mishra, B.: A solution to Kronecker's problem. Appl. Algebra Eng. Commun. Comput. **5**, 343–370 (1994)

[66] Gilmer, R.: Multiplicative Ideal Theory. Queens Paper in Pure and Applied Mathematics, vol. 90. Marcel Dekker, New York (1992)

[67] Glaz, S.: On the weak dimension of coherent group rings. Commun. Algebra **15**, 1841–1858 (1987)

[68] Glaz, S.: Commutative Coherent Rings. Lectures Notes in Mathematics, vol. 1371, 2nd edn. Springer, Berlin/Heidelberg/New York (1990)

[69] Glaz, S.: Finite conductor properties of $\mathbf{R}(X)$ and $\mathbf{R}\langle X \rangle$. In: Ideal Theoretic Methods in Commutative Algebra (Columbia, MO, 1999). Lecture Notes in Pure and Applied Mathematics, vol. 220, pp. 231–249. Marcel Dekker, New York (2001)

[70] Glaz, S., Vasconcelos, W.V.: Flat ideals III. Commun. Algebra **12**, 199–227 (1984)

[71] Gräbe, H.: On lucky primes. J. Symb. Comput. **15**, 199–209 (1994)

[72] Grayson, D.R., Stillman, M.E.: Macaulay2, A Software System for Research in Algebraic Geometry. Available at http://www.math.uiuc.edu/Macaulay2/

[73] Greuel, G.M., Pfister, G.: A Singular Introduction to Commutative Algebra. Springer, Berlin/Heidelberg/New York (2002)

[74] Greuel, G.-M., Seelisch, F., Wienand, O.: The Gröbner basis of the ideal of vanishing polynomials. J. Symb. Comput. **46**, 561–570 (2011)

[75] Gruson, L., Raynaud, M.: Critères de platitude et de projectivité. Techniques de "platification" d'un module. Invent. Math. **13**, 1–89 (1971)

[76] Gupta, S.K., Murthy, M.P.: Suslin'S Work on Linear Groups over Polynomial Rings and Serre Problem. Indian Statistical Institute Lecture Notes Series, vol. 8. Macmillan, New Delhi (1980)

[77] Hadj Kacem, A., Yengui, I.: Dynamical Gröbner bases over Dedekind rings. J. Algebra **324**, 12–24 (2010)

[78] Havas, G., Majewski B.S.: Extended gcd calculation. In: Proceedings of the Twenty-sixth Southeastern International Conference on Combinatorics, Graph Theory and Computing (Boca Raton, FL, 1995). Congr. Numer. **111**, 104–114 (1995)

[79] Heinzer, W., Papick, I.J.: Remarks on a remark of Kaplansky. Proc. Am. Math. Soc. **105**, 1–9 (1989)

[80] Huckaba, J.: Commutative Rings with Zero-Divisors. Marcel Dekker, New York (1988)

[81] Hurwitz, A.: Ueber einen Fundamentalsatz der arithmetischen Theorie der algebraischen Größen, pp. 230–240. Nachr. kön Ges. Wiss., Göttingen (1895) [Werke, vol. 2, pp. 198–207]

[82] Jambor, S.: Computing minimal associated primes in polynomial rings over the integers. J. Symb. Comput. **46**, 1098–1104 (2011)

[83] Joyal, A.: Spectral spaces and distibutive lattices. Not. Am. Math. Soc. **18**, 393 (1971)

[84] Joyal, A.: Le théorème de Chevalley-Tarski. Cahiers de Topologie et Géomérie Différentielle **16**, 256–258 (1975)

[85] Kapur, D., Narendran, P.: An equational approach to theoretical proving in first-order predicate calculus. In: Proceedings of the International Joint Conference on Artificial Intelligence, IJCAI, pp. 1146–1153 (1985)

[86] Kandry-Rody, A., Kapur, D.: Computing a Gröbner basis of a polynomial ideal over a euclidean domain. J. Symb. Comput. **6**, 37–57 (1988)

[87] Kemper, G.: A Course in Commutative Algebra. Graduate Texts in Mathematics. Springer, Berlin (2011)

[88] Kreuzer, M., Robbianno, L.: Computational Commutative Algebra, vol. 2. Springer, Berlin (2005)

[89] Kronecker, L.: Zur Theorie der Formen höherer Stufen Ber, pp. 957–960. K. Akad. Wiss. Berlin (1883) [Werke 2, 417–424]

[90] Kunz, E.: Introduction to Commutative Algebra and Algebraic Geometry. Birkhäuser, Basel (1991)

[91] Lam, T.Y.: Serre's Conjecture. Lecture Notes in Mathematics, vol. 635. Springer, Berlin/New York (1978)

[92] Lam, T.Y.: Serre's Problem on Projective Modules. Springer Monographs in Mathematics. Springer, Berlin (2006)

[93] Laubenbacher, R.C., Woodburn, C.J.: An algorithm for the Quillen-Suslin theorem for monoid rings. J. Pure Appl. Algebra **117/118**, 395–429 (1997)

[94] Laubenbacher, R.C., Woodburn, C.J.: A new algorithm for the Quillen-Suslin theorem. Beiträge Algebra Geom. **41**, 23–31 (2000)

[95] Lenstra, A.K., Lenstra, Jr. H.W., Lovász, L.: Factoring polynomials with rational coefficients. Math. Ann. **261**, 515–534 (1982)

[96] Lequain, Y., Simis, A.: Projective modules over $R[X_1,\ldots,X_n]$, R a Prüfer domain. J. Pure Appl. Algebra 18(2), 165–171 (1980)

[97] Logar, A., Sturmfels, B.: Algorithms for the Quillen-Suslin theorem. J. Algebra 145(1), 231–239 (1992)

[98] Lombardi, H.: Le contenu constructif d'un principe local-global avec une application à la structure d'un module projectif de type fini. Publications Mathématiques de Besançon. Théorie des nombres (1997)

[99] Lombardi, H.: Relecture constructive de la théorie d'Artin-Schreier. Ann. Pure Appl. Logic 91, 59–92 (1998)

[100] Lombardi, H.: Dimension de Krull, Nullstellensätze et Évaluation dynamique. Math. Z. 242, 23–46 (2002)

[101] Lombardi, H.: Platitude, localisation et anneaux de Prüfer, une approche constructive. Publications Mathématiques de Besançon. Théorie des nombres. Années 1998–2001

[102] Lombardi, H.: Hidden constructions in abstract algebra (1) integral dependance relations. J. Pure Appl. Algebra 167, 259–267 (2002)

[103] Lombardi, H.: Constructions cachées en algèbre abstraite (4) La solution du 17ème problème de Hilbert par la théorie d'Artin-Schreier. Publications Mathématiques de Besançon. Théorie des nombres. Années 1998–2001

[104] Lombardi, H.: Constructions cachées en algèbre abstraite (5) Principe local-global de Pfister et variantes. Int. J. Commut. Rings 2(4), 157–176 (2003)

[105] Lombardi, H., Perdry, H.: The Buchberger algorithm as a tool for ideal theory of polynomial rings in constructive mathematics. In: Gröbner Bases and Applications (Proceedings of the Conference 33 Years of Gröbner Bases). Mathematical Society Lecture Notes Series, vol. 251, pp. 393–407. Cambridge University Press, London (1998)

[106] Lombardi, H., Quitté, C.: Constructions cachées en algèbre abstraite (2) Le principe local-global. In: Fontana, M., Kabbaj, S.-E., Wiegand, S. (eds.) Commutative Ring Theory and Applications. Lecture Notes in Pure and Applied Mathematics, vol. 131, pp. 461–476. Marcel Dekker, New York (2002)

[107] Lombardi, H., Quitté, C.: Seminormal rings (following Thierry Coquand). Theor. Comput. Sci. 392, 113–127 (2008)

[108] Lombardi, H., Quitté, C.: Algèbre Commutative. Méthodes Constructives. Modules projectifs de type fini. Cours et exercices. Calvage et Mounet, Paris (2011)

[109] Lombardi, H., Quitté, C.: Commutative Algebra. Constructive Methods. Finite Projective Modules. Springer, New York (2015)

[110] Lombardi, H., Yengui, I.: Suslin's algorithms for reduction of unimodular rows. J. Symb. Comput. 39, 707–717 (2005)

[111] Lombardi H., Quitté C., Diaz-Toca G. M., *Modules sur les anneaux commutatifs. Cours et exercices*. Calvage et Mounet, 2014.

[112] Lombardi, H., Quitté, C., Yengui, I.: Hidden constructions in abstract algebra (6) The theorem of Maroscia, Brewer and Costa. J. Pure Appl. Algebra 212, 1575–1582 (2008)

[113] Lombardi, H., Quitté, C., Yengui, I.: Un algorithme pour le calcul des syzygies sur $V[X]$ dans le cas où V est un domaine de valuation. Commun. Algebra 42(9), 3768–3781 (2014)

[114] Lombardi, H., Schuster, P., Yengui, I.: The Gröbner ring conjecture in one variable. Math. Z. **270**, 1181–1185 (2012)

[115] Macaulay2. A Quillen-Suslin package. http://wiki.macaulay2.com/Macaulay2/ index.php?title=Quillen-Suslin(2011)

[116] Magma (Computational Algebra Group within School of Maths and Statistics of University of Sydney). http://magma.maths.usyd.edu.au/magma (2010)

[117] Maroscia, P.: Modules projectifs sur certains anneaux de polynomes. C. R. Acad. Sci. Paris Sér. A **285**, 183–185 (1977)

[118] Mialebama Bouesso, A., Sow, D.: Non commutative Gröbner bases over rings. Commun. Algebra **43**(2), 541–557 (2015)

[119] Mialebama Bouesso, A., Valibouze, A., Yengui, I.: Gröbner bases over $\mathbb{Z}/p^{\alpha}\mathbb{Z}$, $\mathbb{Z}/m\mathbb{Z}$, $(\mathbb{Z}/p^{\alpha}\mathbb{Z}) \times (\mathbb{Z}/p^{\alpha}\mathbb{Z})$, $\mathbb{F}_2[a,b]/\langle a^2 - a, b^2 - b \rangle$, and $\mathbb{Z}$ as special cases of dynamical Gröbner bases. Preprint (2012)

[120] Mines, R., Richman, F., Ruitenburg, W.: A Course in Constructive Algebra. Universitext. Springer, Heidelberg (1988)

[121] Mnif, A., Yengui, I.: An algorithm for unimodular completion over Noetherian rings. J. Algebra **316**, 483–498 (2007)

[122] Möller, M., Mora, T.: New constructive methods in classical ideal theory. J. Algebra **100**, 138–178 (1986)

[123] Monceur, S., Yengui, I.: On the leading terms ideals of polynomial ideals over a valuation ring. J. Algebra **351**, 382–389 (2012)

[124] Monceur, S., Yengui, I.: Suslin's lemma for rings containing an infinite field. Colloq. Math. (in press)

[125] Mora, T.: Solving Polynomial Equation Systems I: The Kronecker-Duval Philosophy. Cambridge University Press, Cambridge (2003)

[126] Northcott, D.G.: Finite Free Resolutions. Cambridge University Press, Cambridge (1976)

[127] Norton, G.H., Salagean, A.: Strong Gröbner bases and cyclic codes over a finite-chain ring. Appl. Algebra Eng. Commun. Comput. **10**, 489–506 (2000)

[128] Norton, G.H., Salagean, A.: Strong Gröbner bases for polynomials over a principal ideal ring. Bull. Aust. Math. Soc. **64**, 505–528 (2001)

[129] Norton, G.H., Salagean, A.: Gröbner bases and products of coefficient rings. Bull. Aust. Math. Soc. **65**, 145–152 (2002)

[130] Norton, G.H., Salagean, A.: Cyclic codes and minimal strong Gröbner bases over a principal ideal ring. Finite Fields Appl. **9**, 237–249 (2003)

[131] Park, H.: A computational theory of Laurent polynomial rings and multidimensional FIR systems. University of Berkeley (1995)

[132] Park, H.: A realization algorithm for $SL_2(\mathbf{R}[X_1,\ldots,X_m])$ over the euclidean domain. SIAM J. Matrix Anal. Appl. **21**, 178–184 (1999)

[133] Park, H.: Symbolic computations and signal processing. J. Symb. Comput. **37**, 209–226 (2004)

[134] Park, H.: Generalizations and variations of Quillen-Suslin theorem and their applications. In: Work-shop Gröbner Bases in Control Theory and Signal Processing. Special Semester on Gröbner Bases and Related Methods 2006, University of Linz, Linz, 19 May 2006

[135] Park, H., Woodburn, C.: An algorithmic proof of Suslin's stability theorem for polynomial rings. J. Algebra **178**, 277–298 (1995)

[136] Pauer, F.: On lucky ideals for Gröbner basis computations. J. Symb. Comput. **14**, 471–482 (1992)

[137] Pauer, F.: Gröbner bases with coefficients in rings. J. Symb. Comput. **42**, 1003–1011 (2007)

[138] Perdry, H.: Lazy bases: a minimalist constructive theory of Noetherian rings. MLQ Math. Log. Q. **54**, 70–82 (2008)

[139] Perdry, H.: Strongly Noetherian rings and constructive ideal theory. J. Symb. Comput. **37**, 511–535 (2004)

[140] Perdry, H., Schuster, P.: Noetherian orders. Math. Struct. Comput. Sci. **24**(2), 29 (2014)

[141] Perdry, H., Schuster, P.: Constructing Gröbner bases for Noetherian rings. Math. Struct. Comput. Sci. **21**, 111–124 (2011)

[142] Pola, E., Yengui, I.: A negative answer to a question about leading terms ideals of polynomial ideals. J. Pure Appl. Algebra **216**, 2432–2435 (2012)

[143] Pola, E., Yengui, I.: Gröbner rings. Acta Sci. Math. (Szeged) **80**, 363–372 (2014)

[144] Preira, J.-M., Sow, D., Yengui, I.: On Polly Cracker over valuation rings and $\mathbb{Z}_n$. Preprint (2010)

[145] Quillen, D.: Projective modules over polynomial rings. Invent. Math. **36**, 167–171 (1976)

[146] Rao, R.A.: The Bass-Quillen conjecture in dimension three but characteristic $\neq 2, 3$ via a question of A. Suslin. Invent. Math. **93**, 609–618 (1988)

[147] Rao, R.A., Swan, R.: A regenerative property of a fibre of invertible alternating polynomial matrices (in preparation)

[148] Raynaud, M.: Anneaux Locaux Henséliens. Lectures Notes in Mathematics, vol. 169. Springer, Berlin/Heidelberg/New York (1970)

[149] Richman, F.: Constructive aspects of Noetherian rings. Proc. Am. Mat. Soc. **44**, 436–441 (1974)

[150] Richman, F.: Nontrivial use of trivial rings. Proc. Am. Mat. Soc. **103**, 1012–1014 (1988)

[151] Roitman, M.: On projective modules over polynomial rings. J. Algebra **58**, 51–63 (1979)

[152] Roitman, M.: On stably extended projective modules over polynomial rings. Proc. Am. Math. Soc. **97**, 585–589 (1986)

[153] Rosenberg, J.: Algebraic K-Theory and Its Applications. Graduate Texts in Mathematics. Springer, New York (1994)

[154] Sasaki, T., Takeshima, T.: A modular method for Gröbner-basis construction over $\mathbb{Q}$ and solving system of algebraic equations. J. Inform. Process. **12**, 371–379 (1989)

[155] Schreyer, F.-O.: Syzygies of canonical curves and special linear series. Math. Ann. **275**(1), 105–137 (1986)

[156] Schreyer, F.-O.: A standard basis approach to Syzygies of canonical curves. J. Reine Angew. Math. **421**, 83–123 (1991)

[157] Schuster, P.: Induction in algebra: a first case study. In: 2012 27th Annual ACM/IEEE Symposium on Logic in Computer Science, Proceedings LICS 2012, pp. 581–585. IEEE Computer Society Publications, Dubrovnik (June 2012)

[158] Schuster, P.: Induction in algebra: a first case study. Log. Methods Comput. Sci. 9(3), 19 (2013)

[159] Seindeberg, A.: What is Noetherian? Rend. Sem. Mat. Fis. Milano 44, 55–61 (1974)

[160] Serre, J.-P.: Faisceaux algébriques cohérents. Ann. Math. 61, 191–278 (1955)

[161] Serre, J.-P.: Modules projectifs et espaces fibrés à fibre vectorielle. Sém. Dubreil-Pisot, no. 23, Paris (1957/1958)

[162] Shekhar, N., Kalla, P., Enescu, F., Gopalakrishnan, S.: Equivalence verification of polynomial datapaths with fixed-size bit-vectors using finite ring algebra. In: ICCAD '05: Proceedings of the 2005 IEEE/ACM International Conference on Computer-Aided Design, pp. 291–296. IEEE Computer Society, Washington, DC (2005)

[163] Simis, A., Vasconcelos, W.: Projective modules over $\mathbf{R}[X]$, $\mathbf{R}$ a valuation ring are free. Not. Am. Math. Soc. 18(5), 944 (1971)

[164] Suslin, A.A.: Projective modules over a polynomial ring are free. Sov. Math. Dokl. 17, 1160–1164 (1976)

[165] Suslin, A.A.: On the structure of the special linear group over polynomial rings. Math. USSR-Izv. 11, 221–238 (1977)

[166] Suslin, A.A.: On stably free modules. Mat. Sb. (N.S.), 102(144)(4), 537–550 (1977)

[167] Swan, R.: On seminormality. J. Algebra 67, 210–229 (1980)

[168] Traverso, C.: Seminormality and the Picard group. Ann. Scuola Norm. Sup. Pisa 24, 585–595 (1970)

[169] Traverso, C.: Gröbner Trace Algorithms. In: Proceedings of the International Symposium on Symbolic and Algebraic Computation, ISSAC '88. Lecture Notes in Computer Science, vol. 358, pp. 125–138 (1988)

[170] Traverso, C.: Hilbert functions and the Buchberger's algorithm. J. Symb. Comput. 22, 355–376 (1997)

[171] Trinks, W.: Über B. Buchbergers Verfahren, Systeme algebraischer Gleichungen zu lösen. J. Number Theory 10, 475–488 (1978)

[172] Tolhuizen, L., Hollmann, H., Kalker, A.: On the realizability of Bi-orthogonal M-dimensional 2-band filter banks. IEEE Trans. Signal Process. 43, 640–648 (1995)

[173] Valibouze, A., Yengui, I.: On saturations of ideals in finitely-generated commutative rings and Gröbner rings. Preprint (2013)

[174] Vaserstein, L.N.: K_1-theory and the congruence problem. Mat. Zametki 5, 233–244 (1969)

[175] Vaserstein, L.N.: Operations on orbits of unimodular vectors. J. Algebra 100, 456–461 (1986)

[176] von zur Gathen, J., Gerhard, J.: Modern Computer Algebra. Cambridge University Press, Cambridge (2003)

[177] Wienand, O.: Algorithms for symbolic computation and their applications. Ph.D. thesis, Kaiserslautern (2011)

[178] Winkler, F.: A p-adic approach to the computation of Gröbner bases. J. Symb. Comput. **6**, 287–304 (1987)

[179] Yengui, I.: An algorithm for the divisors of monic polynomials over a commutative ring. Math. Nachr. **260**, 1–7 (2003)

[180] Yengui, I.: Making the use of maximal ideals constructive. Theor. Comput. Sci. **392**, 174–178 (2008)

[181] Yengui, I.: Dynamical Gröbner bases. J. Algebra **301**, 447–458 (2006)

[182] Yengui, I.: The Hermite ring conjecture in dimension one. J. Algebra **320**, 437–441 (2008)

[183] Yengui, I.: Corrigendum to dynamical Gröbner bases [J. Algebra 301(2), 447–458 (2006)] and to Dynamical Gröbner bases over Dedekind rings [J. Algebra 324(1), 12–24 (2010)]. J. Algebra **339**, 370–375 (2011)

[184] Yengui, I.: Stably free modules over $\mathbf{R}[X]$ of rank $> \dim \mathbf{R}$ are free. Math. Comput. **80**, 1093–1098 (2011)

[185] Yengui, I.: The Gröbner ring conjecture in the lexicographic order case. Math. Z. **276**, 261–265 (2014)

[186] Youla, D.C., Pickel, P.F.: The Quillen-Suslin theorem and the structure of n-dimentional elementary polynomial matrices. IEEE Trans. Circ. Syst. **31**, 513–518 (1984)

Index

© Springer International Publishing Switzerland 2015
I. Yengui, *Constructive Commutative Algebra*, Lecture Notes in Mathematics 2138,
DOI 10.1007/978-3-319-19494-3

LECTURE NOTES IN MATHEMATICS

 Springer

Editors in Chief: J.-M. Morel, B. Teissier;

Editorial Policy

1. Lecture Notes aim to report new developments in all areas of mathematics and their applications – quickly, informally and at a high level. Mathematical texts analysing new developments in modelling and numerical simulation are welcome.

   Manuscripts should be reasonably self-contained and rounded off. Thus they may, and often will, present not only results of the author but also related work by other people. They may be based on specialised lecture courses. Furthermore, the manuscripts should provide sufficient motivation, examples and applications. This clearly distinguishes Lecture Notes from journal articles or technical reports which normally are very concise. Articles intended for a journal but too long to be accepted by most journals, usually do not have this "lecture notes" character. For similar reasons it is unusual for doctoral theses to be accepted for the Lecture Notes series, though habilitation theses may be appropriate.

2. Besides monographs, multi-author manuscripts resulting from SUMMER SCHOOLS or similar INTENSIVE COURSES are welcome, provided their objective was held to present an active mathematical topic to an audience at the beginning or intermediate graduate level (a list of participants should be provided).

   The resulting manuscript should not be just a collection of course notes, but should require advance planning and coordination among the main lecturers. The subject matter should dictate the structure of the book. This structure should be motivated and explained in a scientific introduction, and the notation, references, index and formulation of results should be, if possible, unified by the editors. Each contribution should have an abstract and an introduction referring to the other contributions. In other words, more preparatory work must go into a multi-authored volume than simply assembling a disparate collection of papers, communicated at the event.

3. Manuscripts should be submitted either online at www.editorialmanager.com/lnm to Springer's mathematics editorial in Heidelberg, or electronically to one of the series editors. Authors should be aware that incomplete or insufficiently close-to-final manuscripts almost always result in longer refereeing times and nevertheless unclear referees' recommendations, making further refereeing of a final draft necessary. The strict minimum amount of material that will be considered should include a detailed outline describing the planned contents of each chapter, a bibliography and several sample chapters. Parallel submission of a manuscript to another publisher while under consideration for LNM is not acceptable and can lead to rejection.

4. In general, **monographs** will be sent out to at least 2 external referees for evaluation.

   A final decision to publish can be made only on the basis of the complete manuscript, however a refereeing process leading to a preliminary decision can be based on a pre-final or incomplete manuscript.

   Volume Editors of **multi-author** works are expected to arrange for the refereeing, to the usual scientific standards, of the individual contributions. If the resulting reports can be forwarded to the LNM Editorial Board, this is very helpful. If no reports are forwarded or if other questions remain unclear in respect of homogeneity etc, the series editors may wish to consult external referees for an overall evaluation of the volume.

5. Manuscripts should in general be submitted in English. Final manuscripts should contain at least 100 pages of mathematical text and should always include

   - a table of contents;
   - an informative introduction, with adequate motivation and perhaps some historical remarks: it should be accessible to a reader not intimately familiar with the topic treated;
   - a subject index: as a rule this is genuinely helpful for the reader.
   - For evaluation purposes, manuscripts should be submitted as pdf files.

6. Careful preparation of the manuscripts will help keep production time short besides ensuring satisfactory appearance of the finished book in print and online. After acceptance of the manuscript authors will be asked to prepare the final LaTeX source files (see LaTeX templates online: https://www.springer.com/gb/authors-editors/book-authors-editors/manuscriptpreparation/5636) plus the corresponding pdf- or zipped ps-file. The LaTeX source files are essential for producing the full-text online version of the book, see http://link.springer.com/bookseries/304 for the existing online volumes of LNM). The technical production of a Lecture Notes volume takes approximately 12 weeks. Additional instructions, if necessary, are available on request from lnm@springer.com.

7. Authors receive a total of 30 free copies of their volume and free access to their book on SpringerLink, but no royalties. They are entitled to a discount of 33.3% on the price of Springer books purchased for their personal use, if ordering directly from Springer.

8. Commitment to publish is made by a *Publishing Agreement*; contributing authors of multiauthor books are requested to sign a *Consent to Publish form*. Springer-Verlag registers the copyright for each volume. Authors are free to reuse material contained in their LNM volumes in later publications: a brief written (or e-mail) request for formal permission is sufficient.

Addresses:
Professor Jean-Michel Morel, CMLA, École Normale Supérieure de Cachan, France
E-mail: moreljeanmichel@gmail.com

Professor Bernard Teissier, Equipe Géométrie et Dynamique,
Institut de Mathématiques de Jussieu – Paris Rive Gauche, Paris, France
E-mail: bernard.teissier@imj-prg.fr

Springer: Ute McCrory, Mathematics, Heidelberg, Germany,
E-mail: lnm@springer.com

Printed in the United States
By Bookmasters